FOUNDATIONS OF
DIGITAL LOGIC DESIGN

FOUNDATIONS OF DIGITAL LOGIC DESIGN

GIDEON LANGHOLZ
Tel-Aviv University, Israel

ABRAHAM KANDEL
University of South Florida, Tampa, USA

JOE L MOTT
Florida State University, Tallahassee, USA

World Scientific
Singapore • New Jersey • London • Hong Kong

Published by

World Scientific Publishing Co. Pte. Ltd.

P O Box 128, Farrer Road, Singapore 912805

USA office: Suite 1B, 1060 Main Street, River Edge, NJ 07661

UK office: 57 Shelton Street, Covent Garden, London WC2H 9HE

British Library Cataloguing-in-Publication Data
A catalogue record for this book is available from the British Library.

FOUNDATIONS OF DIGITAL LOGIC DESIGN

ISBN 981-02-3110-5

Printed in Singapore by Uto-Print

Contents

APPENDIX: LOGIC GRAPHIC SYMBOLS   **539**

BIBLIOGRAPHY   **551**

ANSWERS TO SELECTED PROBLEMS   **553**

INDEX   **577**

Preface

This text is intended for a first course in *digital logic design*, at the sophomore or junior level, for electrical engineering, computer engineering, and computer science programs, as well as for a number of other disciplines such as physics and mathematics. The book can also be used for self-study or for review by practicing engineers and computer scientists not intimately familiar with the subject.

No specific background is prerequisite other than an understanding of the basic material ordinarily covered in a college algebra course. We have assumed that the student will have had little or no knowledge of discrete mathematics, computer hardware, or software, although such knowledge would be advantageous. After completing this text, the student should be prepared for a second (advanced) course in digital design, switching and automata theory, microprocessors, or computer organization.

We believe that the student should study and understand thoroughly the *fundamental concepts* of logic design in a first course. To do so requires a coordinated and integrated treatment that maintains a balance between theory and application. Hence, this text covers practical aspects of digital logic circuits developed from a sound theoretical basis.

The *theoretical aspects* include:

Number systems
Arithmetic operations
Coding of binary information
Boolean algebra
Simplification of Boolean expressions
Feedback and memory

The *practical applications* include:

> Arithmetic circuits
> Comparators
> Multiplexers and demultiplexers
> Decoders and encoders
> Programmable arrays
> Latches and flip-flops
> Read/write memory
> Registers and shift registers
> Data transfer and formatting
> Counters
> Counting, frequency division and measurement, and timing signals generation

We put emphasis on structured and systematic design principles. We expose the reader to a variety of methods and algorithms that can be applied to a broad class of digital design problems and feature many topics relevant to current technologies such as modular design of digital circuits and the use of MSI and LSI devices. The use of these general principles in the design of logic circuits is illustrated by means of numerous examples in the text and a variety of problems at the end of each chapter. Practical applications are emphasized, although, of necessity, our examples are small-scale prototypical models of real-life problems. But small-scale examples convey much the same concepts as large-scale ones, and, from our perspective, *what* we design is secondary to *how* we design it: Our main objective is to teach the *principles* of good logic design. We believe that, once these principles are understood, the student will be able to adapt them to changing technologies and apply them to different problems.

The topics covered in this book include all those described in Subject Area 6 (Logic Design) of the 1983 IEEE Computer Society Model Program in Computer Science and Engineering, with the exception of Module 5 (Register Transfer Logic, RTL). We feel that it is too ambitious to try and cover RTL in a first course, particularly since we have already extended some of the topics included in the other modules of the Model Program.

The book is divided into seven chapter. **Chapter 1** provides some introductory definitions and background to digital systems and presents a brief overview of integrated-circuit (IC) technologies. **Chapter 2** presents the various number systems and codes used to represent information in digital systems. Since digital systems manipulate discrete elements of information represented in binary form, this chapter provides a frame of reference for further developments throughout the text.

Chapter 3 introduces Boolean algebra as the basic mathematical tool essential to the study of digital circuits. Boolean algebra enables the designer to build a bridge from the conceptual world of word description of circuits to the physical world of implementation. The interplay between design specifications of digital circuits, Boolean expressions of switching functions, and implementation of these functions with digital devices is a fundamental concept of digital design. As a consequence, the idea of Boolean algebra, though somewhat abstract, provides the language by which digital circuits are analyzed, designed, documented, and verified. The last two sections of Chapter 3 are devoted to properties of Boolean functions and to the implementation of special circuits such as symmetric logic circuits, decomposable logic circuits, and threshold logic circuits.

The minimization of switching functions of a large number of variables is very important due to the extensive use of programmable arrays in the design of VLSI chips. **Chapter 4** provides the conceptual framework that is prequisite to understanding the minimization problem and process. This chapter presents algorithms for simplifying Boolean expressions so that cost-effective combinational circuit implementations can be obtained, but emphasizes the trade-off between minimality and reliability. We discuss two simplification methods: The Karnaugh map method and the Quine-McCluskey tabular method. The map method is useful for simplifying expressions of up to six variables, while the tabulation technique is applicable to any number of variables and can be fairly easily programmed on a digital computer. We apply these techniques to single-output as well as to multiple-output digital circuits.

Digital circuits can be implemented with a variety of components such as electronic devices, fluidic elements, and electromechanical devices. Most digital circuits, however, are constructed with integrated-circuit chips, and the digital designer must be familiar with those that are most often used. In **Chapter 5**, we introduce a variety of combinational MSI and LSI circuit components and discuss in detail devices such as adders, decoders, multiplexers, ROMs, PLAs, and PALs. The use of these components in the design of digital circuits is illustrated through numerous examples.

Memory is the distinguishing feature between combinational and sequential circuits. The last two chapters introduce the analysis and design of sequential circuits in a systematic and unified manner. **Chapter 6** introduces many of the concepts and techniques common to both asynchronous and synchronous sequential circuits, but also covers the formal procedures for the analysis and design of asynchronous circuits. The same algorithmic approach is used in **Chapter 7** to present the design process of synchronous sequential circuits. Asynchronous circuits are useful in a variety of applications where speed of operation is important, and in systems where synchronization is not viable. Nevertheless, synchronous circuits are the type encountered most frequently in digital systems. In Chapter 7, we discuss a variety of MSI and LSI devices such as registers, shift registers, and counters, and illustrate their applications through numerous examples. These applications include: Parallel and serial information transfers, random access read/write memories, data format conversions, parallel and serial binary addition, counting and frequency division, frequency measurement, and generation of timing sequences and signals.

Unlike many texts that discuss synchronous circuits first, we chose to treat asynchronous circuits first for two main reasons: (1) The operation of sequential circuits depends on proper timing and is best understood in the context of asynchronous circuits, and (2) synchronous circuits are actually a special case of asynchronous circuits, and, consequently, a unified design methodology for both circuit types can be presented and duplication of similar topics can be avoided. Nevertheless, should the instructor wish to introduce synchronous circuits first, then except for Section 6.1, only the first parts of Section 6.4 and Section 6.5 need to be covered before proceeding to Chapter 7.

We have taught numerous courses in digital logic design, based on the material included in this text, at a number of different universities in electrical engineering, computer engineering, and computer science programs. The text is self-contained and can be covered in a single-term course. Some sections are marked with an asterisk (*), indicating that the material in these sections is optional and can be skipped without loss of continuity. In a semester course, almost all the material in the text can be covered. In a quarter course, the instructor must be selective in the material covered and the emphasis placed on some of the topics.

We have retained throughout the text the distinctive-shape graphic symbols for logic functions (ANSI/IEEE Standard No. 91-1973) rather than introduce the new ANSI/IEEE Standard No. 91-1984. Because the new standard is quite involved, we feel that using it in a text for a first course in logic design would overburden the reader unnecessarily. Therefore, we delegate this discussion to the Appendix where the interested reader will find a brief overview of the new standard.

To fully understand the technical material, the student must have the opportunity to work problems. Consequently, over 270 end-of-chapter problems have been included. The problems range from the routine to the moderately difficult, and most of them are designed to promote understanding of, and provoke thought about, digital logic systems. Selected answers appear at the end of the book.

We would like to acknowledge our indebtedness to many people who contributed to this book. We owe much to our colleagues and students from whom we have learned and by whom we have been encouraged. Special thanks to Dr. Yan-Qing Zhang, Vijaya Ramamurthi, and Judy Hyde for their enormous help during the various stages of preparing the manuscript.

Completing this text would not have been possible without the encouragement and support of Professor Horst Bunke, Dean of the Institut fur Informatick und Angewandte Mathematik, Universitat Bern, Switzerland, Professor Michael Kovac, Dean of the College of Engineering at the University of South Florida and Professor Uri Shaked, Dean of the Faculty of Engineering at Tel-Aviv University.

Gideon Langholz
Abraham Kandel
Joe L. Mott

Tampa, Florida, 1998

1 Introduction to Digital Systems

1.1 OVERVIEW

In recent years digital systems have become an integral part of our everyday lives. We wake up to a digital alarm clock's preselected music at the right time, drive in a digitally controlled automobile, or shop for computer-coded grocery items. Digital computers are undoubtedly the best known and most prominent class of digital systems. Computers have made possible many industrial, scientific, commercial, and educational advances that would have been unattainable otherwise. They are found in satellite communication systems, space vehicles, telephone networks, industrial production lines, defense systems, and home appliances.

But digital techniques are also applied to many other areas, some quite unrelated to computers. Digital traffic controllers measure traffic flow on crossroads and adjust the period of traffic lights to maximize flow. Digital antiskid devices measure the speed of a truck's wheels and adjust the braking force to prevent skids. Ignition controllers measure the carburetion and timing of automobile engines to reduce emissions.

The single major force behind the proliferation of digital techniques into virtually every area of technology is undoubtedly the rapid development of *integrated circuits* (*ICs*). Their low costs have greatly expanded the range and area of application of digital techniques, as well as reduced the price of complete digital systems. Every day, ingenious designers discover new ways to apply digital techniques to their problems. As digital applications have grown in importance, efficient analysis and design procedures applicable to almost any type of digital system have been developed.

We begin, therefore, with the following question: What is a digital system? Since this book is intended to teach the principles of designing digital systems, this question is a good starting point. To answer it, we must first establish the meaning of **digital**.

In almost any field of endeavor, we constantly deal with *quantities*. Quantities are used in most physical systems and are observed, measured, manipulated, and recorded. The world we live in, for the most part, is studied and explained by continuously changing phenomena like time, temperature, distance, weight, and speed. Basically, we can represent the numerical values of these quantities in two ways: analog and digital. In **analog representation**, one quantity is represented by another quantitiy proportional to it. For example, a room thermostat operates so that the curvature of the bimetallic strip is proportional to the room temperature. As the temperature changes continuously, the bending of the strip changes proportionally. Thus, analog quantities vary gradually over a continuous range of values.

The term *digital* comes from the use of the word *digit* in describing real numbers. In **digital representation**, quantities are not represented by proportional quantities but by symbols called digits. For example, the digital clock provides the time in the form of decimal digits representing the hours and minutes. While time is changing continuously, the clock reading is not; rather, it changes in *discrete* steps of one per minute. The flight pattern of a plane landing on an aircraft carrier takes on a continuum of values, but a pulse radar tracking system sends out radar pulses periodically so that the returning signals form a finite sequence of numbers that approximate the plane's present position, present velocity, and future position. To represent the thermostat's analog data in digital form, we record only specific quantities in tabular form. The table then contains a column of discrete temperature values and a column of corresponding values indicating the bimetallic strip's curvature.

Thus we see that the process of representing analog quantities in digital form requires that these quantities change in discrete steps. The original variable is *sampled* and is replaced by numerical values whose digits represent the size of the variable at particular points in time. This process is called a **digital (discrete) process**, the variable is said to be **digitized**, and the finitely many numerical observations are **called discrete elements** of information.

To answer the original question, we say, therefore, that a **digital system** is one that processes a finite set of data in digital (numerical) form, rather than processing continuous variables. Thus digital systems are capable of manipulating discrete elements of information. Such discrete elements are represented in a digital system by physical quantities called **signals**.

The signals processed in digital systems are usually represented in **binary** form. Binary quantities can be represented by any device that has only two operating **states**. For example, a switch has only two states, open or closed, and a light bulb can be either on or off. The two distinct states of binary signals are customarily represented by 0 and 1. Hence, we can assign a 0 to designate an open switch and a 1 to represent the closed switch. Note, however, that this assignment of binary values is arbitrary; we could have just as well reversed the assignments.

The binary representation of information will be the focus of our discussions throughout this book. The study of the techniques of digital systems design is sometimes referred to as **switching theory** or **logic design**.

Human reasoning tends to be binary, too. Logic according to Webster's dictionary is "a particular model of reasoning viewed as valid or faulty." Logic tells us that a certain proposition is true if certain conditions are true. Thus, "the light is on" and "the switch is off" are propositions that can be classified as true or false. A logic statement or *function* is a combination of several propositions. For example, the logic statement "the light is on" is true if the two conditions "the switch is on" and "the bulb is not burned out" are true. Since many problems in our daily lives can be formulated in terms of logic functions, digital systems with their two-state characteristics are extremely useful.

For the most part, our intention is to concentrate on the principles of analysis and design of digital systems while remaining relatively independent of the particular technologies presently in vogue. We believe that we can teach the basics of a general methodology that will not become obsolete with technological advances. But remember that digital systems are tools and, like any other practical tool designed by human beings, must be built out of available parts to perform a function in a cost-effective manner.

A design result is not unique as a rule; rather, it is likely to be a result of several compromises influenced by the interplay of various factors. The digital system designer is constrained by things like reliability, speed of operation, and power consumption. Often there is no a priori reason for choosing one compromise over another; but after having obtained a preliminary design, the designer generally can see how to improve it. The design process is evolutionary in nature; some of its aspects are routine and straightforward, while others are quite demanding. There is the challenge; may we all develop our competency to meet it.

1.2 PRELIMINARY CONCEPTS

The purpose of this section is to introduce some fundamental concepts associated with digital circuits and systems. We will introduce them using a simple example and then proceed to discuss logic levels, pulse waveform characteristics, and various other topics related to the performance of digital devices.

An Example

Consider the problem of turning a light bulb on or off from opposite ends of a room. Figure 1.1 shows a schematic diagram of the required electrical

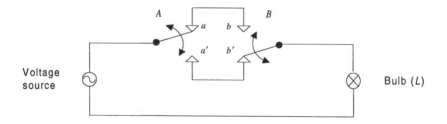

■ **Figure 1.1** Wiring diagram for turning a light bulb on or off from two ends of a room

circuit. The circuit consists of two double-pole, single-throw (DPST) switches, A and B, each located at one end of the room, a voltage source, a light bulb L, and the required electrical wires to connect these components. We will analyze this circuit in some detail and establish how to characterize its operation in mathematical terms.

Each switch can be positioned at either one of its two poles. Switch A can touch either pole a or pole a', and switch B can touch either pole b or pole b'. Clearly, when *not* in a, switch A must be in a' (and similarly for switch B). Hence, a' is actually NOT(a) and can therefore be referred to as the **complement (inverse)** of a. We designate the complement by using a prime symbol; thus, a' is referred to as complemented (or *primed*), while a is called uncomplemented (or *unprimed*).

Let us now examine how this circuit operates. When switch A touches a and switch B touches b', as shown in Figure 1.1, the electrical circuit is open and the light bulb L is OFF. At one end of the room, where A is located, we can turn L ON by switching A to a'. Alternatively, at the other end of the room, where B is located, we can turn L ON by switching B to b. If we turn L ON from A's side of the room, we can clearly turn it OFF from B's side, and vice versa.

The operation of a digital circuit can be expressed by formulating logic expressions. A **logic expression** consists of a *proposition* (statement) and a set of conditions. The proposition can be classified as *true* or *false* depending on whether the conditions are satisfied or not. Applying this idea to our circuit, we can summarize its operation by formulating two logic expressions:

$$L = \text{ON } \textit{if and only if } (A = a \text{ AND } B = b)$$

$$\text{OR } (A = a' \text{ AND } B = b') \tag{1.1a}$$

$$L = \text{OFF } \textit{if and only if } (A = a' \text{ AND } B = b)$$

$$\text{OR } (A = a \text{ AND } B = b') \tag{1.1b}$$

We will formally define the **logic operators** AND and OR in Chapter 3. Nevertheless, loosely interpreting them as we do in our daily use of these words, we can verify that the two expressions *completely* characterize the operation of the circuit. In either of these expressions, the first statement is true (satisfied) only if either of the last two statements is true. For example, in Equation (1.1a), the first statement ("the light bulb is ON") is the proposition and the other two statements are the conditions on which the proposition depends. Note, however, that the two expressions are **complementary** since $L = \text{ON}$ implies that L is NOT OFF. Therefore, either of these expressions would suffice to characterize the operation of the circuit.

Since the status of the bulb (ON or OFF) depends only on the positions of the two switches, L is referred to as the *dependent variable* while A and B are called the *independent variables*. The dependent variables are usually referred to as the *outputs* of the digital system, while the independent variables are the *inputs* to the system. Since the variables can assume only one of two distinct states (switch A can be positioned at either a or a'; L can be either ON or OFF), they are termed *binary variables*.

An alternative description for Equations (1.1a) and (1.1b) can be obtained by representing the operation of the circuit in the form of a **truth table**. In general, a truth table lists *all* the possible binary combinations of the independent variables and displays for each of these combinations the corresponding binary values of the dependent variables of the digital system. If there are n independent binary variables, the total number of possible binary combinations is 2^n, and the truth table contains 2^n rows.

In our example, $n = 2$ and the truth table contains four rows. Since A can be at either a or a', let $A = a$ and $A = a'$ be denoted by 1 and 0, respectively. Similarly for B, let $B = b$ and $B = b'$ be denoted by 1 and 0, respectively. For the dependent variable L, let us designate the condition $L = \text{ON}$ by 1 and the condition $L = \text{OFF}$ by 0. As we have already mentioned in Section 1.1, the assignment of binary values to the variables is arbitrary. We could have assigned 0 to $A = a$ and 1 to $A = a'$. Similarly, we could have denoted $L = \text{ON}$ by 0 and $L = \text{OFF}$ by 1. Nevertheless, keeping with the chosen assignments, we can obtain the truth table for the circuit of Figure 1.1 which is shown in Table 1.1.

Examining the truth table, we see that it completely characterizes the operation of our circuit. The information contained in row 0 OR row 3 is equivalent to Equation (1.1a), while the information contained in row 1 OR row 2 is equivalent to Equation (1.1b). We leave it to the reader to verify the one-to-one correspondence between Table 1.1 and Equations (1.1a) and (1.1b).

■ **Table 1.1** Truth table for the circuit of Figure 1.1

Row Number	A	B	L
0	0	0	1
1	0	1	0
2	1	0	0
3	1	1	1

Let us carry the discussion of this example a step further. We have already described the operation of the circuit using logic expressions and a truth table. We can now attempt to write the **logic functions** (also called **Boolean functions**; see Chapter 3) that characterize its operation. Since A and B are the independent (input) variables and L is the dependent (output) variable, the kind of logic function we are looking for is of the form $L = f(A, B)$; that is, L is ON as a function of A and B. We can also look for the *complementary* logic function $L' = g(A, B)$; that is, L is OFF as another function of A and B.

To derive these functions, recall that when $A = a$, then $A \neq a'$. Let us denote by A the former condition and by A' the latter condition. Hence, we can argue that, since $a' = \text{NOT}(a)$ and since A stands for $A = a$, then A' stands for $A = a'$ and, therefore, $A' = \text{not}\,(A)$. Similar reasoning would lead us to designate by B and B' the conditions $B = b$ and $B = b'$, respectively. Based on these notations and using either Equation (1.1a) or rows 0 and 3 of Table 1.1, we can write the following logic function:

$$L = A \cdot B + A' \cdot B' \qquad (1.2a)$$

where the dot stands for AND and the plus stands for OR. Similarly, using either Equation (1.1b) or rows 1 and 2 of Table 1.1, we can write the following complementary logic function:

$$L' = A' \cdot B + A' \cdot B' \qquad (1.2b)$$

Equation (1.2a) tells us that $L = 1$ (ON) if $\{A = 1\ (A = a)$ and $B = 1$ $(B = b)\}$ or $\{A = 0\ (A = a')$ and $B = 0\ (B = b')\}$. The complementary equation, (1.2b), tells us that $L' = 1$ (OFF, since $L' = 1$ implies $L = 0$) if $\{A' = 1 \quad (A = a') \quad \text{and} \quad B = 1\,(B = b)\}$ or $\{A = 1\,(A = a) \quad \text{and} \quad B' = 1$ $(B = b')\}$. The reader is encouraged to verify the equivalence between Equations (1.1a) and (1.2a) and Equations (1.1b) and (1.2b), respectively, and to study their correspondence with the circuit's truth table.

Logic Levels and Pulse Characteristics

Binary values are designated interchangeably as ON or OFF, TRUE or FALSE, HIGH or LOW, or 1 or 0, and are often referred to as *logic states*. We will use the 1/0 and HIGH/LOW notations to indicate the two states of a variable. If the higher of the two physical values of a variable is represented by 1 and the lower value is represented by 0, the digital system is referred to as a **positive logic** system. If we reverse the notation, designating the higher of the two physical values of a variable by 0 and the lower value by 1, the digital system is referred to as a **negative logic** system. Digital systems can be implemented in either positive or negative logic or both. Positive logic, however, is more common and we will therefore use it almost exclusively throughout the text.

A **pulse** is defined in the 1984 IEEE dictionary as "a wave that departs from an initial level for a limited duration of time and ultimately returns to the original level." A pulse can be either positive-going or negative-going, as shown in Figure 1.2. A **positive-going pulse** occurs when a binary variable makes a transition from its normally LOW (0) state to its HIGH (1) state and then back again to its LOW (0) state. The positive-going transition in this case is termed the **leading edge** of the pulse, and the negative-going transition is called the **trailing edge**. A **negative-going pulse** occurs when the binary variable makes a transition from its normally HIGH (1) state to its LOW (0) state and then back to its HIGH (1) state. In this case, the leading edge of the pulse is the negative-going transition, and the trailing edge is the positive-going transition.

A **train of pulses** can be either periodic or aperiodic. A **periodic** pulse train repeats itself at a fixed interval T, called the *period*, as shown in Figure 1.3(a). The *pulse width* (t_p) designates the duration of a pulse. An **aperiodic** pulse train does not repeat itself at any fixed interval and may contain pulses of varying widths, as shown in Figure 1.3(b).

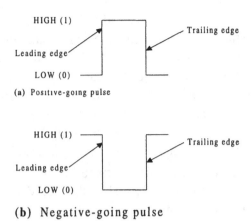

Figure 1.2 Single pulse waveforms

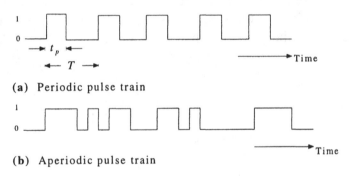

Figure 1.3 Pulse train waveforms

The *frequency* (f) of a periodic waveform is the reciprocal of the period, $f = 1/T$. It is measured in hertz (Hz) or, since the waveform is pulse shaped, in pulses per second, referred to as the *pulse repetition rate* (PRR) of the signal. A periodic pulse train is also characterized by a **duty cycle**, which is defined as the percentage of the pulse width t_p to the period T:

$$\text{duty cycle} = \frac{t_p}{T} \times 100\% \tag{1.3}$$

Figure 1.4 shows some examples of periodic pulse waveforms having the same period (frequency) but different duty cycles.

The pulses shown in Figures 1.2 and 1.3 are *ideal* since the transitions between the two states are assumed to occur instantaneously. Although we will assume ideal pulses for most of our work in analyzing and designing digital systems, this is *not* the case in practice. Variables do not change states instantaneously, but rather in a fashion similar to that depicted in Figure 1.5. The transition from the LOW to the HIGH state of the *nonideal* pulse (assumed to be positive-going here) is characterized by a nonzero *rise time* (t_r), defined as the time required by the pulse to rise from 10% to 90% of the difference between its HIGH and LOW values. Similarly, the transition from the HIGH to the LOW state is characterized by a nonzero *fall time* (t_f), the time required by the pulse to fall from 90% to 10% of the difference between its HIGH and LOW values. The *pulse width* (t_p) is defined as the time interval between certain points on the rising and falling edges. Often, the pulse width is taken as the interval between the 50% points on these edges, as shown in Figure 1.5. Another approach to defining the pulse width will be discussed in the following section.

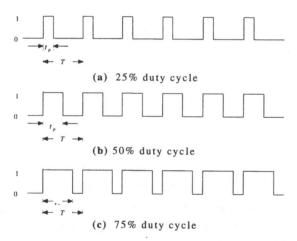

(a) 25% duty cycle

(b) 50% duty cycle

(c) 75% duty cycle

■ **Figure 1.4** Periodic pulse waveforms having different duty cycles

Performance Measures

One key characteristic of a digital circuit is its *speed* of operation, measured by the propagation time delay. The **propagation time delay** is a measure of the speed with which a signal can propagate from the input to the output of a digital (logic) device. To introduce this measure, consider a digital device having an input x and an output y, and let its input and output waveforms be those depicted in Figure 1.6. We assume that the input waveform is an ideal pulse, whereas the output waveform is an actual pulse, similar to the one shown in Figure 1.5.

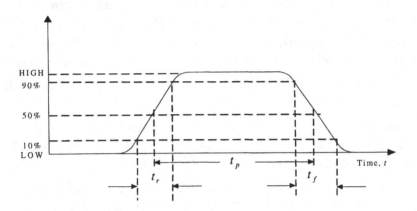

■ **Figure 1.5** Nonideal pulse τ

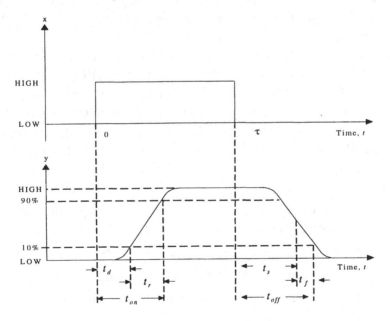

■ **Figure 1.6** Input-output waveforms of a logic device

Figure 1.6 shows that the output does not immediately respond to the input signal. As x makes a transition from LOW to HIGH at $t=0$, y is initially delayed by a *time delay* (t_d), which is defined as the time taken by the output signal to reach 10% of its HIGH value. The output then takes a further amount of time (t_r) to reach 90% of its HIGH value. Similarly, when the input signal returns to its LOW state at $t=\tau$, the output again fails to respond immediately. The interval between the negative-going transition of the input (at $t=\tau$) and the time when the output falls to 90% of its HIGH value is called the *storage time* t_s. The output then takes a further amount of time (t_f) to fall to 10% of its HIGH value.

Based on these definitions, we can characterize a digital device in terms of two parameters: the turn-on time and the turn-off time. The *turn-on* time is defined as the sum of the delay and rise times, $t_{on}=t_d+t_r$. The *turn-off* time is defined as the sum of the storage and fall times, $t_{off}=t_s+t_f$. Unfortunately, these definitions are not very useful in determining the propagation time delay of a digital device because (1) the input is not an ideal pulse, and (2) in practice the input does not have to reach its HIGH value (or 90% of it, if the input pulse is nonideal) before the output of the device can change state.

Let us therefore consider the more realistic situation shown in Figure 1.7. As the input signal rises from LOW toward HIGH, a change in the output state from LOW to HIGH is initiated at some *switching threshold* value TH(0). Similarly, as the input falls from HIGH toward LOW, a change in the output state from HIGH to LOW is initiated at some *other* switching threshold value TH(1). Sometimes, both TH(0) and TH(1) are arbitrarily assumed to be at 50% of the LOW-HIGH swing. The *propagation delay ON time* T_{ON} (which is distinguished from t_{on}) is defined as the interval between the times when the input and output reach TH(0). Similarly, the *propagation delay OFF time* T_{OFF} (which is distinguished from t_{off}) is defined as the interval between the times when the input and output reach TH(1). In general, $T_{ON} \neq T_{OFF}$, and the propagation time delay is defined as the average of the two:

$$T_{\text{PD}} = \frac{1}{2}(T_{\text{ON}} + T_{\text{OFF}}) \tag{1.4}$$

Finally, as shown in Figure 1.7, the pulse width (t_p) is defined as the interval between the TH(0) and TH(1) points of the pulse.

Another important characteristic feature of digital circuits is the **noise immunity**. This parameter is a measure of the ability of a digital device to withstand additive noise at its input without changing the logic state of its output. To understand this parameter, note that the values of both the input and output HIGH and LOW states of a digital device are not specified accurately, but rather within a range of values. In fact, the output HIGH state is actually defined by a value $V_{OH}(\text{min})$ such that (output) values not less than $V_{OH}(\text{min})$ are still considered as HIGH. The output LOW state is similarly defined by a value $V_{OL}(\text{max})$ such that (output) values not more than $V_{OL}(\text{max})$ are still considered as LOW. Likewise, at the input to the digital device, input values not less than $V_{IH}(\text{min})$ and not

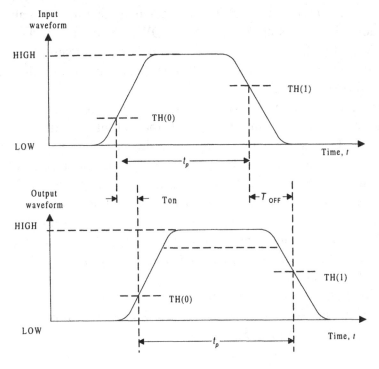

■ **Figure 1.7** Definition of propagation time delay

more than $V_{IL}(\max)$ are considered as HIGH and LOW states, respectively.

Digital devices are designed such that $V_{OH}(\min) > V_{IH}(\min)$ and $V_{OL}(\max) < V_{IL}(\max)$. The noise value at the input that will cause the device to malfunction when its output is in the HIGH (or LOW) state is called the **noise margin**. If we designate the noise margin in the HIGH state by NM(HIGH) and that in the LOW state by NM(LOW), then these values are given by

$$NM(\text{HIGH}) = V_{OH}(\min) - V_{IH}(\min) \qquad (1.5a)$$

$$NM(\text{LOW}) = V_{IL}(\max) - V_{OL}(\max) \qquad (1.5b)$$

The two noise margins indicate the noise immunity capability of the digital device. To illustrate this, consider Figure 1.8 which shows a cascade connection of two digital devices. If the output of device A is LOW and there are fluctuations due to noise on the line between the two devices, the output of device B is not affected as long as the peak value of these fluctuations does not exceed NM(LOW). Similarly, if the output of device A is HIGH, the output of device B is not affected as long as the peak value of the fluctuations on the line between the two devices does not exceed NM(HIGH).

Finally, let us mention two other characteristic features of digital devices: fan-out and fan-in. In general, the output of a digital device is required to drive several inputs of other devices. The **fan-out** M of a digital device indicates the maximum number of input lines of similar devices that the device can drive without affecting its output LOW and HIGH states. For example, a digital device having a fan-out of 12 can drive 12 inputs of similar devices simultaneously (in parallel). If more than 12 inputs are connected to the output of the device, its output logic levels cannot be guaranteed (a HIGH output may be forced to become LOW, and vice versa).

A digital device can have one input or more than one input. The number of inputs N of a device is called the **fan-in**. For example, a 3-input device has a fan-in of 3.

■ **Figure 1.8** Cascade connection of two digital devices

1.3 DIGITAL MICROELECTRONICS TECHNOLOGIES

As mentioned in Section 1.1, our treatment of digital systems will be largely technology independent. Nevertheless, one technology, that of **digital microelectronics**, cannot be dismissed so easily. Its impact on the evolution of digital systems is so outstanding that due consideration, albeit brief, of this technology is necessary. The centerpiece of digital electronics is the **integrated circuit (IC)**. IC technology has rapidly advanced in recent years so that digital systems can now be manufactured with higher speeds, lower power consumptions, smaller sizes, and lower costs than ever before. This is vividly illustrated by considering that the first electronic digital computer, ENIAC, built in 1946, occupied a floor space of 300 square meters but was only approximately comparable in computational power to the 1977 model of Hewlett-Packard's HP-67 hand-held calculator.

In an integrated circuit, all components are placed on a single, small, thin silicon semiconductor sheet. Referred to as a **chip**, the circuit is *integrated* since the components and wiring are all an integral part of the chip and cannot be separated from each other. The different sizes of integration of IC chips are usually defined in terms of the number of logic gates that they contain. (Logic gates are physical devices that implement logic operations such as AND and OR encountered in our introductory example in Section 1.2.) Customarily, IC chips are classified in one of the four following categories:

1. A *small-scale integration* (SSI) device contains less than 10 gates.

2. A *medium-scale integration* (MSI) device contains 10 to 100 gates.
3. A *large-scale integration* (LSI) device contains 100 to 10,000 gates.

   A *very large-scale integration* (VLSI) device contains more than 10,000 gates.

Integrated-circuit technology is rapidly developing, resulting in devices that can implement a large variety of logic functions. From simple gates through single-chip microprocessors, the digital systems designer has a wealth of building blocks from which to choose.

Two different types of transistors are used to fabricate digital IC chips: the *bipolar* transistor and the *MOS* (metal oxide semiconductor) transistor. A number of different basic gate circuits, referred to as **IC logic families**, are manufactured for each type of transistor. All the circuits of a logic family have similar characteristics and can be interconnected without special interfaces. Logic families based on bipolar transistors include DTL (diode-transistor logic), TTL (transistor-transistor logic), ECL (emitter-coupled logic), EFL (emitter-function logic), and IIL (integrated-injection logic). Logic families based on MOS technology include PMOS (*p*-channel MOS), NMOS (*n*-channel MOS), CMOS (complementary MOS), CMOS/SOS (CMOS by silicon on sapphire), DMOS (double-diffused MOS), and VMOS (V-groove MOS).

Other types of logic families include GaAs MOSFET (gallium arsenide metal semiconductor field-effect transistor) and CCD (charge-coupled device) logic. Many of the IC logic families include a number of variations. For example, each MOS logic family has two different realizations, static and dynamic MOS circuits. The TTL logic family includes variations such as Schottky TTL, low-power TTL, and high-speed TTL. The reason for having so many logic families is that no one basic circuit can be used for all applications.

Most digital systems are implemented using chips of only one logic family. Since logic families have different characteristics, the digital system designer is invariably faced with the difficulty of choosing the appropriate family for a given application. Logic families vary in many aspects: electronic performance (e.g., speed, noise immunity, power consumption, and power-supply requirements), logic capability (given logic functions can be implemented with fewer gates using certain logic families than with others), integration size, cost, and design procedures (design and implementation procedures may vary depending on the chosen logic family).

PROBLEMS

1.1   Analyze the operation of the circuit shown in Figure P1.1.
   (a)   Obtain a truth table.
   (b)   Write the logic function relating L to $A, B,$ and C.

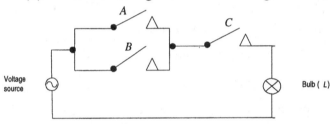

■   **Figure P1.1**

1.2   Repeat Problem 1.1 for the circuit shown in Figure P1.2.

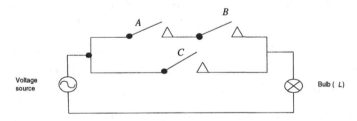

■   **Figure P1.2**

1.3   Designate the two binary levels by V_H and V_L $(V_H > V_L)$ and consider the following truth table:

A	B	F
V_L	V_L	V_H
V_L	V_H	V_H
V_H	V_L	V_H
V_H	V_H	V_L

(a)   Find the logic function F obtained for positive logic.
(b)   Find the logic function F obtained for negative logic.

1.4   Tabulate the truth tables for the following logic functions:
   (a) $L = (A + B) \cdot (A' + B')$ (b) $L = (A + B) \cdot C + (A' + B)C'$

1.5   How many rows will the truth table contain if the number of independent binary variables is (a) 5; (b) 7; (c) 10?

1.6   For a negative pulse, does the leading edge correspond to the negative-going or positive-going transition?

1.7   Repeat Problem 1.6 for a positive pulse.

1.8   Define period and frequency.

1.9   Calculate the pulse repetition rate (PRR) of a periodic train of pulses if the period is (a) 2 microseconds; (b) 4 milliseconds; (c) 0.2 second.

1.10   For a given period, does the duty cycle increase or decrease when the pulse width is increased?

1.11   A periodic pulse train has a pulse width of 15 microseconds and a frequency of 16.67 kHz ($k = 1000$). Find the duty cycle.

1.12   Calculate the duty cycle of a periodic train of pulses whose period is 4 microseconds for a pulse width of (a) 1 microsecond; (b) 2 microseconds; (c) 3 microseconds.

1.13   Verify the duty cycles of the waveforms shown in Figure 1.4.

1.14   The HIGH value of a pulse is 5 and the LOW value is 0. What is the actual change in amplitude during the rise-time interval?

1.15   For the signal shown in Figure P1.15, estimate (a) t_r; (b) t_f; (c)

t_p.

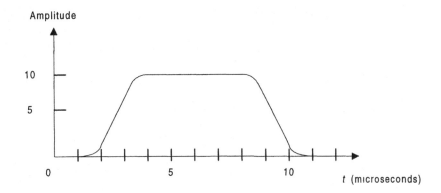

■ **Figure P1.15**

2 Number Systems, Arithmetic Operations and Codes

2.1 INTRODUCTION

Digital systems are characterized by signals restricted to two possible values which are represented by an alphabet of only two characters, commonly denoted by 0 and 1. **Binary** data representations are therefore of fundamental importance to the analysis and design of digital systems.

We begin this chapter by considering numerical data representations. Working with digital systems usually requires familiarity with several number systems:

Decimal
Binary
Octal
Hexadecimal
Binary-coded decimal (BCD)

The binary number system is the most natural to use in a digital system. Sometimes, however, it is more convenient to employ the other number systems, in particular the familiar decimal number system. In such cases, the numbers are manipulated within the number system in use but are, nevertheless, represented as binary numbers within the digital system.

We will introduce arithmetic operations to provide a basis for understanding how digital systems handle numbers in arithmetic computations. We will discuss unsigned and signed binary number representations and consider the four basic arithmetic operations:

addition, subtraction, multiplication, and division.

Binary coding is another important issue. Numbers can be represented in a variety of codes, some of which are more useful than others in particular situations, and can be operated upon. Nevertheless, since arithmetic (numerical) data are not the only data type used, we also discuss codes for representing nonnumerical data, such as letters of the alphabet. Finally, we provide a brief introduction to the fundamentals of protecting against various transmission errors when data are transferred over communication links, and the principles of coding control instructions in a digital system.

2.2 POSITIONAL NUMBER SYSTEMS

A **positional number system** is characterized by a **base** (or **radix**) r, which is an integer greater than 1, and a set of r **digits**. For example, if $r = 10$, the **decimal**, or base 10, number system is comprised of ten digits represented by the characters 0, 1, 2, 3, 4, 5, 6, 7, 8, 9. The **binary** system is base 2 and is comprised of two digits, 0 and 1. An abbreviated word for a binary digit is **bit**. The **octal** system is base 8 and has eight characters, represented by the digits 0 through 7. The **hexadecimal** system is base 16 and is comprised of 16 characters represented by the symbols 0 through 9, A, B, C, D, E, and F. In the hexadecimal system, A stands for decimal 10, B for 11, C for 12, D for 13, E for 14, and F for 15. Another system, the **BCD** number system, will be considered later in this section.

Any integer, $(N)_r$, can be represented by a finite sequence of digits (symbols) concatenated to form a **digit (symbol) string**:

$$(N)_r = (b_{n-1} b_{n-2} \ldots b_1 b_0)_r \tag{2.1}$$

where each b_i is an integer such that $0 \leq b_i \leq r - 1$. The $b_i's$ are the digits of the number N, and n is referred to as the **length** of the string. For example, the digit string $(9A0F)_{16}$ represents a hexadecimal number whose length is $n = 4$ and the value of each symbol cannot exceed $r - 1 = 15$ or be less than zero.

Each digit in the positional notation of an integer N has a special meaning depending on its **position**. In particular, the leftmost digit in the string (2.1), b_{n-1}, is referred to as the **most significant** digit, and the rightmost digit, b_0, is called the **least significant** digit. In the binary number system, these are termed, respectively, the *most significant bit* (abbreviated *msb*) and the *least significant bit* (abbreviated *lsb*). In general, a bit string is referred to as a **word**, but some bit strings have special names: A string of four bits is called a **nibble**, and a string with eight bits is referred to as a **byte**.

The **numerical value** of the integer represented by the digit string (2.1) is given by

$$b_{n-1}r^{n-1} + b_{n-2}r^{n-2} + \ldots + b_1 r^1 + b_0 r^0 = \sum_{i=0}^{n-1} b_i r^i \qquad (2.2)$$

Equation (2.2) is a **weighted sum**. The weights are successive powers of r, and each digit is weighted corresponding to its position in the digit string. The most significant digit weighs most and the least significant digit weighs least. For example, the base 6 number system is comprised of six digits (0 through 5), and the numerical value of the base 6 number 5032 is given by

$$(5032)_6 = 5 \cdot 6^3 + 0 \cdot 6^2 + 3 \cdot 6^1 + 2 \cdot 6^0$$

Thus far we have considered integer numbers only. But what about fractions and mixed integer-fraction numbers? To distinguish between integers and fractions, we use the **radix point** convention. The integer part of a number is located to the left of the radix point, while its fraction appears to the right of the radix point. (If the number is an integer, we can dispose of the radix point.) As an example, consider the digit string $(35.274)_{10}$. The radix (decimal) point separates the integer part of the number (35) from the fractional part (274). The numerical value of this number is given by

$$(35.274)_{10} = 3 \cdot 10^1 + 5 \cdot 10^0 + 2 \cdot 10^{-1} + 7 \cdot 10^{-2} + 4 \cdot 10^{-3} \qquad (2.3)$$

As we see, just as the digits to the left of the radix (decimal) point are the coefficients of a polynomial of *decreasing positive* powers of the base (10 in this example) of the number system, the digits to the right of the radix point are the coefficients of a polynomial of *increasing negative* powers of the base.

We can generalize Equation (2.3) to any number whose integer part consists of n digits and whose fractional part consists of m digits. Consider, therefore, the number represented by the digit string

$$(N)_r = (b_{n-1}b_{n-2}\ldots b_1 b_0 . \, b_{-1}b_{-2}\ldots b_{-m})_r$$

Note that b_{n-1} is still the most significant digit but, now, b_{-m} is the least significant digit. The numerical value of this number is given by

$$(N)_r = b_{n-1}r^{n-1} + b_{n-2}r^{n-2} + \ldots + b_1 r^1 + b_0 r^0 + b_{-1}r^{-1}$$
$$+ b_{-2}r^{-2} + \ldots + b_{-m}r^{-m} \tag{2.4}$$

Or, rewriting Equation (2.4) more compactly, we have

$$(N)_r = \sum_{i=-m}^{n-1} b_i r^i \tag{2.5}$$

In the binary system, $r = 2$ and b_i is either 0 or 1 for every i. We will refer to a binary number whose value is determined in accordance with Equation (2.5) or (2.2) (i.e., by using successive weights that are integer powers of 2) as a **straight** binary number. The use of other kinds of weights is also possible and is introduced in Section 2.5.

Conversion Between Bases

Base r to Base 10 Conversion

The conversion from any base to the base 10 system is easily accomplished by using Equation (2.2) or (2.5). We simply sum the terms of the polynomial representing the base r number to obtain its decimal equivalent.

■ **Example 2.1** Convert $(110101)_2$ to a decimal number.

$$(110101)_2 = 1 \cdot 2^5 + 1 \cdot 2^4 + 0 \cdot 2^3 + 1 \cdot 2^2 + 0 \cdot 2^1 + 1 \cdot 2^0$$
$$= (32)_{10} + (16)_{10} + (0)_{10} + (4)_{10} + (0)_{10} + (1)_{10} = (53)_{10}$$ ■

■ **Example 2.2** Convert $(0.1101)_2$ to a decimal number.

$$(0.1101)_2 = 1 \cdot 2^{-1} + 1 \cdot 2^{-2} + 0 \cdot 2^{-3} + 1 \cdot 2^{-4}$$
$$= (0.5)_{10} + (0.25)_{10} + (0)_{10} + (0.0625)_{10} = (0.8125)_{10}$$ ■

■ **Example 2.3** Convert $(73.452)_8$ to a decimal number.

$$(73.452)_8 = 7 \cdot 8^1 + 3 \cdot 8^0 + 4 \cdot 8^{-1} + 5 \cdot 8^{-2} + 2 \cdot 8^{-3}$$
$$= (56)_{10} + (3)_{10} + (0.5)_{10} + (0.078125)_{10} + (0.0039063)_{10}$$
$$= (59.5820313)_{10}$$ ■

■ **Example 2.4** Convert $(32AF.C4)_{16}$ to a decimal number.

$$(32AF.C4)_{16} = 3 \cdot 16^3 + 2 \cdot 16^2 + A \cdot 16^1 + F \cdot 16^0 + C \cdot 16^{-1} + 4 \cdot 16^{-2}$$
$$= (12288)_{10} + (512)_{10} + (160)_{10} + (15)_{10} + (0.75)_{10}$$
$$+ (0.015625)_{10}$$
$$= (12975.766)_{10}$$ ■

[Recall that $(A)_{16} = (10)_{10}$, $(C)_{16} = (12)_{10}$, and $(F)_{16} = (15)_{10}$.] ■

Base r to Base t Conversion

There are various ways of converting numbers from one base system to another. Probably the most simple and straightforward method is that of **repeated divisions** for integers and **repeated multiplications** for fractionals. To convert a mixed integer-fraction number, we split the process. We convert the integer part first using repeated divisions and then the fractional part by repeated multiplications. We introduce this method in this section and defer to the following section the discussion of conversion between numbers represented in the octal, hexadecimal, and binary systems. As we will see, these particular conversions can be accomplished without resorting to repeated divisions and multiplications.

Let us first consider the conversion of a base r integer $(N)_r$ to its equivalent $(M)_t$ in base t. The conversion is accomplished by repeated divisions by t such that the remainder after each step is one digit of the required base t number, while the quotient is again divided by t. The remainder after the first division by t is the least significant digit of $(M)_t$, while the last remainder is the most significant digit of $(M)_t$. Examples 2.5 and 2.6 illustrate this procedure.

■ **Example 2.5** Convert $(15247)_{10}$ to a hexadecimal number.

$$\frac{15247}{16} = 952 \qquad \text{remainder} = (15)_{10} = (F)_{16}$$

$$\frac{952}{16} = 59 \qquad \text{remainder} = (8)_{10} = (8)_{16}$$

$$\frac{59}{16} = 3 \qquad \text{remainder} = (11)_{10} = (B)_{16}$$

$$\frac{3}{16} = 0 \qquad \text{remainder} = (3)_{10} = (3)_{16}$$

Hence, $(15247)_{10} = (3B8F)_{16}$.   ■

To convert a base r fractional $(.F)_r$ to its equivalent $(.E)_t$ in base t, we repeatedly multiply by t (instead of dividing by t). The integer part after each step is one digit of the required base t fraction, while the remaining fraction is again multiplied by t. The integer resulting from the first multiplication by t is the most significant digit of $(.E)_t$, while the last integer is the least significant digit of $(.E)_t$. Example 2.7 illustrates this procedure.

■ **Example 2.6** Convert $(12A7C)_{16}$ to an octal number. We can simplify the conversion if we first find the decimal equivalent of the hexadecimal number. Doing that, we get $(12A7C)_{16} = (76412)_{10}$. Now converting the decimal number into octal, we have

$$\frac{76412}{8} = 9551 \qquad \text{remainder} = 4$$

$$\frac{9551}{8} = 1193 \qquad \text{remainder} = 7$$

$$\frac{1193}{8} = 149 \qquad \text{remainder} = 1$$

$$\frac{149}{8} = 18 \qquad \text{remainder} = 5$$

$$\frac{18}{8} = 2 \qquad \text{remainder} = 2$$

$$\frac{2}{8} = 0 \qquad \text{remainder} = 2$$

Therefore, $(12A7C)_{16} = (225174)_8$. ■

Example 2.7 shows that the process of converting a fraction might not terminate. In other words, it may not be possible to exactly represent the fraction in the new base with a *finite* number of digits. The number of digits retained after the radix point depends on how *precise* we want the conversion to be. If we terminate the process in Example 2.7 after the first two significant fractional bits, the resulting binary fraction would be $(0.11)_2$, which equals $(0.75)_{10}$. This represents an *error* of about 1.4% from the actual decimal fraction $(0.761)_{10}$. On the other hand, if we terminate the process after ten bits, the resulting binary fraction would be $(0.1100001011)_2$ [$= (0.7607422)_{10}$] and the error would be about 0.03% from the required $(0.761)_{10}$.

■ **Example 2.7** Convert $(0.761)_{10}$ to a binary number.

$$0.761 \times 2 = 1.522 \qquad \text{integer part} = 1$$
$$0.522 \times 2 = 1.044 \qquad \text{integer part} = 1$$
$$0.044 \times 2 = 0.088 \qquad \text{integer part} = 0$$
$$0.088 \times 2 = 0.176 \qquad \text{integer part} = 0$$
$$0.176 \times 2 = 0.352 \qquad \text{integer part} = 0$$
$$0.352 \times 2 = 0.704 \qquad \text{integer part} = 0$$
$$0.704 \times 2 = 1.408 \qquad \text{integer part} = 1$$
$$0.408 \times 2 = 0.816 \qquad \text{integer part} = 0$$
$$0.816 \times 2 = 1.632 \qquad \text{integer part} = 1$$
$$0.632 \times 2 = 1.264 \qquad \text{integer part} = 1$$
$$\vdots \qquad\qquad\qquad \vdots$$

Hence, $(0.761)_{10} = (0.1100001011\ldots)_2$.   ■

Binary, Octal, and Hexadecimal Conversions

Since $8 (= 2^3)$ and $16 (= 2^4)$ are both powers of two, conversions between the binary, octal, and hexadecimal number systems are quite simple. In particular, these relations imply that each octal digit can be binary coded by three bits, as shown in Table 2.1, while each hexadecimal digit can be binary coded by four bits, as shown in Table 2.2. For reference, these tables also include the equivalent decimal numbers.

■ **Table 2.1** Binary-coded octal numbers

Decimal	Octal	Binary
0	0	000
1	1	001
2	2	010
3	3	011
4	4	100
5	5	101
6	6	110
7	7	111

To convert from *binary* to *octal*, we partition the binary number, starting from the radix point, into groups of three bits each. The octal digit corresponding to each group is then assigned using Table 2.1. To convert from *octal* to *binary*, we reverse the process. Each octal digit is assigned a group of three bits using Table 2.1, and the equivalent binary number is obtained by concatenating these groups.

■ **Example 2.8** Convert $(110111010011)_2$ to octal.

Since the given number is an integer, we start from the least significant bit, partition the number into groups of three bits each, and then assign to each group the corresponding octal digit:

$$\underbrace{110}_{6} \quad \underbrace{111}_{7} \quad \underbrace{010}_{2} \quad \underbrace{011}_{3}$$

The equivalent octal number is therefore $(6723)_8$. ■

■ **Table 2.2** Binary-coded hexadecimal numbers

Decimal	Hexadecimal	Binary
0	0	0000
1	1	0001
2	2	0010
3	3	0011
4	4	0100
5	5	0101
6	6	0110
7	7	0111
8	8	1000
9	9	1001
10	A	1010
11	B	1011
12	C	1100
13	D	1101
14	E	1110
15	F	1111

■ **Example 2.9** Convert $(1101000101.01011)_2$ to octal.

Starting from the radix point, we partition the integer and fractional parts and then use Table 2.1 to assign the corresponding octal digits:

$$
\underbrace{1}_{1} \quad \underbrace{101}_{5} \quad \underbrace{000}_{0} \quad \underbrace{101}_{5} \quad . \quad \underbrace{010}_{2} \quad \underbrace{11}_{6}
$$

The equivalent octal number is $(1505.26)_8$.   ■

Notice in Example 2.9 that the leftmost group (of the integer part) contains only one bit and that the rightmost group (of the fractional part) contains only two bits. To qualify for the octal representation, however, each group must contain three bits. To "fill in" the groups, we append *leading* zeros to the integer part and *trailing* zeros to the fractional part. Adding leading or trailing zeros does not change the value of the original number and serves only to facilitate the conversion. In Example 2.9, we appended two leading zeros to the leftmost group to obtain $(001)_2 = (1)_8$ and one trailing zero to the rightmost group to obtain $(110)_2 = (6)_8$.

■ **Example 2.10** Convert $(4736)_8$ to binary.

Using Table 2.1, we have

$$
\underbrace{4}_{100} \quad \underbrace{7}_{111} \quad \underbrace{3}_{011} \quad \underbrace{6}_{110}
$$ ■

Binary to *hexadecimal conversions* are as easily obtained. Starting from the radix point, we partition the binary number into groups of four bits (nibbles) each. Using Table 2.2, each nibble is assigned its corresponding hexadecimal digit. To convert from hexadecimal to binary, we simply reverse the process.

■ **Example 2.11** Convert $(12A7F)_{16}$ to binary.

Using Table 2.2, we have

$$\underline{\lfloor 1 \rfloor} \quad \underline{\lfloor 2 \rfloor} \quad \underline{\lfloor A \rfloor} \quad \underline{\lfloor 7 \rfloor} \quad \underline{\lfloor F \rfloor}$$
$$\;0001 \qquad 0010 \qquad 1010 \qquad 0111 \qquad 1111$$

∎

∎ **Example 2.12** Convert $(111011100010111001)_2$ to hexadecimal.

Grouping the binary number into nibbles and using Table 2.2,

$$\underline{\lfloor 1 \rfloor} \quad \underline{\lfloor 1101 \rfloor} \quad \underline{\lfloor 1100 \rfloor} \quad \underline{\lfloor 0101 \rfloor} \quad \underline{\lfloor 1001 \rfloor}$$
$$\;\;1 \qquad\;\; D \qquad\;\;\; C \qquad\;\;\; 5 \qquad\;\;\; 9$$

Since the leftmost group does not contain four bits, we have to append three zeros to facilitate the conversion into the most significant hexadecimal digit (see discussion following Example 2.9). ∎

∎ **Example 2.13** Convert the binary number of Example 2.12 into octal.

$$\underline{\lfloor 11 \rfloor} \quad \underline{\lfloor 101 \rfloor} \quad \underline{\lfloor 110 \rfloor} \quad \underline{\lfloor 011 \rfloor} \quad \underline{\lfloor 011 \rfloor} \quad \underline{\lfloor 001 \rfloor}$$
$$\;\;3 \qquad\;\; 5 \qquad\;\;\; 6 \qquad\;\;\; 1 \qquad\;\;\; 3 \qquad\;\;\; 1$$

∎

Binary-Coded Decimal (BCD) Representation

The **binary-coded decimal (BCD)** representation is a compromise between the decimal and binary number systems. In BCD, we use four bits (a nibble) to represent each of the decimal digits 0 through 9. To represent a decimal number beyond 9, we need two or more decimal digits and, correspondingly, its BCD representation requires two or more nibbles. Table 2.3 shows the binary and BCD equivalents of decimals 0 through 15. Note that a nibble can actually represent 16 numbers (decimals 0 to 15); however, the BCD system uses only ten of them. It should be understood that BCD is just a code in binary as opposed to a *straight* binary number that represents the actual value.

■ **Table 2.3**   Decimal, binary, and BCD relationships

Decimal	Binary	BCD
0	0000	0000
1	0001	0001
2	0010	0010
3	0011	0011
4	0100	0100
5	0101	0101
6	0110	0110
7	0111	0111
8	1000	1000
9	1001	1001
10	1010	0001 0000
11	1011	0001 0001
12	1100	0001 0010
13	1101	0001 0011
14	1110	0001 0100
15	1111	0001 0101

Decimal-to-BCD Conversion

To find the BCD representation of a decimal number, each decimal digit is independently represented by its corresponding binary nibble.

■ **Example 2.14**   Convert $(3729)_{10}$ to BCD.

Using Table 2.3, each decimal digit is represented by its corresponding BCD nibble, resulting in the BCD number

| 3 | 7 | 2 | 9 |
| 0011 | 0111 | 0010 | 1001 | ■

Note that this conversion is much simpler than the conversion of $(3729)_{10}$ to binary. To convert $(3729)_{10}$ to binary, we have to resort to repeated divisions to get 111010010001. Example 2.14 also points out the inefficiency of the BCD representation. The binary representation of $(3729)_{10}$ requires 12 bits, whereas the BCD equivalent contains 16 bits. Indeed, the larger the decimal number is, the more wasteful its BCD representation becomes. Nevertheless, the convenience of using the BCD system often makes up for its inefficiency.

BCD-to-Decimal Conversion

Converting a BCD number to its decimal equivalent is just as easily accomplished. Starting from the radix point, we partition the binary pattern into groups of four bits each. Each group is then converted to its corresponding decimal number using Table 2.3.

- **Example 2.15** Convert $(1001001101010001)_{BCD}$ to decimal.

  Since the given number is an integer, we start from the least significant bit, partition the number into groups of 4-bit patterns, and then assign to each group the corresponding decimal number:

  $$\underbrace{1001}_{9} \quad \underbrace{0011}_{3} \quad \underbrace{0101}_{5} \quad \underbrace{0001}_{1}$$

  Hence, the equivalent decimal number is 9351. ∎

Note that, if the BCD number in Example 2.15 had been mistakenly interpreted as a binary number, the resulting decimal number would have been 37713. Again, Example 2.15 demonstrates the inefficiency involved in using the BCD system. A 16-bit BCD number can represent up to four decimal digits (decimal numbers up to 9999), while a 16-bit binary number can represent up to five decimal digits (decimal numbers up to 65535).

- **Example 2.16** Convert $(110010011.01101)_{BCD}$ to decimal.

  Grouping into nibbles, we get

$$\underline{|\quad 1\quad|} \quad \underline{|1001|} \quad \underline{|0011|} \quad . \quad \underline{|0110|} \quad \underline{|1\quad|}$$
$$1 \qquad\quad 9 \qquad\quad 3 \qquad\qquad 6 \qquad\quad 8$$

The equivalent decimal number is 193.68. Note that to facilitate the conversion, we have to append leading and trailing zeros to the leftmost and rightmost groups, respectively. (See also Examples 2.9 and 2.12.)

■

BCD-to-Binary Conversion

To convert BCD to binary, and vice versa, it is easiest to make an intermediate conversion to the decimal equivalent.

■ **Example 2.17** Convert $(100101000001)_{BCD}$ to binary.

Converted to decimal, this BCD number becomes $(941)_{10}$. Then, using repeated divisions, we get the binary equivalent (1110101101).

■

■ **Example 2.18** Convert $(101110101101)_2$ to BCD.
Converted to decimal, this binary number becomes $(2989)_{10}$. Then, using Table 2.3, the BCD equivalent is $(0010\ 1001\ 1000\ 1001)$.

■

BCD-to-Hexadecimal Conversion

BCD-to-hexadecimal conversion (as well as to other bases), and vice versa, is best done by first converting either number to decimal.

■ **Example 2.19** Convert $(3F6)_{16}$ to BCD.

Converting $(3F6)_{16}$ to decimal yields $(1014)_{10}$. The equivalent BCD number is then $(1\ 0000\ 0001\ 0100)$. (Notice that we have dropped the first three leading zeros from the leftmost nibble.)

■

■ **Example 2.20** Convert $(1110010111 0101)_{BCD}$ to hexadecimal.

The decimal equivalent of the BCD number is 3975. Converting to hexadecimal, we get F87.

■

2.3 NUMBER REPRESENTATION

Numbers used in scientific calculations are designated by a **sign**, by the **magnitude** of the number, and by the **position** of the radix point. The position of the radix point is required to represent fractions, integers, or mixed integer-fraction numbers. There are two ways of specifying the position of the radix point: by giving it a **fixed** position or by using the **floating-point** representation. In the following, we will limit our discussion to fixed-point number representation. Note, however, that the difference between these two representations has to do with the range of numbers that can be accommodated by a digital system having a given word length.

The two most commonly used positions for the radix point are (1) a radix point to the extreme left, making the number a fraction that is strictly less than one, or (2) a radix point to the extreme right, making the number an integer. In either case, the radix point is not actually present in the digital system; rather, its position is implied by the fact that the number is *predefined* as an integer or as a fraction. Therefore, in integer arithmetic the numbers are lined up to the right as if there were a radix point at the extreme right. In fraction arithmetic, the numbers (regardless of their length) are lined up to the left as if there were a radix point at the extreme left.

Note, however, that these two conventions are essentially equivalent and that it is rather easy to convert between integers and fractions. This conversion process is referred to as **shifting**. Shifting the radix point of a base r number k *places to the left* is equivalent to multiplying the number by r^{-k}, and shifting the radix point k *places to the right* has the effect of multiplying the number by r^k. However, to restore the original value, we must multiply the left-shifted number by r^k and the right-shifted number

by r^{-k}. For example, if we shift the radix point of the binary number
10110.101 three places to the left, we obtain 10.110101 and we must
multiply it by 2^3 to restore the original value; that is,
$10110.101 = 2^3(10.110101)$. If we shift the radix point of the original
number two places to the right, we obtain 1011010.1 and we must multiply
it by 2^{-2} to restore the original value; that is, $10110.101 = 2^{-2}(1011010.1)$.

Thus, any base r, n-digit integer can be considered as a fraction
multiplied by a constant factor r^n, and any m-digit fraction in base r can
be considered as an integer multiplied by a constant factor r^{-m}. Note,
however, that shifting is not limited to integer-fraction conversions. For
example, when we multiply or divide, there are intermediate steps that
require numbers to be shifted to the left or to the right (see Section 2.4).

Sign Notations

To carry out arithmetic operations, we have to specify a number as
either positive or negative. Since the digital system cannot recognize the
conventional sign labels $(+)$ and $(-)$, we have to introduce a *code* to
represent signed numbers. The sign convention commonly employed in
digital systems is to reserve the leftmost digit position for the **sign digit**.

For **positive** fixed-point numbers, the sign digit is set to 0 and the
remaining digits display the *true magnitude* of the number. Let $(N)_r$ be a
number in base r with an integer part of n digits (*including* the sign digit)
and a fractional part of m digits. Then the positive number $(N)_r \geq 0$ is
represented by the digit string

$$(N)_r = (0b_{n-2}\ldots b_1 b_0 . b_{-1} b_{-2}\ldots b_{-m})_r \tag{2.6}$$

and its magnitude equals

$$\sum_{i=-m}^{n-2} b_i r^i \tag{2.7}$$

Three different notations are commonly used to represent a **negative**
fixed-point number: sign-magnitude, $(r-1)'$s complement, and r's
complement. In either representation, *the value of the sign digit is* $(r-1)$.
To emphasize that the leftmost digit *always* represents the sign of the

number, we will set it in **boldface**. To introduce these representations, let $(\overline{N})_r$ be the negative version of the positive number $(N)_r$ defined in Equation (2.6).

Sign-Magnitude Representation

In sign-magnitude representation, the magnitude of the negative number equals that of the positive number and they differ only in the sign digit. Thus

$$(\overline{N})_r = ((r-1)b_{n-2}...b_1b_0 . b_{-1}b_{-2}...b_{-m})_r \tag{2.8}$$

where the digits $b_{n-2},...,b_1,b_0,b_{-1},b_{-2},...,b_{-m}$ represent the true magnitude of the negative number.

- **Example 2.21** The sign-magnitude representation of $(+687)_{10}$, which is a positive number whose magnitude is $(687)_{10}$, is the 4-digit string $(0687)_{10}$. Since $r-1=9$, the sign-magnitude representation of $(-687)_{10}$ is $(9687)_{10}$.
  ∎

- **Example 2.22** The sign-magnitude representation of $(+1101.011)_2$ is $(01101.011)_2$, and the sign-magnitude representation of $(-1101.011)_2$ is $(11101.011)_2$.
  ∎

$(r-1)$'s Complement Representation

The $(r-1)$'s complement of $(N)_r$ is defined as

$$(\overline{N})_r = r^n - r^{-m} - (N)_r \tag{2.9}$$

If we represent $(\overline{N})_r$ by the digit string

$$(\overline{N})_r = ((r-1)\overline{b}_{n-2}...\overline{b}_1\overline{b}_0 . \overline{b}_{-1}\overline{b}_{-2}...\overline{b}_{-m})_r \tag{2.10a}$$

then each digit is given by

$$\overline{b}_i = r-1-b_i \quad \text{for every } i \tag{2.10b}$$

Equation (2.10b) tells us that the $(r-1)$'s complement is obtained by subtracting each digit from $r-1$. Since $r=2$ in the binary system, the application of Equation (2.10b) to generate the *1's complement* is quite simple: *In the given number, the 0's are changed to 1's and the 1's to 0's to obtain its 1's complement.*

■ **Example 2.23**

(a) The 9's complement of $(04857.43)_{10}$ is obtained from Equation (2.9) with $n=5$ and $m=2$:

$$(\overline{N})_r = r^n - r^{-m} - (N)_r = (10^5 - 10^{-2} - 04857.43)_{10}$$
$$= 99999.99 - 04857.43 = 95142.56$$

The reader can verify that the same result can be obtained by using Equation (2.10b).

(b) The 9's complement of $(0.2731)_{10}$ is obtained from Equation (2.9) with $n=1$ and $m=4$:

$$(\overline{N})_r = r^n - r^{-m} - (N)_r = (10 - 10^{-4} - 0.2731)_{10}$$
$$= 9.9999 - 0.2731 = 9.7268$$

(c) The 7's complement of $(0374)_8$ is given by

$$(\overline{N})_r = r^n - r^{-m} - (N)_r = (10^4 - 1 - 0374)_8$$
$$= 7777 - 0374 = 7403$$

(d) The 15's complement of $(04B7)_{16}$ is given by

$$(\overline{N})_r = r^n - r^{-m} - (N)_r = (10^4 - 1 - 04B7)_{16}$$
$$= FFFF - 04B7 = FB48$$

(e) To find the 1's complement of $(01101.101)_2$, we can use Equation (2.9) with $n=5$ and $m=3$. In this case, however, it is simpler to obtain the 1's complement using the procedure outlined following

Equation (2.10b). Hence, if we change all 0's to 1's and all 1's to 0's, we obtain

$$(\overline{N})_2 = (10010.010)$$

r's Complement Representation

The r's complement of $(N)_r$ is defined as

$$(\overline{N})_r = r^n - (N)_r \qquad (2.11)$$

From Equation (2.11), we see that to find the r's complement of a number we leave all the least significant zeros unchanged, subtract the first nonzero least significant digit from r, and then subtract all the other digits from $r-1$. Since $r = 2$ in the binary system, the application of Equation (2.11) to generate the *2's complement* can be stated as follows: *Starting with the least significant bit in the given number, leave all 0's and the first 1 unchanged and then replace 1's by 0's and 0's by 1's in all the remaining bits.*

The r's complement can also be obtained from the $(r-1)$'s complement. By comparing Equations (2.9) and (2.11), we see that adding r^{-m} to the $(r-1)$'s complement results in the r's complement. The following example illustrates these ideas.

■ **Example 2.24**

(a) The 10's complement of $(04857.43)_{10}$ is obtained from Equation (2.11) with $n = 5$:

$$(\overline{N})_r = r^n - (N)_r = (10^5 - 04857.43)_{10}$$
$$= 100,000 - 04857.43 = 95142.57$$

We can also obtain this result from the 9's complement of the given number, which was derived in Example 2.23(a). Since $m = 2$, we have

$$95142.56 + 10^{-2} = 95142.57$$

(b) The 10's complement of $(0.2731)_{10}$ is obtained from Equation

(2.11) with $n = 1$:

$$(\overline{N})_r = r^n - (N)_r = (10 - 0.2731)_{10} = 9.7269$$

Since $m = 4$, this result can also be obtained by adding 10^{-4} to the 9's complement of the given number, which was derived in Example 2.23(b).

(c) The 8's complement of $(0374)_8$ is given by

$$(\overline{N})_r = r^n - (N)_r = (10^4 - 0374)_8$$
$$= (10000 - 0374)_8 = 7404$$

(d) The 16's complement of $(04B7)_{16}$ can be obtained by adding 16^{-0} (since $m = 0$) to the 15's complement derived in Example 2.23(d). Hence

$$\textbf{FB48} + 1 = \textbf{FB49}$$

(e) To find the 2's complement of $(01101.101)_2$, we can use the procedure outlined following Equation (2.11). We leave the least significant bit unchanged (note that there are no least significant zeros) and then replace all 0's by 1's and all 1's by 0's. Hence

$$(\overline{N})_2 = (\textbf{10010.011})$$

■

■ **Example 2.25** Consider a fixed-point number with true magnitude $(9)_{10} = (1001)_2$. We want to represent its positive and negative versions in the three sign notations using eight bits (a byte).

The fixed-point binary byte representation for $(\pm 9)_{10}$ is shown in Table 2.4. The binary representation of $(+9)_{10}$ is the same in all three notations. (Note that we added leading zeros to "fill up" the byte.)

Since $r = 2$, the value of the sign bit of $(-9)_{10}$ in all three notations is $r-1=1$. To obtain the sign-magnitude representation of $(-9)_{10}$, we simply set the sign bit to 1 and leave the remaining bits unchanged [see Equation (2.8)]. To obtain the 1's complement, we use the procedure outlined following Equation (2.10b). The 2's complement is obtained by using the procedure outlined following Equation (2.11).   ■

Table 2.4 Fixed-point representation of $(\pm 9)_{10}$

Fixed-Point Representation	+9	−9
Sign-magnitude	00001001	10001001
1's complement	00001001	11110110
2's complement	00001001	11110111

To conclude this section, note that in either sign notation, taking the negative of a negative number restores the positive number. Thus

$$\overline{[(\overline{N})_r]}_r = (N)_r$$
$(r-1)$'s complement $[(r-1)$'s complement of $(N)_r] = (N)_r$
r's complement$[r$'s complement of $(N)_r] = (N)_r$

Signed Binary Numbers

Consider the sequence of binary numbers 0000 through 1111 that correspond to the decimal numbers 0 through 15. These *same* binary values can be used to represent signed numbers from decimal +7 to −8, as shown in Table 2.5. Several issues arise from the representation of signed binary numbers. To illustrate them, we will use the 4-bit numbers shown in Table 2.5; however, a generalization to any number of bits is straightforward.

Note that all the fixed-point numbers in Table 2.5 have unique representations *except* the zeros in the sign-magnitude and 1's complement notations. We see that *positive* and *negative* zeros have *different* representations in these two notations, while in the 2's complement convention there is a *unique* zero. The reason for this is very simple. A positive zero is represented by 0000 in any of the three notations. (Remember that the leftmost bit is the sign bit.) Since $r = 2$, the sign-magnitude representation of a negative zero is 1000 in accordance with Equation (2.8). To obtain the 1's complement representation of a negative zero, we use the procedure outlined following Equation (2.10b), which results in 1111. The 2's complement representation of a negative zero is obtained using the procedure outlined following Equation (2.11), which results in 0000.

■ **Table 2.5** Four-bit signed binary number representations

Decimal	1's Complement	2's Complement	Sign-Magnitude
+7	0111	0111	0111
+6	0110	0110	0110
+5	0101	0101	0101
+4	0100	0100	0100
+3	0011	0011	0011
+2	0010	0010	0010
+1	0001	0001	0001
+0	**0000**	**0000**	**0000**
−0	**1111**	**0000**	**1000**
−1	1110	1111	1001
−2	1101	1110	1010
−3	1100	1101	1011
−4	1011	1100	1100
−5	1010	1011	1101
−6	1001	1010	1110
−7	1000	1001	1111
−8	Impossible	1000	Impossible

The largest positive 4-bit binary number equals decimal $+7$. No binary equivalent of decimal +8 exists if we use only four bits to represent signed binary numbers. Indeed, if we use the conventional binary notation and assign 1000 to represent 8, then in *signed* binary number notation the 1 in the sign-bit position indicates that this number should be regarded as a negative number.

The smallest negative 4-bit binary number equals decimal (-8) in 2's complement notation. It has no positive counterpart since both +0 and −0 in 2's complement have the same representation (0000) and occupy one of the positive numbers' bit patterns, leaving one fewer for the remaining positive numbers. In the 1's complement notation, (-0) occupies one of the negative numbers' bit patterns, leaving one fewer for the remaining negative numbers. Therefore, the smallest negative 4-bit binary number in 1's complement equals decimal (-7). In sign-magnitude representation, the magnitude of the smallest negative number must equal that of the largest positive number. Since decimal +7 is the largest positive number, the smallest negative sign-magnitude number is decimal (−7).

Range of Signed Binary Numbers

We can generalize the preceding observations to establish the **range** of fixed-point, signed binary integer numbers. For an n-bit number, the largest positive integer that can be represented by n bits *including* the sign bit determines the **upper bound**. Since the largest positive integer in all three representations is $(01...1)_2$, the upper bound equals $(2^{n-1} - 1)_{10}$. The **lower bound** is determined by the smallest negative number. It is $(11...1)_2 = -(2^{n-1} - 1)_{10}$ for sign-magnitude numbers and $(10...0)_2 = -(2^{n-1} - 1)_{10}$ for 1's complement numbers. Since the zero is uniquely represented in 2's complement, the lower bound in this case is $(10...0)_2 = -(2^{n-1})_{10}$.

■ **Example 2.26** What is the range of signed binary numbers represented by a byte of data?

Since $n = 8$, the upper bound is $(2^7 - 1)_{10} = +127$. In sign-magnitude and 1's complement notations, the lower bound is $-(2^7 - 1)_{10} = -127$. In 2's complement, the smallest negative number that can be represented is $-(2^7)_{10} = -128$. ■

Whenever we exceed the range of signed numbers, an **overflow** occurs. As will be seen in Section 2.4, an overflow means that the result of an arithmetic operation appears wrong because we have tried to exceed the allowed range of binary values.

Signed Binary-Coded Numbers

The sign-bit convention is also applicable if we use other binary-coded representations. Consider, for example, the number $(EC)_{16} = (1110\ 1100)_2$. If this number is interpreted as an *unsigned* binary number, it is equivalent to $(236)_{10}$. Interpreted as a *signed* binary number, the binary pattern represents a negative number since the sign bit is 1. Hence, it is equivalent to $(-108)_{10}$ in sign-magnitude representation, to $(-19)_{10}$ in 1's complement notation, and to $(-20)_{10}$ in 2's complement.

We are not as fortunate, however, when we deal with BCD numbers. Consider, for example, decimal $+5$. Its BCD (as well as binary) representation is 0101. The sign-magnitude, 1's, and 2's complements representations of this number are 1101, 1010, and 1011, respectively, and *none* is a *valid* BCD code. To handle signed BCD numbers, we must use either the 10's or 9's complement code. In other words, the hardware generating signed BCD numbers must subtract each BCD digit from either $(10)_{10}$ or $(9)_{10}$. To represent the sign, we must append an extra digit. The coding commonly used for this *sign digit* is $(0000)_2 [= (0)_{10}]$ for positive numbers and for zero, and $(1001)_2 [= (9)_{10}]$ for negative numbers.

Consider, for example, 3-digit BCD numbers with a fourth digit appended to indicate the sign. In sign-magnitude notation, $(+45)_{10}$ is coded as $(0045)_{10} = (\mathbf{0000}\ 0000\ 0100\ 0101)_{BCD}$, while $(-45)_{10}$ is coded as $(\mathbf{1001}\ 1001\ 0100\ 0101)_{BCD}$. The *range* of 3-digit numbers, coded as 2-byte sign-magnitude BCD numbers, is from $(-999)_{10}$ to $(+999)_{10}$.

To find the 10's complement code of $(-45)_{10}$, $(45)_{10}$ must be subtracted from 10^4 [recall Equation (2.11) and remember that the signed numbers in the example have four digits, one of which is the sign digit]. Therefore, $10^4 - 45 = (9955)_{10}$ so that $(-45)_{10}$ is then coded as $(\mathbf{1001}\ 1001\ 0101\ 0101)_{BCD}$. The *range* of numbers in the 10's complement, 2-byte BCD representation is from $(1001\ 0000\ 0000\ 0000)_{BCD} = (-1000)_{10}$ to $(0000\ 1001\ 1001\ 1001)_{BCD} = (+999)_{10}$. To obtain the 9's complement, subtract 1 from the 10's complement representation [see Equation (2.9)] to get $(9954)_{10} = (\mathbf{1001}\ 1001\ 0101\ 0100)_{BCD}$.

2.4 ARITHMETIC OPERATIONS

Data in digital systems are binary numbers and other binary-coded information that are operated on to achieve computational results. Digital system arithmetic differs from real arithmetic in the fundamental issue of number precision. Since the numbers manipulated by a digital system have a finite length, only *finite-precision* computations can be performed. In contrast, real arithmetic may produce results of *arbitrary precision* with no restriction on length. Therefore, digital arithmetic can be considered as approximating real arithmetic, subject to appropriate rounding mechanisms.

The purpose of this section is to present the basic aspects of digital arithmetic. We limit our discussion to fixed-point arithmetic and consider the four basic operations of addition, subtraction, multiplication, and division. Logic circuits that implement some of these operations are introduced in Chapter 5.

Addition

Number **addition** is a *binary operation* on the augend and addend digit pairs, with their column value positions having been aligned. We begin by adding the two least significant digits. If the column sum exceeds the largest symbol value (of the number system used), we carry one base value to the next column and reduce the present column sum by one base value to obtain the particular sum digit.

This procedure applies to any base system, as illustrated in the following examples. In these examples, we assume that the numbers are *unsigned*. We will see later how to handle signed numbers.

- **Example 2.27** Add $(347)_{10}$ and $(679)_{10}$.

  Aligning columns, we have:

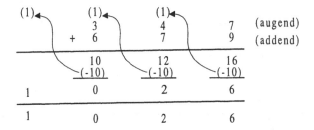

The sum of the least significant digits pair, $7 + 9 = 16$, exceeds the largest symbol value (9). We carry one base value, 10, to the next column and reduce the present column sum by one base value ($16 - 10 = 6$) to obtain the least significant sum digit 6. In the second column we now have $4 + 7 +$ the carried digit $1 = 12$. Again, 10 is carried to the next column while the present column sum is reduced by one base value: $12 - 10 = 2$. In the third column we have a similar situation: $3 + 6 +$ the carried digit $1 = 10$. Again, 10 is carried to the next column, where both the augend and addend digits are 0, while the present column sum is reduced by one base value: $10 - 10 = 0$. Hence, the sum is $(1026)_{10}$.   ■

■ **Example 2.28** Add $(1011)_2$ and $(1101)_2$.
Aligning columns, we have:

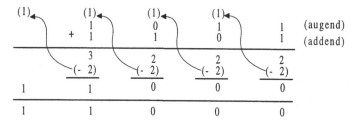

■ **Example 2.29** Add $(CA67)_{16}$ and $(5BC)_{16}$.
Aligning columns, we have:

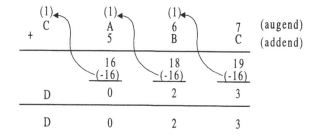

Subtraction

Subtraction is a *binary operation* on the minuend and subtrahend digit pairs, with their column value positions having been aligned. We begin by subtracting the two least significant digits. If the subtrahend exceeds the minuend, we *borrow* one base unit from the next column, add it to the minuend, and carry out the subtraction to obtain the particular difference digit. In the next column subtraction, the borrowed base unit is subtracted from the minuend and then the process continues.

■ **Example 2.30** Subtract $(59)_{10}$ and $(75)_{10}$.
Aligning columns, we have:

$$
\begin{array}{ccc}
 & \nearrow (+10) & \\
7 & 5 & \text{(minuend)} \\
(-1) & & \\
-\quad 5 & 9 & \text{(subtrahend)} \\
\hline
1 & 6 &
\end{array}
$$

■

The same rules apply to any base system but, in general, subtraction is more complicated to implement than addition. To illustrate the additional complications, assume that we want to subtract 75 from 59. This is accomplished by reversing their order, subtracting the number with the smaller magnitude from the larger, and using the sign of the larger number for the sign of the result. That is, $59 - 75 = -(75 - 59) = -16$.

Notice that here, too, we have assumed that the numbers are unsigned. However, in everyday arithmetic calculations we use signed numbers, which are represented in sign-magnitude notation. When we operate on two sign-magnitude numbers, we must first *compare* their signs. If the two signs are the same, we *add* the two magnitudes, and the sign of the result is the same as the signs of the two numbers. But if the two signs differ, we *compare* the two magnitudes, *subtract* the smaller from the larger, and determine the sign of the result by the sign of the larger number. The following example illustrates these processes.

■ **Example 2.31**

$$
\begin{aligned}
+21 + (+45) &= +(21 + 45) = +66 \\
-21 + (-45) &= -(21 + 45) = -66 \\
+21 + (-45) &= -(45 - 21) = -24 \\
-21 + (+45) &= +(45 - 21) = +24
\end{aligned}
$$

■

To implement these processes we are required to make a long sequence of decisions: compare, add, subtract, and determine sign. Although these processes can be implemented in hardware or software, fixed-point addition and subtraction can be simplified if we use the r's or $(r-1)$'s complement representation, as shown in the following section.

Addition and Subtraction Using Complements

r's Complement Arithmetic

If the numbers are represented in r's complement notation, then the key is Equation (2.11), which can be rewritten as

$$(N)_r + (\overline{N})_r = r^n \tag{2.12}$$

and reinterpreted to say that, when two numbers sum to 0 with a carry of 1 (which we *ignore*), the number are **radix complements** of one another.

r's Complement Addition: To add two numbers represented in r's complement, add their corresponding digits (*including* the sign digits) and *ignore* any carry out of the leftmost digit. This carry is referred to as an **end carry**. The sum thus obtained is also represented in r's complement.

The following example illustrates the addition of positive, negative, and oppositely signed binary numbers represented in 2's complement. The leftmost bit represents the sign and is set in **boldface**, and the numbers shown in parentheses are the decimal equivalents of the binary numbers.

■ **Example 2.32**

$$
\begin{array}{rl}
001110 & (+14) \\
+\ 001100 & (+12) \\
\hline
011010 & (+26)
\end{array}
\qquad
\begin{array}{rl}
110010 & (-14) \\
+\ 110100 & (-12) \\
\hline
\text{end carry}\ \nearrow\ 1\ 100110 & (-26)
\end{array}
$$

$$
\begin{array}{rl}
001110 & (+14) \\
+\ 110100 & (-12) \\
\hline
\text{end carry}\ \nearrow\ 1\ 000010 & (+2)
\end{array}
\qquad
\begin{array}{rl}
110010 & (-14) \\
+\ 001100 & (+12) \\
\hline
111110 & (-2)
\end{array}
$$

$$
\begin{array}{rl}
110100 & (-12) \\
+\ 001110 & (+14) \\
\hline
\text{end carry}\ \nearrow\ 1\ 000010 & (+2)
\end{array}
\qquad
\begin{array}{rl}
001100 & (+12) \\
+\ 110010 & (-14) \\
\hline
111110 & (-2)
\end{array}
\qquad ■
$$

The following example demonstrates that addition of signed numbers must be handled carefully.

■ **Example 2.33**

(a) Add $(9D)_{16}$ and $(A4)_{16}$, both represented in 16's complement.

$$
\begin{array}{r}
9D \\
+ \quad \underline{A4} \\
(1) \ 41
\end{array}
$$

Notice that both the augend and addend are negative numbers. (To see this, convert the hexadecimal numbers to binary.) However, their sum turned out to be a positive number, which cannot be correct. The reason for this is that we exceeded the range of negative 2-digit hexadecimal numbers and an *overflow* has occurred. To see this, note that the smallest negative number that can be represented in signed notation by two hexadecimal digits is $-(128)_{10}$. Now, the 16's complements of $(9D)_{16}$ and $(A4)_{16}$ are $(63)_{16}$ and $(5C)_{16}$, respectively, and their decimal equivalents are -99 and -92. Hence, their sum is -191, which is out of range.

(b) Add $(00111001)_2$ and $(01001011)_2$, both represented in 2's complement.

$$
\begin{array}{r}
00111001 \\
+ \quad \underline{01001011} \\
10000100
\end{array}
$$

The augend and addend are positive numbers, but their sum is a negative number, which is obviously incorrect. Once again, we exceeded the range of positive 8-bit binary numbers and an *overflow* has occurred.

■

Since digital arithmetic involves numbers of finite length, an overflow can occur whenever we exceed the range of signed numbers, as shown in Example 2.33. When the signs of the two numbers to be added are opposite, an overflow cannot occur. If the signs of the two numbers to be added are the same, an overflow may occur, depending on the magnitudes of the two numbers, and can be detected by examining the sign of the result. If the sign of the result is opposite the signs of the operands, an overflow has occurred and the result is incorrect. The digital system must, therefore, check for an overflow and provide an indication if it has occurred.

r's **Complement Subtraction:** To subtract two numbers using r's complements, add the minuend to the r's complement of the subtrahend. If there is an end carry out of the sign digit, ignore it. The resulting difference is also represented in r's complement.

The following example illustrates the subtraction of positive, negative, and oppositely signed binary numbers represented in 2's complement. The numbers shown in parentheses are the decimal equivalents of the binary numbers.

■ **Example 2.34**

```
   001110   (+ 14)  ─────▶              001110
 - 001100   (+ 12)  ─────▶           +  110010    (2's complement of subtrahend)
                           end carry ╱▶ 1 000010   (+ 2)


   001100   (+ 12)  ─────▶              001100
 - 001110   (+ 14)  ─────▶           +  110010    (2's complement of subtrahend)
                                        111110    (- 2)


   001110   (+ 14)  ─────▶              001110
 - 110100   (- 12)  ─────▶           +  001100    (2's complement of subtrahend)
                                        011010    (+ 26)


   110010   (- 14)  ─────▶              110010
 - 001100   (+ 12)  ─────▶           +  110100    (2's complement of subtrahend)
                           end carry ╱▶ 1 100110   (- 26)


   110010   (- 14)  ─────▶              110010
 - 110100   (- 12)  ─────▶           +  001100    (2's complement of subtrahend)
                                        111110    (- 2)


   110100   (- 12)  ─────▶              110100
 - 110010   (- 14)  ─────▶           +  001110    (2's complement of subtrahend)
                           end carry ╱▶ 1 000010   (+ 2)
```

Subtraction of oppositely signed numbers may produce a result that exceeds the range of $r's$ complement numbers, so an *overflow* occurs. To illustrate this, consider the subtraction of 10110101 $[(-75)_{10}]$ from 00111001 $[(+57)_{10}]$. If we add the minuend to the 2's complement of the subtrahend, we get 10000100, which is obviously incorrect. The overflow has occurred because the correct result [which is $(132)_{10}$] exceeds the upper bound of the range of positive 8-bit binary numbers $[(127)_{10}]$. As we can see, a subtraction overflow, similarly to addition overflow (see Example 2.33), can be detected by examining the most significant bit of the result.

(r–1)'s Complement Arithmetic

We can also add and subtract numbers represented in $(r-1)$'s complement notation. The key here is Equation (2.9).

(r–1)'s Complement Addition: To add two numbers, add their corresponding digits (*including* the sign digits). If there is a carry out of the leftmost digit, increment the sum by 1 and ignore the end carry. An alternative way of describing this process is the following: If there is an end carry, then (take it and) add it to the result. This process is referred to as **end-around carry**. The obtained sum is also represented in $(r-1)$'s complement.

Example 2.35 illustrates the addition of positive, negative, and oppositely signed binary numbers represented in 1's complement. (The leftmost bit represents the sign and is set in **boldface.**)

(r – 1)'s Complement Subtraction: To subtract in $(r-1)$'s complement, add the $(r-1)$'s complement of the subtrahend to the minuend. If an end carry is obtained, add it to the result (end-around carry). The resultive difference is also represented in $(r-1)$'s complement.

Example 2.36 illustrates the subtraction of positive, negative, and oppositely signed binary numbers represented in 1's complement.

Addition and subtraction with numbers in $(r-1)$'s complement notation may result in an overflow if we exceed the range of $(r-1)$'s complement representation. Similarly to r's complement numbers, an overflow may occur when we add two numbers having the same signs or when we subtract two oppositely signed numbers. In either case, we detect the overflow by monitoring the most significant digit of the result.

■ **Example 2.35**

```
  001110   (+ 14)                        110001   (- 14)
+ 001100   (+ 12)                      + 110011   (- 12)
  ──────                        end carry  ⌐⌐  1 100100
  011010   (+ 26)                        + ────────▶1   (end-around carry)
                                           ──────
                                           100101   (- 26)
```

```
          001110   (+ 14)                  110001   (- 14)
        + 110011   (- 12)                + 001100   (+ 12)
end carry  ⌐⌐ 1 000001                     ──────
        + ───────▶1   (end-around carry)   111101   (- 2)
          ──────
          000010   (+ 2)
```

```
          110011   (- 12)                  0 01100   (+ 12)
       ⌐⌐ + 001110   (+ 14)              + 110001   (- 14)
end carry   1 000001                       ──────
        + ──────▶1   (end-around carry)    111101   (- 2)
          ──────
          000010   (+ 2)
```

■

Multiplication and Division

Let us conclude our introduction to arithmetic operations by considering multiplication and division of fixed-point signed numbers. In the following, we will restrict our discussion to binary numbers. Note, however, that the procedures outlined apply to any base system.

Multiplication is a repeated process of *left-shift* and *add* operations. Starting with the least significant bit (*lsb*) of the multiplier, if it is 1, the multiplicand is copied to form the first partial product; if it is 0, an all-zero sequence forms the first partial product. Next, the second bit of the multiplier is examined. If it is 1, the second partial product is a copy of the multiplier, *shifted* one place to the left relative to the first partial product; if it is 0, an all-zero sequence forms the second partial product. This process continues until all the bits of the multiplier have been exhausted. Then, all the partial products are summed up to form the final product.

The sign of the product is determined from the signs of the multiplicand and multiplier. If they are the same, the product is positive; if not, the product is negative.

The multiplication process is illustrated in Example 2.37. The signs of

■ **Example 2.36**

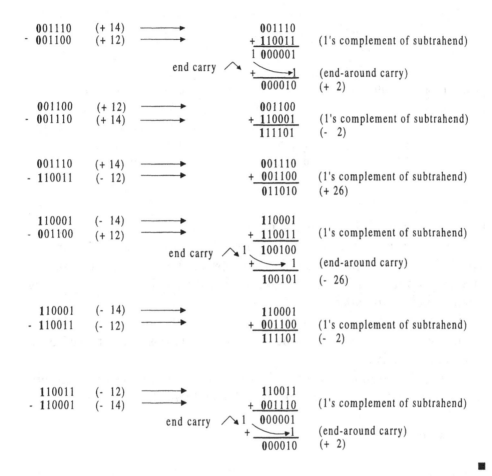

the numbers in this example are assumed to be the same and have been omitted.

Division is a repeated process of *compare, right-shift,* and *subtract* operations. (Notice that subtraction is usually carried out in 2's complement.) Assume that the dividend X is n bits long and that the divisor Y is m bits long. Let X_m denote the m most significant bits of X, $m \leq n$. The division process starts by comparing Y with X_m. If $Y > X_m$, then we compare Y with X_{m+1} and continue doing so until $X_{m+i} \geq Y$. At this point, we enter 1 in the most significant bit (*msb*) of the quotient Q. Y is then right-shifted i places and subtracted from X_{m+i}. The $m+i+1$ *msb* of X is appended to the partial remainder. If the partial remainder is greater than

■ **Example 2.37**

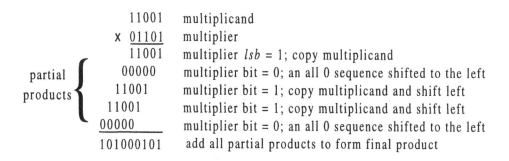

```
                11001    multiplicand
              x 01101    multiplier
                11001    multiplier lsb = 1; copy multiplicand
                00000    multiplier bit = 0; an all 0 sequence shifted to the left
  partial        11001    multiplier bit = 1; copy multiplicand and shift left
  products       11001    multiplier bit = 1; copy multiplicand and shift left
                00000    multiplier bit = 0; an all 0 sequence shifted to the left
              101000101   add all partial products to form final product
```

                                                                      ■

Y, the next *msb* of Q is 1, and Y is shifted right one place and subtracted from the partial remainder. Otherwise, the initial process outlined is repeated, with 0's placed in the appropriate bits of Q. This process continues until all n bits of X have been exhausted, at which point we obtain the last remainder.

     The sign of the quotient is determined from the signs of the dividend and divisor. If they are the same, the quotient is positive; otherwise, the quotient is negative.

     Example 2.38 illustrates the division process. The signs of the number in this example are assumed to be the same and have been omitted.

■ **Example 2.38**

	Dividend	Divisor	Quotient
	X	Y	Q
(a) Compare five most significant bits of X with Y; $X < Y$.	0110110110 / 10001 = 01101		11001
(b) Compare six bits of X; $X > Y$; shift right Y and subtract; enter *msb* of Q = 1.	011011 − 10001 01010		
(c) Partial remainder > Y; shift right Y and subtract; enter 1 in Q.	010100 − 10001 00011		

(d) Partial remainder < Y;   000111
shift right; enter 0 in Q.

(e) Partial remainder < Y;   001111
shift right; enter 0 in Q.

(f) Partial remainder > Y;   011110
shift right Y and subtract; $-\underline{\,10001}$
enter 1 in Q.   01101   (last remainder)
∎

Note that the division operation may result in a quotient with an overflow. An overflow can occur in several situations, but certainly if the divisor is zero.

2.5 BINARY CODES

We have already encountered some numerical codes. We interpreted binary patterns as representing binary, decimal, octal, hexadecimal, BCD, sign-magnitude, and 1's or 2's complement numbers. In this section we consider some additional numerical codes but also non-numerical codes, instruction codes, and error-detecting codes.

Binary codes can be established for *any* set of discrete elements. If the set contains 2^m discrete elements, then *at least m* bits are required to code them. For example, if the set contains 12 elements, then at least four bits are required to encode uniquely each of these elements. Of course, with four bits we can encode up to a maximum of 16 distinct elements. Notice, however, that there is no maximum number of bits that may be used for a binary code. For example, we can use 12 bits to encode the 12 elements. In this code, the first element can be assigned the bit pattern 000000000001, the second element is encoded by 000000000010, and so on, with the twelfth element being encoded by 100000000000.

Decimal Codes

In the BCD code, each decimal digit is encoded as a straight 4-bit binary number. The BCD code is also called the **8421** code because of the successive values of powers of 2 $(2^3, 2^2, 2^1,$ and $2^0)$ that are used to convert the binary bit pattern to its equivalent decimal digit. Since each bit is multiplied by a corresponding weight and the sum of the weighted bits gives the decimal digit, the BCD code is referred to as a **weighted code**.

A large number of codes can be formulated by permuting four bits to represent the ten decimal digits. Some examples of the more common 4-bit decimal codes are shown in Table 2.6. Note that some of the code weights can be positive or negative. Note also that some codes (the 2421 code, for example) use similar weights for different bit positions. Hence, the representations of some decimal digits are not unique in these codes.

- **Table 2.6** Examples of 4-bit decimal codes

Decimal Digit	BCD 8421	2421	84(−2)(−1)			Excess 3
0	0000	0000	00	0	0	0011
1	0001	0001	01	1	1	0100
2	0010	0010	01	1	0	0101
3	0011	0011	01	0	1	0110
4	0100	0100	01	0	0	0111
5	0101	1011	10	1	1	1000
6	0110	1100	10	1	0	1001
7	0111	1101	10	0	1	1010
8	1000	1110	10	0	0	1011
9	1001	1111	11	1	1	1100

To obtain the decimal equivalent of a weighted code word, we multiply each bit by its respective weight. For example, the code word 1101 in the positively weighted code **2421** is equivalent to $1 \cdot 2 + 1 \cdot 4 + 0 \cdot 2 + 1 \cdot 1 = (7)_{10}$, while 1011 in the **84(−2)(−1)** code is equivalent to $1 \cdot 8 + 0 \cdot 4 + 1 \cdot (-2) + 1 \cdot (-1) = (5)_{10}$. Note that some code words may be *valid* under one coding scheme but *invalid* under another. For example, 1101 is valid in the 2421 code, but invalid in the $84(-2)(-1)$ code since $1\cdot8+1\cdot4+0\cdot(-2)+1\cdot(-1) = (11)_{10}$. Some code words, like 1011, are valid under both schemes. Since we are using only ten of the 16 code words possible with four bits, six code words are always invalid with any decimal coding scheme. For the 2421 code, the invalid code words are 0101, 0110, 1000, 1001, and 1010; for the $84(-2)(-1)$ code, the invalid code words are 0001, 0010, 0011, 1100, 1101, and 1110.

Other sets of weights are possible. We can construct 4-bit decimal codes using the weights (6, 3, 2, 1), (6, 4, 2, -3), (4, 2, 2, 1), (7, 4, 2, 1) or (5, 2, 1, 1), to cite but a few examples. Nevertheless, codes do not have to be weighted. Table 2.6 shows an example of an **unweighted code**, the **excess 3** code. Here, each decimal digit is encoded by adding $(3)_{10} = (0011)_2$ to its BCD equivalent. For example, decimal 8 is encoded as 1011, which is decimal 11 $(= 8 + 3)$ in the 8421 code.

One disadvantage of using the BCD code is the difficulty of computing its 9's complement. In contrast, the 2421, $84(-2)(-1)$, and excess 3 codes are a **self-complementing**, meaning that the 9's complement of any decimal digit is obtained by interchanging the 1's and 0's in its binary code. To be self-complementing, the code must be *symmetrical about the center*. For a weighted code this means that the sum of the weights must be equal to 9, while for an unweighted code the binary sum of the code words of the decimal digit and its 9's complement must be equal to the code word of 9. The reader can easily verify that, except for the BCD code, the codes listed in Table 2.6 satisfy these conditions and is encouraged to show that the weighted codes (5, 2, 1, 1), (4, 2, 2, 1), and (6, 4, 2, -3) are also self-complementing. Note, however, that codes consisting of some equal weights (e.g., 2421, 4221) result in nonunique code words. Therefore, we must verify that the generated set of code words is indeed self-complementing (see Problem 2.21).

The self-complementing property, demonstrated in the following example, is very useful when arithmetic operations are performed with binary-coded decimal numbers and subtraction is accomplished by means of the 9's complement.

■ **Example 2.39** Obtain the 9's complement of $(38)_{10}$ in the four coding schemes of Table 2.6. (For simplicity, discard the sign digit.)

BCD: To find the BCD representation of the 9's complement of 38, we have to use Equation (2.9), which yields 61, and then apply Table 2.6. Hence, $(38)_{10} = (0011\ 1000)$ and $(61)_{10} = (0110\ 0001)$.

2421: $(38)_{10} = (0011\ 1110)$. Complementing each bit, we obtain $(1100\ 0001) = (61)_{10}$.

84(−2)(−1): $(38)_{10} = (0101\ 1000)$. Complementing each bit, we obtain $(1010\ 0111) = (61)_{10}$.

Excess 3: $(38)_{10} = (0110\ 1011)$. Complementing each bit, we obtain $(1001 0100)\ = (61)_{10}$.

■

As seen in this example, self-complementing decimal codes enable us to obtain the 9's complement in a simple, straightforward manner, thus relieving us of the necessity to use Equation (2.9). This property enables us to implement a very simple logic circuit that generates 9's complements.

We mentioned earlier that four bits is the *minimum* number of bits required to encode the ten decimal digits. Nevertheless, sometimes it is expedient to use more than four bits. Two such examples are shown in Table 2.7. Both of these codes are useful for error detection, as will be explained in a subsequent secction. The **biquinary** code is a 7-bit weighted code, while the **2-out-of-5** code is a 5-bit unweighted code. Both are characterized by having only two 1's in each code word.

Notice how the biquinary code is constructed: A 01 under the most significant weights $\{5,\ 0\}$ indicates that the decimal digit is in the range from 0 to 4, whereas a 10 under these weights indicates that the decimal digit is in the range from 5 to 9. A 1 under any of the remaining weights indicates the position of the digit within the range; that is, a 1 under the least significant weight $\{0\}$ indicates that the decimal digit is the first within the range, a 1 under the weight $\{1\}$ indicates that the decimal digit is the second within the range, and so on.

The derivation of the 2-out-of-5 code words is quite simple. Each code word must have only two 1 bits with the remaining three bits being 0. Notice, however, that the encoding scheme is not unique. Rather, the code words can be permuted with respect to the decimal digits, with each permutation producing a valid 2-out-of-5 code.

■ **Table 2.7** Examples of decimal codes having more than four bits

Decimal Digit	Biquinary 5043210	2-out-of-5
0	0100001	00011
1	0100010	00101
2	0100100	00110
3	0101000	01001
4	0110000	01010
5	1000001	01100
6	1000010	10001
7	1000100	10010
8	1001000	10100
9	1010000	11000

Gray Code

Many practical applications require codes in which successive code words differ in only one bit. These codes are referred to as *cyclic codes*, of which the **Gray code** is an important member. Table 2.8 shows an example of a 4-bit Gray code. As can be seen, each binary code word differs from either its successor or predecessor by a change of only one bit, from 1 to 0 or from 0 to 1. Note though that the code shown in Table 2.8 is not the only possible cyclic code, and that many other codes with similar characteristics can be devised.

The Gray code is often used in situations where other binary codes might produce erroneous or ambiguous results during transitions from one code word to another in which more than one bit of the code is changing. Consider, for example, using the 8421 code and requiring a transition from 0111 to 1000. Such transition requires that all four bits change simultaneously. Depending on the logic circuit that generates the bits, there may be a significant difference in the transition times of the different bits. For example, if the most significant bit changes faster than the rest, then in going from 0111 to 1000 we will momentarily obtain the (erroneous) code word 1111. The occurrence of this code word, albeit brief, can produce an erroneous operation, whose outcome depends on the relative transition times of the changing bits, in a circuit whose inputs are these bits. A situation like this, in which the outcome depends on the relative transition times of the changing bits, is referred to as a **race**.

■ **Table 2.8**   Four-bit Gray code

Decimal	Gray Code
0	0000
1	0001
2	0011
3	0010
4	0110
5	0111
6	0101
7	0100
8	1100
9	1101
10	1111
11	1110
12	1010
13	1011
14	1001
15	1000

Using the Gray code would eliminate this problem because only one bit changes per transition, and any race between the bits will have no impact on the result.

The Gray code belongs to a class of codes called **reflected codes**. The term reflected code originates from the method used to derive it. As shown in Figure 2.1, an n-bit code is generated by reflecting the $(n-1)$-bit code. At first, the bits 0 and 1 are written in a column [Figure 2.1(a)]. A reflecting line is then drawn below the 1 and the column is reflected about it [Figure 2.1(b)]. Two 0's are added above the reflecting line and two 1's are added below it, as shown in the top part of Figure 2.1(c). Another reflecting line is drawn [Figure 2.1 (c)], and the process continues as shown in Figures 2.1(d) and 2.1(e). This process can be continued to obtain any desired number of combinations.

Alphanumeric Codes

Many applications use data that include not only numbers but also letters of the alphabet, as well as other special characters. Such data are called **alphanumeric** data and may be represented by numeric codes. When numbers are included in the data, they are also represented by *special* codes.

An *alphanumeric character set* typically includes the 26 letters of the alphabet (possibly providing for both upper- and lowercase letters), the ten decimal digits, and a number of special symbols, such as +, =, *, $, ;, and !. The two most common alphanumeric codes are **ASCII** (American Standard Code for Information Interchange), shown in Table 2.9, and **EBCDIC** (Extended Binary-Coded Decimal Interchange Code), shown in Table 2.10. ASCII is a 7-bit code and EBCDIC is an 8-bit code. Since many digital systems handle 8-bit (byte) codes more efficiently, an 8-bit

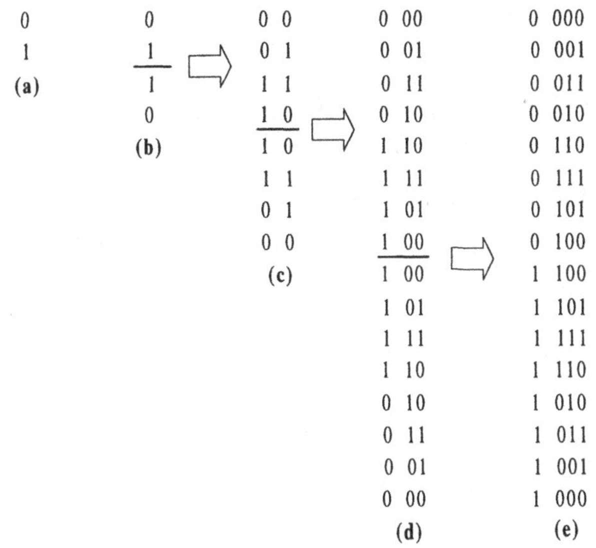

0	0	0 0	0 00	0 000
1	1	0 1	0 01	0 001
(a)	1	1 1	0 11	0 011
	0	1 0	0 10	0 010
	(b)	1 0	1 10	0 110
		1 1	1 11	0 111
		0 1	1 01	0 101
		0 0	1 00	0 100
		(c)	1 00	1 100
			1 01	1 101
			1 11	1 111
			1 10	1 110
			0 10	1 010
			0 11	1 011
			0 01	1 001
			0 00	1 000
			(d)	(e)

■ **Figure 2.1**   Construction of Gray codes

version, called **ASCII-8** (or **USASCII-8**), has also been developed and is shown in Table 2.11.

As we can see from these tables, some special codes are also provided in addition to the alphanumeric character set. These reserved codes are used as communication signals in applications where data transfers take place between digital systems that are connected through communication lines. For example, LF (line feed) and CR (carriage return) are used in conjunction with a printer; BEL is used to activate a bell; ACK (acknowledge), NAK (negative acknowledge), and DLE (data link escape) are exchanged signals over the communication lines.

Any code can be read from the tables in either a binary or hexadecimal format. Hexadecimal is used only for shorthand; the data are handled in binary in the digital system. To find the code corresponding to any character, locate its position in the appropriate table, read across the row to find the most significant hexadecimal digit (binary nibble), and then read its column to find the least significant hexadecimal digit (binary nibble). To find the character associated with a binary or hexadecimal code, we do the opposite. We use the most significant digit to locate the character's row and the least significant digit to locate its column. (Notice that, although ASCII is a 7-bit code, Table 2.9 depicts it in *byte* form by adding a leading zero to the binary representation of the most significant digit.)

- **Example 2.40** Find the hexadecimal and binary codes for lowercase k in the three alphanumeric coding schemes.

  ASCII : Locating k in Table 2.9, its row value is $(6)_{16} = (0110)_2$ and its column value is $(B)_{16} = (1011)_2$. Hence, the ASCII code for k is $(6B)_{16}$ or $(01101011)_2$.

  EBCDIC : Using Table 2.10, the EBCDIC code for k is $(92)_{16}$ or $(10010010)_2$.

  ASCII-8: Using Table 2.11, the ASCII-8 code for k is $(EB)_{16}$ or $(11101011)_2$.   ∎

- **Example 2.41** Find the character represented by $(01011011)_2$ [or equivalently by $(5B)_{16}$] in the three alphanumeric coding schemes.

■ Table 2.9 ASCII code

Least Significant Digit (nibble)

Binary Hex	0000 0	0001 1	0010 2	0011 3	0100 4	0101 5	0110 6	0111 7	1000 8	1001 9	1010 A	1011 B	1100 C	1101 D	1110 E	1111 F
0	NUL	SOH	STX	ETX	EOT	ENQ	ACK	BEL	BS	HT	LF	VT	FF	CR	SO	SI
1	DLE	DC1	DC2	DC3	DC4	NAK	SYN	ETB	CAN	EM	SUB	ESC	FS	GS	RS	US
2	SP	!	"	#	$	%	&	'	(	)	*	+	,	-	.	/
3	0	1	2	3	4	5	6	7	8	9	:	;	<	=	>	?
4	@	A	B	C	D	E	F	G	H	I	J	K	L	M	N	O
5	P	Q	R	S	T	U	V	W	X	Y	Z	[	\	]	^	_
6	`	a	b	c	d	e	f	g	h	i	j	k	l	m	n	o
7	p	q	r	s	t	u	v	w	x	y	z	{	|	}	~	DEL

Most Significant Digit (nibble)

■ Table 2.10 EBCDIC code

Least Significant Digit (nibble)

Binary	0000	0001	0010	0011	0100	0101	0110	0111	1000	1001	1010	1011	1100	1101	1110
Hex	0	1	2	3	4	5	6	7	8	9	A	B	C	D	E
0000 0	NUL	SOH	STX	ETX	PF	HT	LC	DEL			SMM	VT	FF	CR	SO
0001 1	DLE	DC1	DC2	TM	RES	NL	BS	IL	CAN	EM	CC	CU1	IFS	IGS	IRS
0010 2	DS	SOS	FS		BYP	LF	ETB	ESC			SM	CU2		ENQ	ACK
0011 3			SYN		PN	RS	UC	EOT				CU3	DC4	NAK	
0100 4	SP										¢	.	<	(	+
0101 5	&										!	$	*	)	;
0110 6	—	/									¦	,	%	_	>
0111 7											:	#	@	'	=
1000 8		a	b	c	d	e	f	g	h	i					
1001 9		j	k	l	m	n	o	p	q	r					
1010 A		~	s	t	u	v	w	x	y	z					
1011 B															
1100 C	{	A	B	C	D	E	F	G	H	I					
1101 D	}	J	K	L	M	N	O	P	Q	R					
1110 E	\		S	T	U	V	W	X	Y	Z					
1111 F	0	1	2	3	4	5	6	7	8	9					

Most Significant Digit (nibble)

Least Significant Digit (nibble)

Binary		0000	0001	0010	0011	0100	0101	0110	0111	1000	1001	1010	1011	1100	1101	1110	1111
Binary	Hex	0	1	2	3	4	5	6	7	8	9	A	B	C	D	E	F
0000	0	NUL	SOH	STX	ETX	EOT	ENQ	ACK	BEL	BS	HT	LF	VT	FF	CR	SO	SI
0001	1	DLE	DC1	DC2	DC3	DC4	NAK	SYN	ETB	CAN	EM	SUB	ESC	FS	GS	RS	US
0010	2	SP	!	"	#	$	%	&	'	(	)	*	+	,	–	.	/
0011	3	0	1	2	3	4	5	6	7	8	9	:	;	<	=	>	?
0100	4	@	A	B	C	D	E	F	G	H	I	J	K	L	M	N	O
0101	5	P	Q	R	S	T	U	V	W	X	Y	Z	[	\	]	^	–
0110	6	`	a	b	c	d	e	f	g	h	i	j	k	l	m	n	o
0111	7	p	q	r	s	t	u	v	w	x	y	z	{	:	}	~	DEL
1000	8																
1001	9																
1010	A																
1011	B																
1100	C																
1101	D																
1110	E																
1111	F																

Most Significant Digit (nibble)

ASCII : The character is located in Table 2.9 at the intersection of the $(5)_{16}$ row and the $(B)_{16}$ column and is the symbol [.

EBCDIC : Using Table 2.10, the symbol is $.

ASCII-8: Using Table 2.11, the character is ;.   ■

■ **Example 2.42** ASCII and EBCDIC codes for some character strings are shown in Table 2.12. We left spaces between the bytes to make the codes more readable; however, such spaces do not exist in actual data communication. Note that the numerals shown in the table (04, 12, and 84) have codes that are different from their usual BCD representations.   ■

Error-Detection Codes

Digital data are typically prepared on an input device (e.g., a keyboard) and then sent over some communication link and read by a digital system (e.g., a computer). The digital system, in turn, manipulates the data and then may be required to send the results over a communication link to an output device (e.g., a display).

■ **Table 2.12** Character string codes for Example 2.42

Character String	ASCII								EBCDIC							
COMPUTER	43	4F	4D	50	55	54	45	52	C3	D6	D4	D7	E4	E3	C5	D9
$Integer	24	49	6E	74	65	67	65	72	5B	C9	95	A3	85	87	85	99
04/12/84	30	34	2F	31	32	2F	38	34	F0	F4	61	F1	F2	61	F8	F4

The process of transferring digital information is subject to error. Although modern equipment is designed to reduce errors, even relatively infrequent errors can cause serious problems. It is therefore essential to *detect* errors whenever possible. If errors occur infrequently, retransmission of the data can be attempted. However, in modern digital communication systems, this is not considered efficient; retransmission of data for every detected error is liable to be quite costly. For this reason, techniques have been developed to enable the *correction* of detected errors. Employing these techniques usually implies a more complex system. The optimal solution, therefore, lies somewhere between the two approaches - between using a simple and relatively cheap system that does not correct detected errors and using a sophisticated and expensive system capable of correcting many errors. In this section we will consider error-detecting codes only.

An **error-detecting code** is a binary code that detects digital errors during data transmission. One of the most widely used schemes for error detection is the **parity** method. It entails the addition of an extra bit, the **parity bit**, to a binary message that is being transferred from one location to another. In the following discussion, the parity bit is assumed to be appended to the binary message in the rightmost position. (It can also be appended in the leftmost position.)

The parity bit can designate either even or odd parity. In the **even-parity** method, the parity bit is set to 1 so that the *total number* of 1's in the binary message (*including* the parity bit) is an *even number*; otherwise, it is set to 0. Assume, for example, that we want to transmit the EBCDIC character $, whose binary code is 01011011. Since it contains an odd number of 1's, the added parity bit is set to 1 to make the total number of 1's an even number. Therefore, the code that will actually be transmitted is 010110111. For the EBCDIC character C, the code that will be transmitted is 110000110. The **odd-parity** method is used in exactly the same way except that the parity bit is set to 1 so that the *total number* of 1's (*including* the parity bit) is an *odd number*.

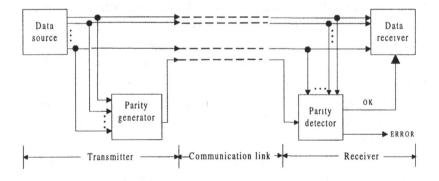

■ **Figure 2.2** Digital data transmission using parity checking

Regardless of whether we use even or odd parity, the parity bit becomes an integral part of the code. Consider Figure 2.2 which shows a block diagram of a digital data-transmission system that uses parity checking. At the sending end, the binary message is supplied first to a **parity generator** that produces the appropriate parity bit. The message, including the parity bit, is then transmitted over the communication link to its destination. At the receiving end, the incoming message is applied to a **parity detector** that verifies the integrity of the parity information. If the number of 1's in the received message matches the generated parity, the message is considered correct and the parity detector issues an OK signal that enables the data receiver. Otherwise, the message is declared incorrect and the parity detector issues an ERROR signal that can be used to request retransmission.

The parity method is limited in that it can only detect errors in the message but cannot detect which bit is in error. Furthermore, regardless of whether we use even or odd parity, the parity method can only detect an *odd* number of errors; an *even* number of errors cannot be detected (see Problem 2.28). More elaborate schemes exist that not only check for multiple errors, but also detect where the errors are and correct them.

Another example of an error-detecting code is the 2-out-of-5 code shown in Table 2.7. This code is constructed such that each decimal digit is represented by a binary code word in which two bits are 1 and all the others are 0. Therefore, if a digit arriving at the receiver has three 1's, an error will be indicated and the digit will be rejected as incorrect. The 2-out-of-5 code is capable of detecting any error involving a change of one or more bits that causes the number of 1's to differ from 2. However, there are occasions when two simultaneous bit changes may occur and a different digit, also having two 1's, is produced and accepted with the code failing to detect the error. For example, if the number 4, represented by the code 01010, is transmitted incorrectly and the least significant bit changes to 1, giving 01011, the error will be immediately detected. However, if the second least significant bit is also affected at the same time, then the resulting code word 01001, representing decimal 3, will be accepted with the errors undetected.

The biquinary code shown in Table 2.7 is yet another example of a code with error-detecting properties. Structured like the 2-out-of-5 code, the biquinary code can be similarly employed for error detection, but will equally fail if two bits change simultaneously and produce a valid code word.

Instruction Codes

Apart from data representation, binary codes are also used to formulate control instructions, referred to as instruction codes. An **instruction code** is a group of bits that tells the digital system to perform a specific operation. The instruction code is usually divided into parts, each having its own particular interpretation. The most basic part of an instruction code is the operation code. An **operation code** (or **op-code**) is a group of bits that define such operations as add, subtract, shift, and complement. It is the instruction's *verb* that tells the system what to do. The other parts of the instruction consist of one or more operands. An **operand** is the *name* used for the object and may be data or an address that tells where the data are.

To execute the instructions formulated for a digital system, they must be binary coded. Consider, for example, the operations *load* and *store*. To load means to copy data from a memory location into a register, and to store means to copy data from a register into a memory location. (Memory and registers will be discussed in Chapter 7.) For example, Motorola's 6800 microprocessor uses the hexadecimal op-code **86** to *LOAD register A with a value.* The same microprocessor uses the hexadecimal op-code B7 to *STORE the contents of register A in a memory location.* (Note that we use the hexadecimal notation only as a convenient shorthand representation; in a digital system, op-codes are represented in binary form.) Instructions containing these op-codes can be, for example, 86 F2 (in hexadecimal), which means *LOAD register A with the value F2,* or B7 00F2, which translates into *STORE the value in register A in the memory location (whose address is) 00F2.* Again, F2 is internally represented in the microprocessor as (11110010).

2.6 INTERPRETATION OF BINARY DATA

It is appropriate to conclude the discussion of binary representations by reflecting on the interpretation of binary data. As we realize by now, binary data can have various meanings depending on the context in which they are used. To illustrate the various interpretations, let us consider the data byte (1001 0101) as an example. Any of the following interpretations is valid for this byte:

1. An unsigned binary number, equivalent to decimal 149, octal 225, or hexadecimal 95.
2. An unsigned binary number (equivalent to decimal 21) with even parity.
3. A sign-magnitude representation of $-(21)_{10}$.
4. A 1's complement representation of $-(106)_{10}$.
5. A 2's complement representation of $-(107)_{10}$.
6. An unsigned BCD number, equivalent to $(95)_{10}$.
7. A sign-magnitude BCD number, equivalent to $-(5)_{10}$.
8. ASCII for the character NAK, with even parity.
9. An op-code in the instruction set of Motorola's 6800 microprocessor that performs the logical AND between the contents of register A and the contents of memory location M.
10. Excess 3-coded decimal 62.
11. Gray-coded decimal 226.

You may wonder how a digital system decides which of these interpretations a given binary data might have. It does not. It is the digital system designer who decides on the meanings of the various binary data within the system.

PROBLEMS

2.1   Convert the following numbers to decimal:

         (a) $(101011)_2$, $(11011.1101)_2$, $(.01101)_2$

    (b) $(476)_8$, $(365.27)_8$, $(.7105)_8$

    (c) $(F23A)_{16}$, $(7A41.C8)_{16}$, $(.FD21)_{16}$

    (d) $(121.201)_3$, $(1302.12)_4$, $(.354)_6$, $(298)_{12}$

2.2   Convert the decimal number 427.5 to base 3, base 6, base 8, and base 16.

2.3   Convert the following numbers from the given base to the bases indicated:

    (a) $(11010.110)_2$ to decimal, octal, and hexadecimal

    (b) $(3DA9.DD)_{16}$ to decimal, octal, and binary

    (c) $(478.12)_{10}$ to binary, octal, and hexadecimal

    (d) $(372.16)_8$ to binary, decimal, and hexadecimal

2.4   Convert the following numbers to BCD:

    (a) $(47.28)_{10}$

    (b) $(362)_8$

    (c) $(7AF.2C)_{16}$

2.5   Convert the BCD code (11101011001.10011) to binary, decimal, and hexadecimal.

2.6   Convert the following numbers to BCD:

    (a) $(10110111011)_2$ (b) $(1986)_{10}$

    (c) $(72AF1)_{16}$

2.7   Convert $(11010111011)_2$ to octal and hexadecimal.

2.8   Obtain the 16's and 15's complements of the hexadecimal numbers 027A, F8B6, and 07AF.4C.

2.9   Obtain the 10's and 9's complements of the decimal numbers 01398, 9239.72, and 9471.

2.10   Obtain the 8's and 7's complements of the octal numbers 0361, 043.216, and 756.

2.11   Obtain the 2's and 1's complements of the binary numbers 01101101, 1010111.1100, and 10101101011.

2.12 Obtain the 10's complement of $((11)197)_{12}$ and $(0BA45)_{16}$, which are represented in sign-magnitude notation.

2.13 Use the 10's and 9's complements to perform the indicated subtractions with decimal numbers (assume positive numbers):

   (a) $375 - 42$ **(b)** $725 - 1956$
   (c) $7240 - 471$

2.14 Use the 2's and 1's complements to perform the indicated subtractions with binary numbers (assume positive numbers):
   (a) $11010 - 1101$ **(b)** $101 - 11001$
   (c) $11010 - 10000$

2.15 Add the following positive numbers:
   (a) $(45B7)_{16} + (15)_{16}$

   (b) $(10111)_2 + (101)_2$

   (c) $(925)_{10} + (634)_8$

2.16 The positive and negative versions of a signed fixed-point number with true magnitude $A = (547)_{10}$ are stored in 16-bit words. List their sign-magnitude, 1's complement, and 2's complement binary fixed-point representations.

2.17 What are the integer range representations in sign-magnitude, 1's complement, and 2's complement in a fixed-point binary arithmetic system consisting of **(a)** 16 bits, and **(b)** 64 bits?

2.18 Obtain the indicated multiplication of the following positive binary numbers: **(a)** 110101×1011 **(b)** 1101101×1101

2.19 Obtain the indicated division of the following positive binary numbers: **(a)** $11010101 / 11011$ **(b)** $010111101 / 1001$

2.20 Represent the decimal number 729 in the BCD, excess 3, and 2421 codes.

2.21 Generate the binary codes for the decimal digits using the weights (3, 2, 3, 1). List a set of self-complementing code words and a set that is not self-complementing.

2.22 Construct self-complementing binary codes for the decimal digits using the following weighted codes:

(a) $(6, 4, 2, -3)$ **(b)** $(4, 2, 2, 1)$
(c) $(5, 2, 1, 1)$

2.23   Obtain the 9's complement of $(+59)_{10}$ in the three coding schemes of Problem 2.22. (Do not code the sign digit.)

2.24   Obtain code words for the base 12 digits using the weights $(5,4,2,1)$.

2.25   Show two alternatives for a Gray code other than the one listed in Table 2.8.

2.26   Find the hexadecimal and binary codes for the following:

(a)   ASCII U   **(b)** EBCDIC $

2.27   Write the ASCII and EBCDIC hexadecimal codes for the following character strings:
(a)   base -2   **(b)** Architecture
(c)   03 / 24 / 86

2.28   Why is it impossible to detect an even number of errors when using either even-parity or odd-parity code?

2.29   Determine the odd-parity bit appended to a message consisting of the ten decimal digits in the $(6, 4, 2, -3)$ code. Assume that each digit is sent separately.

2.30   Generate odd- and even-parity codes for the following messages:
(a)   Binary (10111010)
(b)   Decimal 8 in the $(8, 4, -2, -1)$ code
(c)   ASCII %
(d)   Hexadecimal A4F, with each character sent separately.

2.31   Write your first name and last name in ASCII, appending to each letter an even-parity bit in the most significant bit position.

3   Boolean Switching Algebra

3.1 INTRODUCTION

Digital systems manipulate *discrete elements* of information represented by physical quantities called *signals*. The signals are usually restricted to two possible values and are said to be *binary*. As we saw in Chapter 2, two levels are enough because any desired message, no matter how complex, can be coded in the *binary system* using strings of the symbols 0 and 1. Therefore, two-state devices like switches, diodes, magnetic cores, and transistors can be used to process information because the two states (on versus off, conducting versus nonconducting, positively magnetized versus negatively magnetized, and high potential versus low potential) can represent the two binary symbols 0 and 1.

At the most basic level, digital systems are really little more than a maze of two-state devices that reduce even the most complex information to simple patterns of 1's and 0's. From this perspective, Claude Shannon showed in 1938 how to analyze and design digital circuits by using algebraic expressions involving variables that can take on only two values. He based his approach on the concept of *Boolean algebra*, developed originally by the nineteenth-century mathematician George Boole. Boole was interested in discovering the laws governing the working of the human mind and what Shannon observed was that these same laws govern the behavior of digital circuits. The guiding principle behind Shannon's approach is: *Reduce the problem of digital circuit design and analysis to the study of expressions in a Boolean algebra.*

We begin our discussion of Shannon's system by listing in the next section a basic formulation of the fundamental axioms of Boolean algebra. Then in Section 3.3 we discuss some immediate and useful consequences of these axioms. But the reader should be forewarned: The laws of ordinary algebra we learned in school and the laws of Boolean algebra we list here are similar in some, but dramatically different in other respects. The reader, therefore, will need to observe these differences and commit them to memory.

Basically, we can use Boolean algebra as a tool to analyze and design digital circuits because the set of all n-variable Boolean functions form a Boolean algebra. Moreover, for this Boolean algebra, we can derive (in Section 3.6) one major theoretical result: Every n-variable Boolean function can be expressed in terms of three operators (AND, OR, and NOT) in either one of two canonical forms. These canonical forms correspond in a one-to-one fashion to digital circuit implementations for the function, where the circuit building blocks are AND, OR, and NOT gates. But then we show in Section 3.5 that there are also other types of operators that can be used to express any Boolean function. Consequently, we can implement a digital circuit for the function using other types of gates. We introduce in Section 3.7 the standard gates that are used extensively in digital systems design.

But let us hasten to say that Boolean functions are not always implemented in their canonical forms. In fact, this is seldom the case. The existence of a canonical form just guarantees that we *can* implement all functions with AND, OR, and NOT gates, but the canonical form rarely is the *best* implementation of a function. Instead, we can use Boolean algebra laws to convert an algebraic expression to a simpler equivalent form in order to construct a simpler circuit. We discuss these simplifications in Chapter 4.

3.2 AXIOMS OF BOOLEAN ALGEBRA

We will use *capital* letters to denote sets of well-defined objects and *lowercase* letters to designate the objects. The phrase "is an element (object) of" is denoted by the symbol $\in$. Thus we write $a \in A$ for "a is an element of the set A."

A **Boolean algebra** is an algebraic structure consisting of a set of elements B, together with two *binary* operations $\{+\}$ and $\{\cdot\}$ and a *unary* operation $\{'\}$, such that the following axioms hold:

1. The set B contains at least two elements a, b such that $a \neq b$.

2. **Closure properties**: For every a, $b \in B$
    (i) $a + b \in B$
    (ii) $a \cdot b \in B$

3. **Commutative laws:** For every a, $b \in B$
    (i) $a + b = b + a$
    (ii) $a \cdot b = b \cdot a$

4. **Existence of identities:**
    (i) There exists an identity element with respect to $\{+\}$, designated by 0, such that $a + 0 = a$, for every $a \in B$.
    (ii) There exists an identity element with respect to $\{\cdot\}$, designated by 1, such that $a \cdot 1 = a$, for every $a \in B$.

5. **Distributive laws:** For every a, b, $c \in B$
    (i) $a + (b \cdot c) = (a + b) \cdot (a + c)$
    (ii) $a \cdot (b + c) = a \cdot b + a \cdot c$

6. **Existence of complement:** For each $a \in B$, there exists an element $a' \in B$ (the *complement* of a) such that
    (i) $a + a' = 1$
    (ii) $a \cdot a' = 0$

It is known that this list of axioms is both *consistent* and *independent*; that is, none of the postulates contradicts any other axiom and neither can any postulate be proved from other axioms in the list.

The notations $\{+\}$ and $\{\cdot\}$ that we have adopted to designate the binary operations are those frequently used by digital systems designers. The terms *sum*, *join*, and *disjunction* are often used interchangeably in Boolean algebra to denote the $\{+\}$ operation, as are *product*, *meet*, and *conjunction* to denote the $\{\cdot\}$ operation.

Operator priorities in Boolean algebra are such that an expression inside parentheses must be evaluated before all other operations; the next operation that holds precedence is the complement; then follows the $\{\cdot\}$

and, finally, the $\{+\}$. For example, to evaluate the expression $(a+b)'$, we must first evaluate $a+b$ and then complement the result. Likewise, the expression $a' \cdot b'$ is evaluated by first complementing a and b and then applying the $\{\cdot\}$ operation. In the following, when parentheses are not used, it will be understood that $\{\cdot\}$ operations are performed before $\{+\}$ operations. Also, the $\{\cdot\}$ operation will be omitted for brevity so that ab will be written instead of $a \cdot b$.

Note that the axioms of Boolean algebra are arranged in pairs. Each statement can be obtained from the other by interchanging the $\{+\}$ and $\{\cdot\}$ operations and the identity elements 0 and 1. This is called the **principle of duality** and is illustrated in the following example.

■ **Example 3.1** (a) Axiom 5(i) can be obtained from axiom 5(ii), and vice versa, as follows:

$$a+(b \cdot c) = (a+b) \cdot (a+c)$$
$$\updownarrow \ \updownarrow \qquad \updownarrow \ \updownarrow \ \updownarrow$$
$$a \cdot (b+c) = (a \cdot b) + (a \cdot c)$$

(b) Axiom 6(i) can be obtained from axiom 6(ii), and vice versa, as follows:

$$a+a' = 1$$
$$\updownarrow \qquad \updownarrow$$
$$a \cdot a' = 0$$

■

This important property of Boolean algebra holds for any algebraic expression (theorem) deducible from the axioms. Such an algebraic expression remains valid when the operators and the identity elements are interchanged. In general, every theorem that can be proved for Boolean algebra has a *dual* statement that is also true.

3.3 BASIC THEOREMS OF BOOLEAN ALGEBRA

The purpose of this section is to introduce some of the most basic relationships in Boolean algebra. We will use these results frequently in the analysis and design of digital circuits, and it is imperative that the reader become familiar with them. The proofs for some theorems are shown in detail to illustrate the use of the axioms and the way in which algebraic expressions can be manipulated. But proofs of Boolean algebra laws will not be our main emphasis; *utility* is our interest, not theory alone. Therefore, we are content to leave the proofs of other results as exercises to the reader.

Throughout this section, we will state the dual results in one theorem but only prove one result. The proof of the other result can be obtained by using the principle of duality.

□ **Theorem 3.1** *Idempotent Laws* For every $a \in B$, $a + a = a$ and $aa = a$.

Proof

$$
\begin{aligned}
a + a &= (a + a)1 & &\text{Axiom 4(ii)} \\
&= (a + a)(a + a') & &\text{Axiom 6(i)} \\
&= a + aa' & &\text{Axiom 5(i)} \\
&= a + 0 & &\text{Axiom 6(ii)} \\
&= a & &\text{Axiom 4(i)}
\end{aligned}
$$

□

□ **Theorem 3.2** For every $a \in B$, $a + 1 = 1$ and $a0 = 0$.

Proof

$$
\begin{aligned}
a + 1 &= (a + 1)1 & &\text{Axiom 4(ii)} \\
&= (a + 1)(a + a') & &\text{Axiom 6(i)} \\
&= a + 1a' & &\text{Axiom 5(i)} \\
&= a + a' & &\text{Axioms 4(ii) and 2(ii)} \\
&= 1 & &\text{Axiom 6(i)}
\end{aligned}
$$

□

□ **Theorem 3.3** The elements 0 and 1 are unique.   □

□ **Theorem 3.4** The elements 0 and 1 are distinct; namely, $1' = 0$ and $0' = 1$.

□

□ **Theorem 3.5** For every $a \in B$, there exists a unique complement a'.

□

□ **Theorem 3.6** *Involution Law* For every $a \in B$, $(a')' = a$.   □

□ **Theorem 3.7** *Absorption Laws* Let $a, b \in B$, then $a + ab = a$ and $a(a + b) = a$.

Proof

$$
\begin{aligned}
a + ab &= a1 + ab &&\text{Axiom 4(ii)} \\
&= a1 &&\text{Theorem 3.2} \\
&= a &&\text{Axiom 4(ii)}
\end{aligned}
$$
□

□ **Theorem 3.8** Boolean algebra is *associative* under the operations of $\{+\}$ and $\{\cdot\}$; that is, for all $a, b, c \in B$

$$a + (b + c) = (a + b) + c$$

and

$$a(bc) = (ab)c$$

Proof Let

$$A = [(a + b) + c][a + (b + c)]$$

Then

$$
\begin{aligned}
A &= [(a + b) + c]a + [(a + b) + c](b + c) \\
&= [(a + b)a + ca] + [(a + b) + c](b + c) \\
&= a + [(a + b) + c](b + c) \\
&= a + [(a + b) + c]b + [(a + b) + c]c \\
&= a + (b + c)
\end{aligned}
$$

But, also,

$$
\begin{aligned}
A &= (a + b)[a + (b + c)] + c[(a + (b + c)] \\
&= (a + b)[a + (b + c)] + c \\
&= a[a + (b + c)] + b[a + (b + c)] + c \\
&= (a + b) + c
\end{aligned}
$$

Thus, $a + (b + c) = (a + b) + c$.
□

□ **Theorem 3.9** *DeMorgan's Theorems* For any $a, b \in B$, $(a + b)' =$

$a'b'$ and $(ab)' = a' + b'$.   □

Table 3.1 lists some of the axioms and theorems discussed previously, as well as other important identities of Boolean algebra. They are listed in *dual pairs* designated by part (a) and part (b). Notice that some of the identities involve two or three variables, while others involve an arbitrary number (n) of variables.

We can use the identities to prove other identities or to manipulate Boolean algebraic expressions into some other forms, as illustrated in the following example. At each stage in this example, reference is made to Table 3.1 by noting the number of the identity used. Any proven identity renders its dual true by the principle of duality.

■ **Example 3.2** **(a)** Prove identity (7b): $xy' + y = x + y$.

$$xy' + y = (x + y)(y + y') \qquad\qquad\qquad (2b)$$
$$= (x + y)1 \qquad\qquad\qquad\qquad\quad (3a)$$
$$= x + y \qquad\qquad\qquad\qquad\qquad (1b)$$

(b) Prove identity (9b): $xy + x'z + yz = xy + x'z$.

$$xy + x'z + yz = xy + x'z + yz(x + x') \qquad\qquad (3a), (1b)$$
$$= xy + x'z + xyz + x'yz$$
$$= xy(1 + z) + x'z(1 + y) \qquad\qquad\qquad (2a)$$
$$= xy1 + x'z1 \qquad\qquad\qquad\qquad\quad (5a)$$
$$= xy + x'z \qquad\qquad\qquad\qquad\qquad (1b)$$

(c) Show that $x'y'z + yz + xz = z$.

$$x'y'z + yz + xz = (x'y' + y + x)z$$
$$= (x' + y + x)z \qquad\qquad\qquad (7b)$$
$$= (1 + y)z \qquad\qquad\qquad\quad (3a)$$
$$= 1z \qquad\qquad\qquad\qquad\quad (5a)$$
$$= z \qquad\qquad\qquad\qquad\qquad (1b)$$

■ **Table 3.1** Some important identities of Boolean algebra

(1a) $x + 0 = x$	(1b) $x \cdot 1 = x$	Axiom 4
(2a) $x(y + z) = xy + xz$	(2b) $x + yz = (x + y)(x + z)$	Distributivity
(3a) $x + x' = 1$	(3b) $xx' = 0$	Axiom 6
(4a) $x + x = x$	(4b) $xx = x$	Idempotency
(5a) $x + 1 = 1$	(5b) $x \cdot 0 = 0$	Theorem 3.2
(6a) $x + xy = x$	(6b) $x(x + y) = x$	Absorption
(7a) $(x + y')y = xy$	(7b) $xy' + y = x + y$	
(8a) $(x + y)(x + y') = x$	(8b) $xy + xy' = x$	Logical adjacency
(9a) $(x + y)(x' + z)(y + z) = (x + y)(x' + z)$	(9b) $xy + x'z + yz = xy + x'z$	Consensus
(10a) $(x_1 + x_2 + \dots + x_n)' = x_1' x_2' \cdots x_n'$		DeMorgan's
(10b) $(x_1 x_2 \cdots x_n)' = x_1' + x_2' + \dots x_n'$		theorems
(11a) $f(x_1, x_2, \dots, x_n) = x_i f(x_1, x_2, \dots, x_{i-1}, 1, x_{i+1}, \dots, x_n)$ $+ x_i' \, f(x_1, x_2, \dots, x_{i-1}, 0, x_{i+1}, \dots, x_n)$		Shannon's expansion
(11b) $f(x_1, x_2, \dots, x_n) = [x_i + f(x_1, x_2, \dots, x_{i-1}, 0, x_{i+1}, \dots, x_n)]$ $\cdot [x_i' + f(x_1, x_2, \dots, x_{i-1}, 1, x_{i+1}, \dots, x_n)]$		theorems

(d) Show that $(x + y)[x'(y' + z')]' + x'y' + x'z' = 1$.

$$(x + y)[x'(y' + z')]' + x'y' + x'z'$$
$$= (x + y)[x + (y' + z')'] + x'y' + x'z' \tag{10b}$$
$$= (x + y)(x + yz) + x' y' + x' z' \tag{10a}$$
$$= x + xy + xyz + yz + x'y' + x'z'$$
$$= x(1 + y + yz) + yz + x'y' + x'z' \tag{2a}$$
$$= x + yz + x'y' + x'z' \tag{5a, 1b}$$
$$= x + x'(y' + z') + yz \tag{2a}$$
$$= x + y' + z' + yz \tag{7b}$$
$$= x + y' + z + z' \tag{7b}$$
$$= x + y' + 1 \tag{3a}$$
$$= 1 \tag{5a}$$

(e) Show that $w'x + wxz + wx'yz' + xy = x(w' + z) + wyz'$.

$$w'x + wxz + wx'yz' + xy = x(w' + wz) + y(x + x'wz') \qquad (2a)$$
$$= x(w' + z) + y(x + wz') \qquad (7b)$$
$$= w'x + xz + xy + wyz'$$
$$= w'x + xz + xy(w + w') + wyz' \qquad (3a), (1b)$$
$$= w'x + xz + wxy + w'xy + wyz'$$
$$= w'x(1 + y) + xz + wxy + wyz' \qquad (2a)$$
$$= w'x + xz + wxy + wyz' \qquad (5a), (1b)$$
$$= w'x + xz + wyz' \qquad (9b)$$
$$= x(w' + z) + wyz' \qquad (2a)$$

∎

3.4 SWITCHING ALGEBRA

The simplest example of a Boolean algebra is one in which the domain B is just the set $\{0,1\}$. Since each element of B can only assume either one of two values, 0 or 1, we refer to this two-valued Boolean algebra as **switching algebra.**

The operations of switching algebra are defined in Table 3.2, and an alternative representation in *truth table* form (see Section 1.2) is given in Table 3.3. In the context of digital systems, the $\{+\}$, $\{\cdot\}$, and $\{'\}$ operations are commonly referred to as OR, AND, and NOT, respectively. Thus, for example, $x\text{AND}y$ and xy represent the same expression and can be used interchangeably. As we will see later, each of these operations can be implemented by a physical device known as a *gate*. By using AND, OR, and NOT gates, we can implement Boolean expressions in the form of *logic circuits*. Other types of gates exist, and they too can be used interchangeably to denote a variety of logical operations. These will be considered in Section 3.5 and 3.7.

We can verify that switching algebra satisfies all the axioms and relationships of Boolean algebra by using algebraic manipulations, as in Example 3.2. Alternatively, we can construct a truth table and evaluate both sides of the identity for all possible combinations of values of the variables. We illustrate this latter approach, known as **perfect induction,** in Example 3.3.

■ **Table 3.2** Operator definitions in Boolean algebra over two values

OR Operator			AND operator			NOT operator	
+	0	1	·	0	1	′	
0	0	1	0	0	0	0	1
1	1	1	1	0	1	1	0

■ **Table 3.3** Alternative representation of OR, AND, and NOT

	OR			AND		NOT	
x	y	$x+y$	x	y	xy	x	x'
0	0	0	0	0	0	0	1
0	1	1	0	1	0	1	0
1	0	1	1	0	0		
1	1	1	1	1	1		

■ **Example 3.3** Prove identity 2(b) of Table 3.1: $x + yz = (x + y)(x + z)$. Since we have three variables, each of which can assume one of two values (0 or 1), the number of possible combinations is $2^3 = 8$. The truth table has, therefore, eight rows, as shown in Table 3.4. For each combination of the variables, the truth table shows the corresponding values of the left side and right side of the identity. To evaluate these values, we use the operator definitions of Table 3.3. By comparing the columns for $x + yz$ and $(x + y)(x + z)$, we see that they have the same value for each combination. ■

Boolean algebra resembles ordinary algebra in some respects, but as we said in Section 3.1, there are some differences between them. In particular, axioms 5(i) and 6 (the distributivity of $\{+\}$ over $\{\cdot\}$ and the definition of a complement, respectively) do not hold for ordinary algebra. Besides, some theorems of ordinary algebra are not true for Boolean algebra. For example, the cancellation law for addition in ordinary algebra is

$$\text{If } x + y = x + z, \quad \text{then } y = z$$

But this cancellation law is not true for Boolean algebra. To see this, let $x = 1$, $y = 0$, and $z = 1$. Then $x + y = 1 + 0 = 1$ and $x + z = 1 + 1 = 1$, but $y \neq z$. On the other hand, the statement

$$\text{If } y = z, \quad \text{then } x + y = x + z$$

is true for both algebras, and its verification is left as an exercise for the reader.

■ **Table 3.4** Truth table for Example 3.3

x	y	z	yz	$x+yz$	$x+y$	$x+z$	$(x+y)(x+z)$
0	0	0	0	0	0	0	0
0	0	1	0	0	0	1	0
0	1	0	0	0	1	0	0
0	1	1	1	1	1	1	1
1	0	0	0	1	1	1	1
1	0	1	0	1	1	1	1
1	1	0	0	1	1	1	1
1	1	1	1	1	1	1	1

3.5 BOOLEAN SWITCHING FUNCTIONS

The concept of a polynomial is undoubtedly well known to those familiar with ordinary algebra, but Boolean (algebraic) expressions, representing the analogous concept for Boolean algebra, may not be as familiar. To introduce the concept of a Boolean expression, let $\{x_{n-1}, \ldots, x_1, x_0\}$ be a set of symbols, called *variables*. Then a **Boolean (algebraic) expression** in $\{x_{n-1}, \ldots, x_1, x_0\}$ is an expression formed with the variables, the constants 0 and 1, the binary operators AND and OR, and the unary operator NOT. In other words, we can obtain Boolean expressions by applying the following two rules any finite number of times:

1.   The constants 0 and 1 and the Boolean variables $\{x_{n-1}, \ldots, x_1, x_0\}$ are Boolean expressions in $\{x_{n-1}, \ldots, x_1, x_0\}$.
2.   Moreover, if E_1 and E_2 are Boolean expressions in $\{x_{n-1}, \ldots, x_1, x_0\}$, then so are E_1', E_2', $E_1 + E_2$, and $E_1 E_2$.

For example, the following expressions

$$E_1 = [(x_1 x_2)' + (x_1 + x_3)]'$$
$$E_2 = x_1 x_2 + x_1' x_3 + x_1 x_3'$$
$$E_3 = x_1' x_2 (x_3 + x_1 x_2') + x_2' x_3$$

are all Boolean expressions in x_1, x_2, and x_3.

The preceding definition of a Boolean expression is an example of a *recursive* (or *inductive*) definition. Rule 1 gives us the basis step; it points to some definite objects that are Boolean expressions. Then rule 2 describes how to build more complex expressions from already existing expressions. (Many constructs in computer science, including elements of many programming languages, are defined recursively.)

But now comes a crucial observation. Just as polynomial expressions in ordinary algebra may be regarded as functions, so also Boolean algebraic expressions may be regarded as **Boolean** (or **switching**) **functions**. What we mean is this: Let B and B^n respectively denote the set $\{0, 1\}$ and the set of 2^n possible n-tuples (*words* of length n) of 0's and 1's. Then, for a Boolean expression F in the variables $\{x_{n-1}, \ldots, x_1, x_0\}$, the value of F at $(b_{n-1}, \ldots, b_1, b_0) \in B^n$ is an element of B obtained by assigning each value b_i to the corresponding variable x_i. Since a Boolean variable can only take on the two values 0 or 1, a Boolean (switching) function is just a function that accepts an n-tuple of 0's and 1's and produces either 0 or 1 as its value at the n-tuple. Any switching function is, therefore, completely determined by the list of its functional values at the 2^n n-tuples. Consequently, there are 2^{2^n} such switching functions.

The truth table is a column of functional values together with the n-

tuples viewed as binary numbers starting from 0 to $2^n - 1$. However, while a truth table defines only *one* switching function, many *different* Boolean expressions in $\{x_{n-1}, \ldots, x_1, x_0\}$ can determine the *same* truth table. For example, the expressions

$$E_1 = x_1' x_2' x_3 + x_1' x_2 x_3 + x_1 x_2 x_3' + x_1 x_2 x_3$$
$$E_2 = x_1 x_2 + x_1' x_3 + x_2 x_3$$
$$E_3 = x_1 x_2 + x_1' x_3$$

all specify the same truth table in which the functional values at the 3-tuples $(0,0,1)$, $(0,1,1)$, $(1,1,0)$, and $(1,1,1)$ are 1, and are 0 at all the other 3-tuples. In fact, two Boolean expressions will specify the same truth table if one expression can be transformed into the other by the axioms and laws of Boolean algebra.

In Section 3.6, we will show, conversely, that every Boolean function determines a Boolean expression in canonical form. Then, among other things, we will conclude that every Boolean function can be realized with variables and the operations AND, OR, and NOT.

If $F(x_{n-1}, \ldots, x_1, x_0)$ is a switching function in n variables, then the complement $F'(x_{n-1}, \ldots, x_1, x_0)$ is a function whose value is 1 whenever the value of F is 0, and 0 whenever the value of F is 1. Hence, the truth table of F' can be obtained from that of F simply by complementing each entry in the column of the functional values of F. The *sum* (OR) $F + G$ of two functions F and G is a function whose value is 1 for every n-tuple for which either F or G (or both) equals 1, while the *product* (AND) FG is 1 if and only if both F and G equal 1. Therefore, the truth tables for $F + G$ and FG are obtained by adding (ORing) or multiplying (ANDing) the corresponding entries from the truth tables of F and G. The following example illustrates these procedures.

■ **Example 3.4** Two switching functions in the variables $\{x_3, x_2, x_1\}$ are specified in columns F and G of Table 3.5. The resulting complement F', sum $F + G$, and product FG are specified in their respective columns. ■

With the definitions of $\{+\}$, $\{\cdot\}$, and $\{'\}$, we can conclude that the set of switching functions in n variables forms a Boolean algebra. In particular, let us consider in the following section the Boolean algebra of all switching functions in two variables.

■ **Table 3.5** Complementation, addition, and multiplication of switching functions

x_3	x_2	x_1	F	G	F'	$F+G$	FG
0	0	0	1	1	0	1	1
0	0	1	0	0	1	0	0
0	1	0	1	1	0	1	1
0	1	1	0	0	1	0	0
1	0	0	1	0	0	1	0
1	0	1	0	1	1	1	0
1	1	0	0	1	1	1	0
1	1	1	1	1	0	1	1

Binary Logic Operations

When the number of variables is $n = 2$, there are 16 (2^{2^2}) possible Boolean functions. The truth tables for the 16 Boolean functions of two variables are grouped together and listed in Table 3.6. Each of the 16 columns, F_0 to F_{15}, together with the four binary combinations of the independent variables A and B, represents a truth table of one possible function of these variables. Among these, we can identify F_1, F_7, F_{10}, and F_{12} as representing, respectively, the AND operation, the OR operation, and the two (unary) complement (NOT) operations. Table 3.6 also lists the Boolean expressions for each of the 16 functions.

Scanning the list of functions, we see two constant functions, F_0 and F_{15}, four functions with unary operations, F_3, F_5, F_{10}, and F_{12}, and ten functions with binary operators. All these functions are formed with the Boolean operators OR, AND, and NOT, but some are often denoted by special names. Two of the binary operators, *inhibition* and *implication*, are seldom used by digital systems designers. Therefore, let us consider the four binary operators, other than AND and OR, that are used extensively in the design of digital systems.

From the truth table of F_{14}, we see that **NAND** is the complement of AND (its name stands for not-AND). By virtue of DeMorgan's theorem [(10b) in Table 3.1], NAND can also be expressed as $A' + B'$. Sometimes, a special symbol, called the *Sheffer stroke*, is used to denote the NAND operator.

- **Table 3.6** Boolean functions of two variables

A	B	F_0	F_1	F_2	F_3	F_4	F_5	F_6	F_7	F_8	F_9	F_{10}	F_{11}	F_{12}	F_{13}	F_{14}	F_{15}
0	1	0	0	0	0	0	0	0	0	1	1	1	1	1	1	1	1
0	1	0	0	0	0	1	1	1	1	0	0	0	0	1	1	1	1
1	0	0	0	1	1	0	0	1	1	0	0	1	1	0	0	1	1
1	1	0	1	0	1	0	1	0	1	0	1	0	1	0	1	0	1

$F_0 = 0$	Binary constant 0
$F_1 = AB$	AND
$F_2 = AB'$	Inhibition (A but not B)
$F_3 = A$	Transfer
$F_4 = A'B$	Inhibition (B but not A)
$F_5 = B$	Transfer
$F_6 = AB' + A'B$ $= A \oplus B$	Exclusive-OR (XOR)
$F_7 = A + B$	OR
$F_8 = (A + B)'$ $= A'B'$	NOR (not-OR; in mathematical logic it is referred to as Peirce arrow and denoted $A \downarrow B$)
$F_9 = A'B' + AB$ $= A \odot B$	Equivalence (exclusive-NOR)
$F_{10} = B'$	Complement (NOT)
$F_{11} = A + B'$	Implication (B implies A)
$F_{12} = A'$	Complement (NOT)
$F_{13} = A' + B$	Implication (A implies B)
$F_{14} = (AB)'$ $= A' + B'$	NAND (not-AND; in mathematical logic it is referred to as Sheffer stroke and denoted $A|B$)
$F_{15} = 1$	Binary constant 1

By comparing the truth tables for F_7 and F_8, we see that the **NOR** function is the complement of OR (its name stands for not-OR). The NOR function can be expressed also as $A'B'$ by virtue of DeMorgan's theorem (10a), and sometimes, it is denoted by a special symbol called the *Peirce arrow*. (Both the Sheffer stroke and Peirce arrow are symbols used most often by mathematicians but hardly ever by digital systems designers.)

The **exclusive-OR** (F_6), designated by the operator symbol $\oplus$ and often referred to as **XOR**, is similar to OR except that $F_6 = 0$ when $A = B = 1$. The **equivalence** function (F_9), designated by the operator symbol $\odot$, derives its name from the fact that it attains a value of 1 if and only if $A = B$. The truth tables of F_6 and F_9 show that the 2-variable XOR and equivalence functions are *complementary*. In fact, F_9 is the exclusive-NOR function.

Extension to Multiple Variables

A binary logic operator can be extended to accommodate *multiple variables* if the operation is both commutative and associative. We have already mentioned that inhibition and implication are hardly ever used in digital systems design. Therefore, we have to consider only the extensions to multiple variables of the binary operators AND, OR, XOR, equivalence, NAND, and NOR.

By axiom 3 and Theorem 3.8 we know that the binary operators OR and AND are both commutative

$$x + y = y + x \quad \text{and} \quad xy = yx$$

and associative

$$x + (y + z) = (x + y) + z = x + y + z \quad \text{and} \quad x(yz) = (xy)z = xyz$$

Therefore, both operators can be extended to more than two variables. To illustrate this, Table 3.7 lists the truth tables for AND and OR with three independent variables.

We can also show that the binary XOR operation is both commutative and associative; that is,

$$A \oplus B = B \oplus A$$

and

$$(A \oplus B) \oplus C = A \oplus (B \oplus C) = A \oplus B \oplus C$$

Similarly, the binary equivalence operation is both commutative and associative since it is the complement of the XOR operator. Hence, both operators can be extended to multiple inputs. However, unlike the functions AND and OR, the extensions of XOR and equivalence to more than two inputs are not straightforward. The reason for this is that XOR is an *odd* function, while equivalence is an *even* function. An odd function is equal to 1 if an odd number of the independent variables are equal to 1. In contrast, an even function is equal to 1 if the independent variables have an even number of 0's.

To illustrate the problem, consider in Table 3.8 the extensions of the XOR and equivalence operators to three and four variables. To obtain the truth tables, we first have to express the Boolean functions in terms of OR, AND, and NOT operators. For example,

$$\begin{aligned} A \oplus B \oplus C &= (A \oplus B) \oplus C = (A'B + AB') \oplus C \\ &= (A'B + AB')'C + (A'B + AB')C' \\ &= ABC + A'B'C + A'BC' + AB'C' \end{aligned} \tag{3.1}$$

From expression (3.1) we can get the corresponding truth table shown in Table 3.8. The other truth tables are derived in a similar way.

■ **Table 3.7** Extensions of AND and OR to three variables

x	y	z	xyz	$x+y+z$
0	0	0	0	0
0	0	1	0	1
0	1	0	0	1
0	1	1	0	1
1	0	0	0	1
1	0	1	0	1
1	1	0	0	1
1	1	1	1	1

■ **Table 3.8** Extensions of XOR and equivalence to three and four variables

A	B	C	XOR $A \oplus B \oplus C$	Equivalence $A \odot B \odot C$
0	0	0	0	0
0	0	1	1	1
0	1	0	1	1
0	1	1	0	0
1	0	0	1	1
1	0	1	0	0
1	1	0	0	0
1	1	1	1	1

A	B	C	D	XOR $A \oplus B \oplus C \oplus D$	Equivalence $A \odot B \odot C \odot D$
0	0	0	0	0	1
0	0	0	1	1	0
0	0	1	0	1	0
0	0	1	1	0	1
0	1	0	0	1	0
0	1	0	1	0	1
0	1	1	0	0	1
0	1	1	1	1	0
1	0	0	0	1	0
1	0	0	1	0	1
1	0	1	0	0	1
1	0	1	1	1	0
1	1	0	0	0	1
1	1	0	1	1	0
1	1	1	0	1	0
1	1	1	1	0	1

Next, we make the following observations. If we have n variables, then $2^n / 2$ of the possible 2^n binary combinations have an odd number of

1's, and $2^n / 2$ combinations have an even number of 0's. But if n is *odd*, then the combinations having an odd number of 1's are the *same* combinations having an even number of 0's. On the other hand, if n is *even*, then the combinations having an odd number of 1's form the *complementing* set to the combinations having an even number of 0's. You can verify these observations by inspecting the binary combinations for $n = 3$ and $n = 4$ variables in Table 3.8.

If now we compare the columns headed by XOR and Equivalence in Table 3.8, we see that for $n = 3$ (odd), the two functions are equal, while for $n = 4$ (even) they are complementary. These observations are also true in general. If the number of variables n is even, then XOR and equivalence are complementary; but if n is odd, then XOR and equivalence are equal.

Let us turn finally to the NOR and NAND operators. Both these binary operators are commutative, but neither is associative. Therefore, strictly speaking, neither operator can be extended to multiple variables.

■ **Table 3.9** Extensions of NOR and NAND to three variables

A	B	C	NOR $(A+B+C)'$	NAND $(ABC)'$
0	0	0	1	1
0	0	1	0	1
0	1	0	0	1
0	1	1	0	1
1	0	0	0	1
1	0	1	0	1
1	1	0	0	1
1	1	1	0	0

Nevertheless, we can circumvent this difficulty by recalling that NOR is not-OR and NAND is not-AND. We can, therefore, define the multiple-variable NOR operator simply as

$$(x_1 + x_2 + \cdots + x_n)'$$

and the multiple-variable NAND operator as

$$(x_1 x_2 \cdots x_n)'$$

As an example, Table 3.9 lists the extensions of NOR and NAND to three variables based on these definitions.

Functional Completeness

A set of operations is **functionally complete** (or **universal**) if and only if every Boolean function can be expressed in terms of the operators in the set. Since Boolean functions are formed using the operations OR, AND, and NOT, the set $\{+, \cdot, '\}$ is therefore functionally complete. But due to DeMorgan's law (10a) (see Table 3.1), the set $\{\cdot, '\}$ is also functionally complete since $x + y = (x'y')'$ and, therefore, AND and NOT can replace the OR operation in any Boolean function. Similarly, by DeMorgan's law (10b), the set $\{+, '\}$ is also functionally complete.

There are many functionally complete sets of operations. A common technique to prove that a given set of operations is functionally complete is to show that the given operators can *generate* each of the operators of a *known* functionally complete set (such as $\{+, '\}$ or $\{\cdot, '\}$, for example).

Some functionally complete sets contain a *single* operator. For example, the NOR operator is functionally complete because we can use it to generate the functionally complete set $\{+, '\}$ as follows:

$$x \downarrow x = x'x' = x' \quad \text{(NOT)}$$
$$(x \downarrow y) \downarrow (x \downarrow y) = [(x+y)' + (x+y)']' = x + y \quad \text{(OR)}$$

Similarly, the NAND operation is functionally complete because it generates the functionally complete set $\{\cdot, '\}$ as follows:

$$x \mid x = x' \quad \text{(NOT)}$$
$$(x \mid y) \mid (x \mid y) = [(xy)'(xy)']' = xy \quad \text{(AND)}$$

Since we can implement any Boolean function with either of these single-operator sets, NOR and NAND gates are very popular in digital systems design and are available on integrated-circuit (IC) chips in any of the logic families.

Other functionally complete sets of operations are illustrated in the following examples.

■ **Example 3.5**   The set $\{f\}$, where $f(x, y, z) = x'yz + y'z' + xy'$, is functionally complete since

$$f(x, x, y) = x'xy + x'y' + xx' = x \downarrow y \quad (\text{NOR})$$ ■

■ **Example 3.6**   The set $\{f, 1\}$, where $f(x, y) = x'y$, is functionally complete since

$$f(x, 1) = x' \quad (\text{NOT})$$
$$f(x', y) = xy \quad (\text{AND})$$ ■

■ **Example 3.7**   The set $\{f, 1\}$, where $f(x, y, z) = x'y' + x'z' + y'z'$, is functionally complete since

$$f(x, x, x) = x' \quad (\text{NOT})$$
$$f(x', y', 1) = xy \quad (\text{AND})$$ ■

We encourage the reader to verify that neither $\{\text{XOR}\}$ nor $\{\text{equivalence}\}$ is a functionally complete set, but, on the other hand, the sets $\{\text{XOR, OR, 1}\}$ and $\{\text{equivalence, OR, 0}\}$ are functionally complete.

Complements

The complement f' of a switching function f is obtained by interchanging the 0 and 1 values of f. We can obtain f' from the truth table of f as illustrated in Example 3.4. We can also derive the complement algebraically by using DeMorgan's theorems (10a) and (10b) (see Table 3.1), as shown in Example 3.8.

Alternatively, the complement of a function can be obtained by first deriving the dual of the function and then complementing each variable.

■   **Example 3.8   (a)** Find the complement of $f = (x' + y)(xz + yz')$.

$$
\begin{aligned}
f' &= [(x' + y)(xz + yz')]' \\
&= (x' + y)' + (xz + yz')' &&\text{(10b)} \\
&= xy' + (xz)'(yz')' &&\text{(10a)} \\
&= xy' + (x' + z')(y' + z) &&\text{(10b)}
\end{aligned}
$$

(b) Find the complement of $f = xy'z + x'yz'$.

$$
\begin{aligned}
f' &= (xy'z + x'yz')' \\
&= (xy'z)'(x'yz')' &&\text{(10a)} \\
&= (x' + y + z')(x + y' + z) &&\text{(10b)}
\end{aligned}
$$

■

(Recall that the dual f^d of f is obtained by interchanging ANDs and ORs, and 0's and 1's; see Example 3.1.) We illustrate this procedure in the following example.

■   **Example 3.9**   Repeat Example 3.8 by finding the dual and then complementing each variable.

(a) $f^d = x'y + (x + z)(y + z')$. Complementing each variable, we obtain

$$
f' = xy' + (x' + z')(y' + z)
$$

(b) $f^d = (x + y' + z)(x' + y + z')$. Complementing each variable, we obtain

$$
f' = (x' + y + z')(x + y' + z)
$$

■

3.6 BOOLEAN CANONICAL FORMS

Thus far, we have emphasized the conversion from a Boolean expression to a switching function in a truth table form. Now let us change our perspective and consider the reverse process of deriving an algebraic expression for a switching function from its truth table representation.

Two binary variables, x and y, operated on by the AND operator result in four combinations of *product* terms: $x'y'$, $x'y$, xy', and xy. Similarly, using the OR operation results in four combinations of *sum* terms: $x+y$, $x+y'$, $x'+y$, and $x'+y'$. Note that, in each of the terms generated, a variable appears either complemented or uncomplemented, but *not* in both forms. But the two symbols x and x' (or y and y') are not two different variables because they involve only x (or y). To distinguish them, we designate both possibilities as $x*$ (or $y*$) and refer to either as a **literal**. A literal, therefore, defines a variable with or without a prime. Hence, x and x' refer to the *same* variable but are two *different* literals. With this definition in mind, we see that each product or sum term is formed with *distinct* literals.

In general, a *product term* formed with n distinct literals is called a **minterm** of n variables, whereas a *sum term* formed with n distinct literals is called a **maxterm** of n variables. As we will see shortly, minterms and maxterms provide the key for obtaining a Boolean expression from its functional truth table representation.

Table 3.10 lists in truth table form the minterms and maxterms of three binary variables, x, y, and z. Note that each binary combination of the three variables *uniquely* defines a minterm (maxterm). The converse is also true; that is, each minterm (maxterm) is associated with only one binary combination. But since there is a one-to-one correspondence between the binary combinations and their decimal equivalents, there is a one-to-one correspondence between each decimal and a minterm (maxterm). For example, decimal 6 uniquely corresponds to binary 110, to minterm xyz' (designated by m_6), and to maxterm $x'+y'+z$ (designated by M_6).

Notice that if the number of variables is specified there is no ambiguity in this chain of one-to-one correspondences. For example, m_6 for four variables, w, x, y, and z, is $w'xyz'$. In contrast, if the number of variables is *not* specified and we want to express m_6, then all that we can say is that a *minimum* of three variables is required. We cannot specify the maximum number of variables, nor can we deduce the *actual* number of variables.

■ **Table 3.10** Three-variable minterms and maxterms

Decimal	x	y	z	Minterm		Maxterm	
0	0	0	0	$x'y'z'$	(m_0)	$x+y+z$	(M_0)
1	0	0	1	$x'y'z$	(m_1)	$x+y+z'$	(M_1)
2	0	1	0	$x'yz'$	(m_2)	$x+y'+z$	(M_2)
3	0	1	1	$x'yz$	(m_3)	$x+y'+z'$	(M_3)
4	1	0	0	$xy'z'$	(m_4)	$x'+y+z$	(M_4)
5	1	0	1	$xy'z$	(m_5)	$x'+y+z'$	(M_5)
6	1	1	0	xyz'	(m_6)	$x'+y'+z$	(M_6)
7	1	1	1	xyz	(m_7)	$x'+y'+z'$	(M_7)

To obtain the minterm that corresponds to a given binary combination, we form a product term in which a variable is complemented if its indicated binary value is 0 or is uncomplemented if the binary value is 1. Maxterms are generated by applying the principle of duality; a variable is complemented if its indicated binary value is 1 or is uncomplemented if the binary value is 0. You can verify these rules by generating the 3-variable minterms and maxterms of Table 3.10. But these rules are applicable to any number of variables. For example, consider the binary combination 01101 (decimal 13). We require at least five variables to specify its corresponding minterm and maxterm. If we designate these variables as v, w, x, y, and z, where v is the most significant bit, then $m_{13} = v'wxy'z$ and $M_{13} = v+w'+x'+y+z'$.

In general, there are 2^n minterms (maxterms) generated by n binary variables $\{x_{n-1},...,x_1,x_0\}$. To characterize the structure of a minterm (maxterm) let $(K)_{10}$ denote the decimal equivalent of the Kth binary combination $(k_{n-1},...,k_1,k_0)_2$ of the n variables; hence, $K = 0, 1, 2, ..., 2^n - 1$. If we denote by $m_K = x^*_{n-1} \cdots x^*_1 x^*_0$ the Kth minterm, then

$$x_i^* = x_i' \quad \text{if} \quad k_i = 0$$
$$\quad = x_i \quad \text{if} \quad k_i = 1$$

Similarly, if we let $M_K = x_{n-1}^* + \cdots + x_1^* + x_0^*$ denote the Kth maxterm, then

$$x_i^* = x_i \quad \text{if} \quad k_i = 0$$
$$\quad = x_i' \quad \text{if} \quad k_i = 1$$

Note that the value of any minterm is 1 *if and only if* the variables assume the values of its corresponding binary combination, and it is 0 for any other combination. Similarly, the value of any maxterm is 0 *if and only if* the variables assume the values of its corresponding binary combination, and it is 1 for any other combination. Therefore, for the *same* binary combination, the corresponding minterm and maxterm are complements of one another. For example, with three variables, $m_6 = 1$ ($M_6 = 0$) if and only if $x = 1$, $y = 1$, and $z = 0$, and $m_6 = 0$ ($M_6 = 1$) for any other combination. These properties hold true for any number of variables and can be summarized as follows:

$$m_i m_j = 0 \quad \text{if} \quad i \neq j$$
$$\quad = m_i \quad \text{if} \quad i = j \tag{3.2a}$$

$$M_i + M_j = 1 \quad \text{if} \quad i \neq j$$
$$\quad = M_i \quad \text{if} \quad i = j \tag{3.2b}$$

$$m_i = M_i' \quad \text{and} \quad M_i = m_i' \quad \text{for every } i \tag{3.2c}$$

We can now use these concepts to derive the algebraic expression of a Boolean function from a given truth table. We can do it in two ways:

1. By summing (ORing) those minterms for which the function takes a value 1.
2. By multiplying (ANDing) those maxterms for which the function assumes a value 0.

We illustrate these procedures in the following example.

■ **Example 3.10**   Consider the following truth table of a Boolean function of three variables $f(x_2, x_1, x_0)$:

Decimal	x_2	x_1	x_0	f
0	0	0	0	0
1	0	0	1	1
2	0	1	0	0
3	0	1	1	0
4	1	0	0	1
5	1	0	1	1
6	1	1	0	1
7	1	1	1	1

(a) Since $f = 1$ for the binary combinations whose decimal equivalents are 1, 4, 5, 6, and 7, the function can be expressed as

$$f(x_2, x_1, x_0) = m_1 + m_4 + m_5 + m_6 + m_7$$

Each of these minterms can be explicitly specified in terms of the three variables to yield the following Boolean algebraic expression for f :

$$f(x_2, x_1, x_0) = x_2'x_1'x_0 + x_2x_1'x_0' + x_2x_1'x_0 + x_2x_1x_0' + x_2x_1x_0$$

(b) Since $f = 0$ for the binary combinations whose decimal equivalents are 0, 2, and 3, the function can be alternatively expressed as

$$f(x_2, x_1, x_0) = M_0 M_2 M_3$$

Each of these maxterms can be explicitly specified in terms of the three variables to yield the following algebraic expression for f:

$$f(x_2, x_1, x_0) = (x_2 + x_1 + x_0)(x_2 + x_1' + x_0)(x_2 + x_1' + x_0')$$

Later, we will show how to convert from one form to the other. In the meantime, however, we ask the reader to verify, using the identities of Table 3.1, that the expressions in parts (a) and (b) are equal and that f can be represented in a simplified form as $x_2 + x_1' x_0$ [or, equivalently, as $(x_2 + x_1')(x_2 + x_0)$]. ∎

In Example 3.10, the function f is represented as a sum (OR) of the five minterms corresponding to those binary combinations of values of x_2, x_1, x_0 for which f takes on the value 1. Because there is a one-to-one correspondence between these binary combinations, their decimal equivalents, and their minterms, we can use a shorthand notation and express f as

$$f(x_2, x_1, x_0) = \sum(1, 4, 5, 6, 7) \tag{3.3}$$

The symbol $\sum$ in Equation (3.3) stands for "the logical sum (OR) of the minterms indicated by the decimals." Likewise, since f can be also expressed as a product (AND) of those maxterms that correspond to values of x_2, x_1, x_0 for which f is equal to 0, we may represent f as

$$f(x_2, x_1, x_0) = \prod(0, 2, 3) \tag{3.4}$$

The symbol $\prod$ in Equation (3.4) stands for "the logical product (AND) of the maxterms indicated by the decimals." Note that there is no ambiguity in using either one of these formats to express any Boolean function. Rather than carry long expressions in terms of the variables, we can conveniently represent the function using these shorthand forms.

We call the sum of minterms representation of a Boolean function the **canonical sum-of-products (SOP)** form. (Sometimes, the canonical SOP is called the *minterm form* or the *disjunctive normal form*). The product of maxterms representation of a Boolean function is called the **canonical product-of-sums (POS)** form. (The canonical POS is also called the *maxterm form* or the *conjunctive normal form*).

Any Boolean function can be expressed either as a canonical SOP or as a canonical POS. We summarize this statement in the following theorem.

☐ **Theorem 3.10** Every Boolean function $f(x_{n-1}, \ldots, x_1, x_0)$ of n variables can be written in either of the following forms:

$$\text{Canonical sum-of-products}: \; f(x_{n-1}, \cdots, x_1, x_0) = \sum_{K=0}^{2^n-1} \alpha_K m_K \qquad (3.5)$$

$$\text{Canonical product-of-sums}: \; f(x_{n-1}, \ldots, x_1, x_0) = \prod_{K=0}^{2^n-1} (\beta_K + M_K) \qquad (3.6)$$

The coefficients α_K and β_K can each assume a value of either 1 or 0. If $f = 1$ for the Kth minterm or maxterm, then these coefficients will be 1; otherwise, they will be 0. In other words,

$$\alpha_K = \beta_K = f(k_{n-1}, \ldots, k_1, k_0) \qquad (3.7)$$

$\square$

Recall that there are 2^{2^n} functions of n binary variables. Equations (3.5) and (3.6) provide two alternatives for expressing each of these Boolean functions, either as a canonical SOP or as a canonical POS. We can derive these forms by repeated applications of Shannon's expansion theorems (11a) and (11b), respectively (see Table 3.1). To illustrate this, assume that $n = 3$. Hence, there are $2^{2^3} = 2^8 = 256$ Boolean functions of three variables. In the following example, we indicate how to derive all 256 canonical SOP representations by repeated applications of Shannon's expansion theorem (11a).

■ **Example 3.11** Through repeated applications of Shannon's expansion theorem (11a), we obtain

$$\begin{aligned}
f&(x, y, z) \\
&= xf(1, y, z) + x'f(0, y, z) \\
&= x[yf(1,1,z) + y'f(1,0,z)] + x'[yf(0,1,z) + y'f(0,0,z)] \\
&= xy[zf(1,1,1) + z'f(1,1,0)] + xy'[zf(1,0,1) + z'f(1,0,0)] \\
&\quad + x'y[zf(0,1,1) + z'f(0,1,0)] + x'y'[zf(0,0,1) + z'f(0,0,0)]
\end{aligned}$$

Multiplying through and rearranging terms, we get

$$\begin{aligned}
f(x, y, z) = {}& f(0,0,0)x'y'z' + f(0,0,1)x'y'z + f(0,1,0)x'yz' \\
& + f(0,1,1)x'yz + f(1,0,0)xy'z' + f(1,0,1)xy'z \\
& + f(1,1,0)xyz' + f(1,1,1)xyz \qquad (3.8)
\end{aligned}$$

Each of the multipliers $f(\cdot,\cdot,\cdot)$ is either 0 or 1 depending on whether f is 0 or 1, respectively, for the specified binary combination. The multipliers are, therefore, the α_K's of Equation (3.7), and Equation (3.8) is just a repetition of Equation (3.5) for $n = 3$. Since each of the eight multipliers can be either 0 or 1, the total number of binary combinations that they can assume is $2^8 = 256$, the total number of functions of three variables.

We leave it to the reader to derive the canonical POS for three variables by using either Shannon's expansion theorem (11b) or the principle of duality. ∎

Relations between Canonical Forms

Since the two canonical forms provide two alternatives for expressing the same Boolean function, one form can be derived from the other. To see this, recall that the function in Example 3.10 was described by Equation (3.3) as

$$f(x_2, x_1, x_0) = \sum(1, 4, 5, 6, 7) \tag{3.9}$$

Its complement is given by

$$f'(x_2, x_1, x_0) = \sum(0, 2, 3)$$

Taking the complement of the complement and using DeMorgan's law (10a) of Table 3.1, we obtain

$$f(x_2, x_1, x_0) = \left[\sum(0, 2, 3)\right]' = m_0' m_2' m_3' = M_0 M_2 M_3$$

where the last expression is obtained by using Equation (3.2c). Hence,

$$f(x_2, x_1, x_0) = \sum(1, 4, 5, 6, 7) = \prod(0, 2, 3) \tag{3.10}$$

Alternatively, we can view the process of converting from one canonical form to the other as follows: Denote by U the set of decimals associated with the function. If A is the set of decimals of the canonical

SOP, then $U - A$ is the set of decimals of the canonical POS, and vice versa. We illustrate this procedure in the following example.

- **Example 3.12** The set of decimals associated with any 3-variable Boolean function is $U = \{0, 1, 2, 3, 4, 5, 6, 7\}$. The set of decimals of the canonical SOP of f in Equation (3.9) is $A = \{1, 4, 5, 6, 7\}$. Therefore, $U - A = \{0, 2, 3\}$ is the set of decimals corresponding to the canonical POS of f, in agreement with Equation (3.10). ■

Complements and Duals

If an n-variable function f is specified as a canonical SOP, then its complement is given by

$$f' = \left[\sum_{i=0}^{2^n-1} \alpha_i m_i \right]' = \prod_{i=0}^{2^n-1} (\alpha_i' + m_i') = \prod_{i=0}^{2^n-1} (\alpha_i' + M_i) \tag{3.11}$$

The second part of Equation (3.11) is obtained by using DeMorgan's law (10a) of Table 3.1, and the third part is obtained by using Equation (3.2c).

If the function is specified as a canonical POS, then its complement is given by

$$f' = \left[\prod_{i=0}^{2^n-1} (\beta_i + M_i) \right]' = \sum_{i=0}^{2^n-1} \beta_i' M_i' = \sum_{i=0}^{2^n-1} \beta_i' m_i' \tag{3.12}$$

The second part of Equation (3.12) is obtained from DeMorgan's law (10b), and the third part is obtained from Equation (3.2c).

To obtain the dual of an n-variable function f, use Equation (3.13) if f is in canonical SOP, or Equation (3.14) if f is in canonical POS.

$$f^d = \left[\sum_{i=0}^{2^n-1} \alpha_i m_i \right]^d = \prod_{i=0}^{2^n-1} (\alpha_i^d + m_i^d) \tag{3.13}$$

$$f^d = \left[\prod_{i=0}^{2^n-1} (\beta_i + M_i) \right]^d = \sum_{i=0}^{2^n-1} \beta_i^d M_i^d \tag{3.14}$$

The following example illustrates the applications of Equations (3.11) through (3.14).

■ **Example 3.13** Consider the function

$$f = \sum(1, 2, 3, 4) = \prod(0, 5, 6, 7)$$

Therefore, the α's for the canonical SOP form are

$$\alpha_0 = \alpha_5 = \alpha_6 = \alpha_7 = 0, \qquad \alpha_1 = \alpha_2 = \alpha_3 = \alpha_4 = 1$$

and the β's for the canonical POS form are

$$\beta_0 = \beta_5 = \beta_6 = \beta_7 = 0, \qquad \beta_1 = \beta_2 = \beta_3 = \beta_4 = 1$$

(a) With f in canonical SOP form, we obtain its complement from Equation (3.11):

$$f' = M_1 M_2 M_3 M_4$$

(b) If f is in canonical POS form, we obtain from Equation (3.12)

$$f' = m_0 + m_5 + m_6 + m_7$$

(c) With f in canonical SOP form, we obtain its dual from Equation (3.13):

$$f^d = m_1^d m_2^d m_3^d m_4^d = M_6 M_5 M_4 M_3$$

(d) If f is in canonical POS form, we obtain from Equation (3.14)

$$f^d = M_0^d + M_5^d + M_6^d + M_7^d = m_7 + m_2 + m_1 + m_0 \qquad\blacksquare$$

As we mentioned in Section 3.5, the complement can be obtained also by finding the dual first and then complementing the variables. Hence, if f is in canonical SOP, complement the variables in Equation (3.13) to obtain Equation (3.11). Similarly, if f is in canonical POS, complement the variables in Equation (3.14) to obtain Equation (3.12).

Noncanonical Forms

Recall that in Example 3.10 we obtained the two canonical forms

[Equations (3.3) and (3.4)] for the given function $f(x_2, x_1, x_0)$. As we mentioned there, either canonical form can be algebraically manipulated to obtain f in a simplified form, either as

$$f(x_2,\ x_1,\ x_0) = x_2 + x_1'x_0$$

or as

$$f(x_2, x_1, x_0) = (x_2 + x_1')(x_2 + x_0)$$

The first expression is in sum-of-products (SOP) form, and the second expression is in product-of-sums (POS) form, but neither expression is canonical because the product terms (sum terms) are not minterms (maxterms). We will refer to them by simply omitting the word canonical and using the abbreviations SOP and POS. In the following section, we show how to convert from noncanonical to canonical forms.

Conversion to Canonical Forms

To convert a sum-of-products expression to its canonical SOP form, we use the identity $x + x' = 1$. If one or more variables are missing from any product, we AND them to it by using this identify and then remove the parentheses using the distributive law. The procedure is illustrated in the following example.

■ **Example 3.14** Convert $f(x, y, z) = x'y + z + xy'$ to canonical SOP.

$$f(x, y, z) = x'y(z + z') + (x' + x)(y' + y)z + xy'(z + z')$$
$$= x'yz + x'yz' + x'y'z + xy'z + x'yz + xyz + xy'z + xy'z'$$
$$= x'yz + x'yz' + x'y'z + xy'z + xyz + xy'z'$$

In this derivation, notice that both $x'yz$ and $xy'z$ appear twice. We eliminate this *redundancy* by using identity (4a) of Table 3.1.   ■

To convert a sum-of-products expression to its canonical POS form, we may find it easier to convert the expression first to the canonical SOP form and then to the canonical POS form.

■ **Example 3.15** Repeat Example 3.14 by converting the function to canonical POS. Since $f(x, y, z) = \sum(1, 2, 3, 4, 5, 7)$, then f is equal to $\prod(0, 6)$. Therefore,

$$f(x, y, z) = (x + y + z)(x' + y' + z)$$ ■

To convert a product-of-sums expression to its canonical POS form, we use the identity $xx' = 0$. In this conversion, if one or more variables are missing from any product, we OR them to it by using this identity and then repeatedly use identity (2b) of Table 3.1 to obtain the canonical POS. This procedure is clarified by the following example.

■ **Example 3.16** Convert $f(x, y, z) = x(y' + z)$ to canonical POS.
In the first product, x, both y and z are absent. By ORing yy' to it and using the distributive law (2b), we get

$$x = x + yy' = (x + y)(x + y')$$

But z is absent from each of these products; hence,

$$x + y = x + y + zz' = (x + y + z)(x + y + z')$$
$$x + y' = x + y' + zz' = (x + y' + z)(x + y' + z')$$

The variable x is missing from the second product term of f; therefore,

$$y' + z = xx' + y' + z = (x + y' + z)(x' + y' + z)$$

By ANDing all these terms and removing those that are redundant (appear more than once), we obtain the canonical POS:

$$f(x, y, z) = (x + y + z)$$
$$\cdot (x + y + z')(x + y' + z)(x + y' + z')(x' + y' + z)$$ ■

To convert a product-of-sums expression to canonical SOP form, it is simpler to convert first to canonical POS and then to the canonical SOP.

■ **Example 3.17** Repeat Example 3.16 by converting the function to canonical SOP. Since $f(x, y, z) = \prod(0, 1, 2, 3, 6)$, it follows that f is equal to $\sum(4, 5, 7)$; hence,

$$f(x, y, z) = xy'z' + xy'z + xyz \qquad \blacksquare$$

3.7 DIGITAL LOGIC IMPLEMENTATION

The most important application of Boolean algebra lies in the realm of digital systems design. The various functional units of a digital system are **logic** (or **switching**) **circuits** that accept a collection of inputs and generate a collection of outputs. Each input and output signal represents a variable that assumes only one of two distinct values and carries one bit of information. Any information can be operated on by passing binary signals through various combinations of logic circuits.

In subsequent chapters we will discuss in considerable detail the analysis and design of a variety of digital circuits. Nevertheless, with the ideas presented thus far we already possess some basic design tools that can be applied to a number of design problems. The purpose of this section, therefore, is twofold: (1) to introduce the standard building blocks, called *gates*, that are used to implement a digital design, and (2) to introduce the design process by means of a nontrivial example. Other design problems can be found as problems at the end of this chapter.

Logic Gates

Logic circuits that perform logical operations such as AND, OR, and NOT are called **gates**. A gate is a block of hardware that produces a logic 0 or a logic 1 output signal in response to binary signals applied to its inputs.

Of the 16 Boolean functions defined in Table 3.6, two are constants and four (complement, transfer, inhibition, and implication) are repeated twice. Inhibition and implication are not used as standard gates because they are not commutative and associative. Hence, only eight functions are implemented as standard logic gates: AND, OR, NOT (complement), transfer, NAND, NOR, exclusive-OR, and equivalence.

The eight standard gates are shown in Figure 3.1. The graphic symbols, referred to as the *distinctive-shape* symbols, correspond to ANSI/IEEE Standard No. 91-1973. Recently, however, a new standard of graphic symbols for logic functions has emerged (ANSI/IEEE Standard No. 91-1984). The new standard is a very powerful symbolic language that describes the relationship between each input and each output of a digital circuit, without showing explicitly the internal construction of the circuit. Nevertheless, rather than overburden the reader unnecessarily with a fairly complex new standard, we chose to use the distinctive-shape symbols throughout the book and to provide a brief overview of the new standard in the Appendix at the end of the book.

In Figure 3.1, the small circle at the output of a gate indicates a complement (NOT) operation. The NOT gate itself is often called an *inverter* since it inverts (complements) the binary signal at its input. The transfer function does not operate on the input variable and produces at the output the same value of the input. The transfer gate is used mostly as a **buffer** in digital circuits to increase the *fan-out* (see Section 1.2) of a signal source. For example, assume that a certain device has to drive 12 inputs but that its fan-out is only 5. Therefore, the device can drive up to five buffers (the number of buffers necessary depends on their fan-out capabilities), and we can use their outputs to drive the 12 inputs.

Notice that, except for the single-input gates, all the other gates shown in Figure 3.1 have two inputs. But we have already established in Section 3.5 that the binary operators AND, OR, XOR, equivalence, NAND, and NOR can be extended to multiple variables. Therefore, their corresponding gates can be extended to multiple inputs. As far as their graphic symbols are concerned, we can modify each by simply adding the appropriate number of extra input lines.

Wired Logic

If the outputs of two or more gates are connected together directly, the operation of the circuit may be indeterminate. There are, however, numerous applications where two or more outputs of gates or devices are tied together on a common line, called a **bus line** (or simply a **bus**). For example, a digital computer may be required to send data to, or receive data from, a large number of input/output devices. Since it is impractical for each device to have its own (separate) set of lines, most computers communicate with input/output devices through a bus onto which all the devices are connected in *parallel*.

Function	Graphic symbol	Truth table

Function	Truth table
AND	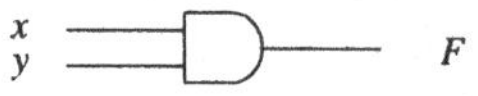 x y → F	x y F 0 0 0 0 1 0 1 0 0 1 1 1
OR	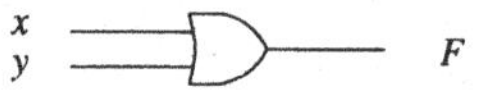 x y → F	x y F 0 0 0 0 1 1 1 0 1 1 1 1
Inverter (NOT)	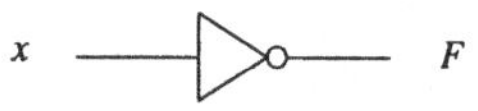 x → F	x F 0 1 1 0
Buffer (transfer)	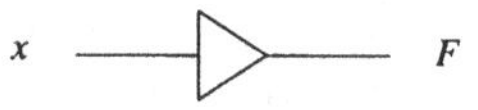 x → F	x F 0 0 1 1
NAND	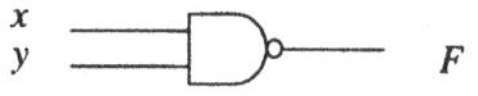 x y → F	x y F 0 0 1 0 1 1 1 0 1 1 1 0
NOR	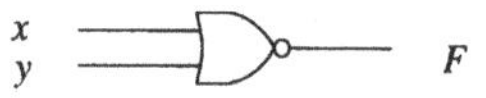 x y → F	x y F 0 0 1 0 1 0 1 0 0 1 1 0
Exclusive-OR (XOR)	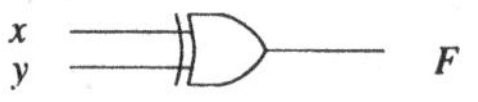 x y → F	x y F 0 0 0 0 1 1 1 0 1 1 1 0
Equivalence (exclusive-NOR)	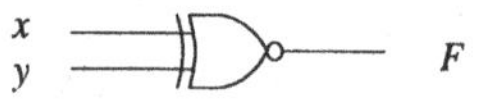 x y → F	x y F 0 0 1 0 1 0 1 0 0 1 1 1

■ **Figure 3.1** Standard logic gates

One type of logic used to implement bus-oriented digital systems is called **wired logic** and is supported by some IC logic families that allow two or more outputs to be interconnected to produce a meaningful signal.

For example, Figure 3.2(a) shows that the logic function generated by interconnecting the outputs of two NOR gates is the OR function. This type of connection is often called a **wired-OR**. Similarly, Figure 3.2(b) shows the implementation of a **wired-AND** function. In each figure, the lines are drawn through the center of the gate to distinguish it from a regular gate. Note that the wired-AND or wired-OR gate is not a physical gate but a symbol to designate the function implemented by the indicated wired connections.

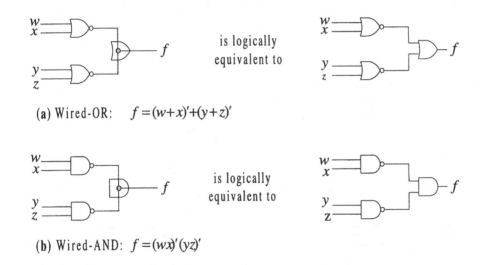

(a) Wired-OR: $f = (w+x)' + (y+z)'$

(b) Wired-AND: $f = (wx)'(yz)'$

■ **Figure 3.2** Wired logic

Logic Design Example

In subsequent chapters we will introduce the design process of digital circuits in a systematic way and discuss a number of algorithms that can be used to produce good designs. Nevertheless, the concepts introduced thus far can already be integrated to illustrate how the design of digital switching circuits can be carried out from the *specifications* of the required circuit to the derivation of a *circuit logic diagram*.

The first step in the design process is to formulate the problem concisely. This is done by mapping the general description of the problem into either a Boolean equation or a truth table form. To appreciate how this is done, let us consider a device, called *seven-segment display* (*SSD*), that is used to display the decimal digits. Figure 3.3(a) shows the seven segments labeled with letters from *a* through *g*. Each segment can be separately illuminated (turned on) so that by appropriately selecting segments, we can display the decimal digits 0 through 9 as shown in Figure 3.3(b).

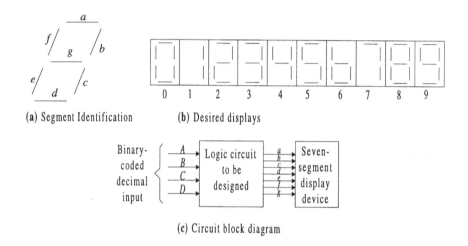

(a) Segment Identification   (b) Desired displays

(c) Circuit block diagram

Decimal	Inputs					Outputs						
displayed	*A*	*B*	*C*	*D*		*a*	*b*	*c*	*d*	*e*	*f*	*g*
0	0	0	0	0		1	1	1	1	1	1	0
1	0	0	0	1		0	1	1	0	0	0	0
2	0	0	1	0		1	1	0	1	1	0	1
3	0	0	1	1		1	1	1	1	0	0	1
4	0	1	0	0		0	1	1	0	0	1	1
5	0	1	0	1		1	0	1	1	0	1	1
6	0	1	1	0		0	0	1	1	1	1	1
7	0	1	1	1		1	1	1	0	0	0	0
8	1	0	0	0		1	1	1	1	1	1	1
9	1	0	0	1		1	1	1	0	0	1	1

(d) Truth table

■ **Figure 3.3** Seven-segment logic circuit design

Our problem, therefore, is to design a logic circuit that will turn on the correct segments in response to a binary-coded decimal (BCD) input (see Section 2.2). Since the BCD code is represented by four binary variables (bits), the circuit will have four inputs, one for each bit. With seven segments, the circuit must have seven outputs, each of which will switch the corresponding segment either on or off. Hence, the required circuit block diagram is the one shown in Figure 3.3(c).

We can tabulate the outputs corresponding to each *valid* binary combination of the inputs as shown in the truth table of Figure 3.3(d). The input codes are BCD representations of the displayed digits, and the output columns are filled in by inspecting the displays shown in Figure 3.3(b): A 1 is placed in the appropriate column if the corresponding segment should be on and a 0 if it should be off. For example, digit 7 requires only segments *a*, *b*, and *c*, and we therefore write 1110000 in the corresponding output section of the truth table.

Notice that the truth table of Figure 3.3(d) is incomplete. With four input variables, we can generate 16 binary combinations. Therefore, the complete truth table should have contained 16 rows. But only ten rows are actually required for the specification of our problem, so what about the remaining six combinations? Since the input code is BCD, we can assume that the binary combinations corresponding to decimals 10 through 15 will never occur. If they never occur, no corresponding output will be generated. In other words, *we do not really care what the outputs are for these inputs*. In the complete truth table, this can be indicated by placing X's in the output columns opposite the rows that correspond to decimals 10 to 15. The X's, referred to as **don't-cares**, can be either 0 or 1, and it makes no difference which value each assumes because the corresponding input combination is assumed never to occur. Hence, their omission from the table in Figure 3.3(d) implies the same thing as "don't care." The notion of don't-cares will be considered in more detail in Section 4.6.

Having represented the operation of the circuit in truth table form, we can now obtain the Boolean functions that govern its behavior. Consider, for example, segment *d*, which should be on ($d = 1$) when displaying the digit 0, 2, 3, 5, 6, or 8. Expressing these conditions in canonical SOP form, we obtain from the truth table

$$d = \sum(0, 2, 3, 5, 6, 8)$$

Similar expressions can be obtained for the remaining outputs:

$$a = \sum(0,2,3,5,7,8,9)$$
$$b = \sum(0,1,2,3,4,7,8,9)$$
$$c = \sum(0,1,3,4,5,6,7,8,9)$$
$$e = \sum(0,2,6,8)$$
$$f = \sum(0,4,5,6,8,9)$$
$$g = \sum(2,3,4,5,6,8,9)$$

We can also express these output functions in the alternative canonical POS form. For example, d expressed in this form would be given by:

$$d = \prod(1,4,7,9)$$

We can substitute the corresponding minterms or maxterms for the decimals and implement each output as a function of the four input variables, using AND, OR, and NOT gates. Of course, we can also use any other functionally complete set of operators to implement the circuit. Figure 3.4 shows two examples of implementing the canonical POS function of segment d; in part (a), AND, OR, and NOT gates are used and in part (b), an all NOR implementation is shown. (Note how a NOR gate is used to implement an inverter.)

*3.8   SOME PROPERTIES OF BOOLEAN FUNCTIONS

In this section we address some important properties of Boolean functions. Most of these properties will be exploited in the following Section 3.9 and in Chapter 4. Throughout this section we will refer to n-variable Boolean switching functions $f(x_1, x_2, \ldots, x_n)$ but omit the argument whenever it is understood.

---

* This section is optional.

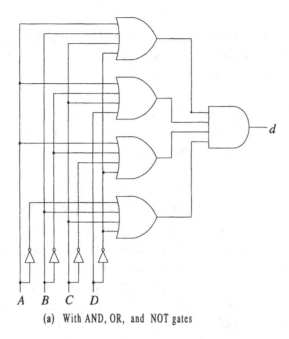

(a)   With AND, OR, and NOT gates

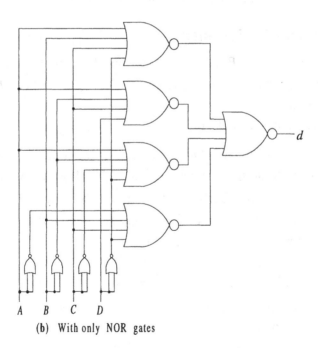

(b)   With only NOR gates

■   **Figure 3.4**   Implementation of segment $d = \prod(1, 4, 7, 9)$

□ **Definition 3.1** Two Boolean expressions f and g in the same variables are said to be **(logically) equivalent** if they determine the same truth table, in which case we write $f = g$.   □

Hence, logical equivalence of Boolean expressions is an *equivalence relation* on the set of all Boolean expressions in n variables. Thus, the set of Boolean expressions in n variables may be partitioned into *disjoint* equivalence classes, where each equivalence class is associated with a *unique* truth table, the one specified by any expression in that class. The number of *distinct* equivalence classes is, therefore, at most the number 2^{2^n} of different truth tables of switching functions in n variables. In fact, Theorem 3.10 of Section 3.6 shows that any switching function in n variables determines a Boolean expression in canonical form. Thus, the number of equivalence classes of Boolean expressions in n variables is exactly the same as the number of switching functions in n variables. It follows then that two Boolean expressions are equivalent if and only if their canonical SOP (canonical POS) forms are identical.

Implication and Equivalence

□ **Definition 3.2** A Boolean expression f is said to **imply** another Boolean expression g, designated as $f \Rightarrow g$, if, when f and g are considered as functions, g has the value 1 at least at every combination at which f has the value 1.   □

Thus, $f \Rightarrow g$ whenever the associated truth tables are such that the rows of the truth table of g, where g has value 1, contain all the rows of the truth table of f for which f has value 1. Note that, by the principle of duality, $f \Rightarrow g$ if and only if f has the value 0 at least at every combination at which g has the value 0.

■ **Example 3.18** Let $f = y'$ and $g = x + x'y'$. Then, $g = 1$ if $x = 1$ or $x = 0$ and $y = 0$. Since $f = 1$ if $y = 0$ and x is either 0 or 1 (note that f is independent of x), then $f \Rightarrow g$.   ■

The following result is an immediate consequence of Definition 3.2.

□ **Theorem 3.11** Two Boolean expressions f and g are equivalent if and only if $f \Rightarrow g$ and $g \Rightarrow f$. □

We can test implication by comparing the truth tables of the two expressions or by using the following result.

□ **Theorem 3.12** If f and g are Boolean expressions, then $f \Rightarrow g$ if and only if $f + g = g$ and $fg = f$. □

■ **Example 3.19** Reconsidering Example 3.18, we observe that $f + g = y' + x + x'y' = x + y' = g$ [since $x + y' = x + x'y'$ by identity (7b) of Table 3.1], and $fg = y'(x + x'y') = y' = f$. ■

Finally, we list one additional useful conclusion about equivalence of Boolean expressions.

□ **Theorem 3.13** Two Boolean expressions f and g are equivalent if and only if one expression can be transformed into the other using the axioms and laws of Boolean algebra. □

Informally speaking, what this last observation implies is that the axioms of Boolean algebra are sufficient to determine whether or not two Boolean expressions yield the same truth table. Therefore, we can conclude that the axioms and laws of Boolean algebra are all that is needed to analyze what may be two dramatically different digital circuit configurations.

□ **Definition 3.3** Let p be a product (sum) of literals and f be a function. If $p \Rightarrow f$, then p is said to be an **implicant (implicate)** of f. □

■ **Example 3.20** Let $p = wxy'$ and $f = wx + yz$. By Theorem 3.12, since $p + f = wxy' + wx + yz = wx + yz = f$ and $pf = wxy' = p$, then $p \Rightarrow f$ and wxy' is an implicant of f. ■

□ **Definition 3.4** A product (sum) of literals p is a **prime implicant (prime implicate)** of f if $p \Rightarrow f$ and if deleting any literal from p results in a new product (sum) term that does not imply f. □

■ **Example 3.21** If $p_1 = wxy'$ and $f = wx + yz$, then p_1 is clearly not a prime implicant of f, because deleting y' from it results in a new product term, $p_2 = wx$, which is also an implicant of f. However, p_2 is a prime implicant of f since $p_2 \Rightarrow f$, and deleting either w or x from p_2 results in a product term that does not imply f.   ∎

Classification of Variables

□ **Definition 3.5**   A function f is said to be **positive (negative)** in x_k if and only if it is possible to express f in a SOP or POS form without $x'_k (x_k)$.   □

∎ **Example 3.22   (a)** The function $f(x,y,z) = xy + z'$ is positive in x and y and negative in z.

   (b) The function $f(x,y) = x \oplus y = x'y + xy'$ is neither positive nor negative in x and y.

   (c) The function $f(x,y) = xy' + y$ seems to be positive in x and neither positive nor negative in y. However, using identity (7b) of Table 3.1, $f(x,y) = x + y$ and is, therefore, positive in both x and y.   ∎

A function f is positive in x_i if f either changes from 0 to 1 or remains unchanged whenever x_i changes from 0 to 1. It is negative in x_i if f either changes from 1 to 0 or remains unchanged whenever x_i changes from 0 to 1. As we saw in Example 3.22(c), we must take care in determining whether a function is positive or negative in any of its variables.

□ **Definition 3.6**   A function is **unate** in x_i if it is either positive or negative in x_i.   □

If f is neither positive nor negative in x_i, then both x_i and x'_i must appear in its SOP or POS expression. If neither x_i nor x'_i appears in its SOP or POS expression, then f is **independent** of x_i and is then trivially both positive and negative in x_i. In this case, f is often referred to as **vacuous** in x_i, and x_i is said to be **redundant**. All the *nonredundant* variables are called **essential** for f. If *all* the variables are redundant, then f must be a constant, either $f = 0$ or $f = 1$. The following example illustrates these concepts.

■ **Example 3.23** Consider the function $f(w,x,y,z) = xy + x'z'$. The variables x, y, and z are essential for f. The function is *independent* of w (w is *redundant*), and it is *positive* in y and w, *negative* in z and w, and *unate* in y and z.   ■

Classification of Functions

□ **Definition 3.7** A function is **monotonic nondecreasing** if and only if it is positive in all its variables.   □

□ **Definition 3.8** A function is **monotonic nonincreasing** if and only if it is negative in all its variables.   □

■ **Example 3.24** Of the 16 functions of two variables (x and y), there are six monotonic non-decreasing functions, 0, x, y, xy, $x+y$, and 1, and six monotonic nonincreasing functions, 1, x', y', $x'y'$, $x'+y'$, and 0.   ■

Unate Functions

□ **Definition 3.9** A function f is called **unate** if and only if it is unate in each of its variables.   □

Using Definition 3.6, we can state alternatively that f is unate if and only if it is positive or negative in each of its variables. In other words, f is unate if and only if each variable appears either complemented or uncomplemented, but not both. For this reason, a unate function is sometimes referred to as a **mixed monotonic function**. Hence, if

$X = \{x_1, x_2, \cdots, x_n\}$ denotes the set of variables, then f is a unate function if and only if X can be partitioned into two subsets, Y and Z, such that $X = Y \cup Z$ and $Y \cap Z = \varnothing$ (where $\cup$ and $\cap$ designate the union and intersection of the two sets, respectively, and $\varnothing$ denotes the empty set), so that f is a monotonic nondecreasing (positive) function on Y and a monotonic nonincreasing (negative) function on Z.

■ **Example 3.25** **(a)** The function $f(x,y,z) = x'y + x'z + yz$ is unate because we can partition $X = \{x, y, z\}$ into $Y = \{y, z\}$ and $Z = \{x\}$ so that $Y \cup Z = X$, $Y \cap Z = \varnothing$, and f is positive on Y and negative on Z.

(b) The function $f(x, y) = x'y + xy'$ is clearly not unate.   ■

Symmetric Functions

□ **Definition 3.10** A function f is **totally symmetric** if and only if it remains unchanged under any permutation of its variables.   □

■ **Example 3.26 (a)** The function $f(x, y) = xy' + x'y$ is totally symmetric since x and y play identical roles in determining the value of f, and thus f is unchanged under their permutation.

(b) The function $f(x, y, z) = xy + xz + yz$ is clearly totally symmetric. ■

□ **Definition 3.11** A function f is **partially symmetric** if and only if there exists at least one subset of at least two variables in which f is totally symmetric.   □

■ **Example 3.27** **(a)** $f(x,y,z) = xy + z'$ is partially symmetric in x and y.

(b) $f(x,y,z) = x'y'z + xy'z'$ is not totally symmetric because interchanging x and y yields $x'y'z + x'yz'$ which is not equal to f. However, f is partially symmetric in x and z since interchanging them yields the same expression $xy'z' + x'y'z$. ∎

□ **Definition 3.12** A function f is **mixed symmetric** if and only if it is not totally symmetric, but can be changed into a totally symmetric function by replacing some of its variables by their complements.

□

∎ **Example 3.28** The function $f(x,y,z) = xyz'$ is not totally symmetric. However, by replacing z' by z, we get xyz which is totally symmetric. ∎

Decomposable Functions

Consider the process of forming a function of n variables by taking the composition of several functions, each depending on fewer than n variables. If a function f is so formed, then f is said to be a **decomposable function** and the process is referred to as **functional decomposition**. In this section we consider only one class of decomposable functions, referred to as simple disjunctive decomposable functions.

□ **Definition 3.13** Let $X = \{x_1, x_2, \ldots, x_n\}$ be a set of n variables. Let $Y = \{y_1, y_2, \ldots, y_s\}$, $1 \le s \le n-1$, and $Z = \{z_1, z_2, \ldots, z_r\}$ be two subsets of X such that $Y \cup Z = X$. Then a function $f(X)$ is said to be a **simple decomposable function** if and only if there exist two functions G and H such that

$$f(X) = G[H(Y), Z] \tag{3.15}$$

(Notice that the y's and z's are merely new names for the x's.) □

□ **Definition 3.14** If $r = n - s$, G and H represent a **simple disjunctive decomposition** of f. □

Observe that, if $r = n - s$, then Y and Z are disjoint; that is, $Y \cap Z = \varnothing$ (where $\varnothing$ denotes the empty set). Figure 3.5 shows the simple disjunctive decomposition of f in block diagram form. Note that when $s = 1$ or $s = n - 1$ the decomposition is trivial. Also note that 0 and 1 are decomposable by convention.

■ **Example 3.29 (a)** Consider the function $f(w, x, y, z) = wy'z + x'y'z + w'xyz$. It can be re-written as

$$f(w, x, y, z) = (w + x')y'z + (w + x')'yz$$

If we select $H(w, x) = w + x'$, then

$$f(w, x, y, z) = Hy'z + H'yz$$

In view of Equation (3.15), $Y = \{w, x\}$, $Z = \{y, z\}$, and f is represented by

$$f(w, x, y, z) = G[H(w, x), y, z]$$

The circuit logic diagram of f is shown in Figure 3.6(a).

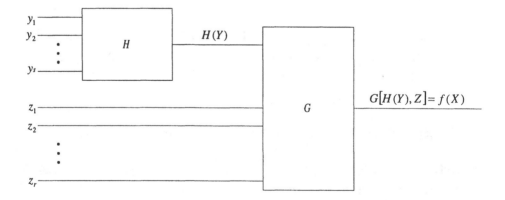

■ **Figure 3.5** Block diagram representation of disjunctive decomposition

(b) Consider the function $f(w, x, y, z) = \sum(0, 5, 7, 9, 11, 12)$. Expanded in SOP form, f can be written as

$$f(w, x, y, z) = w'x'y'z' + w'xy'z + w'xyz + wx'y'z + wx'yz + wxy'z'$$
$$= (w'x' + wx)y'z' + (w'x + wx')yz + (w'x + wx')y'z$$
$$= (w'x' + wx)y'z' + (w'x + wx')(yz + y'z)$$
$$= (w'x' + wx)y'z' + (w'x + wx')z$$

Let $H(w,x) = w'x + wx' = w \oplus x$; then

$$f(w, x, y, z) = (H + y + z)' + Hz = G[H(w,x), y, z]$$

The circuit logic diagram of f is shown in Figure 3.6(b). ∎

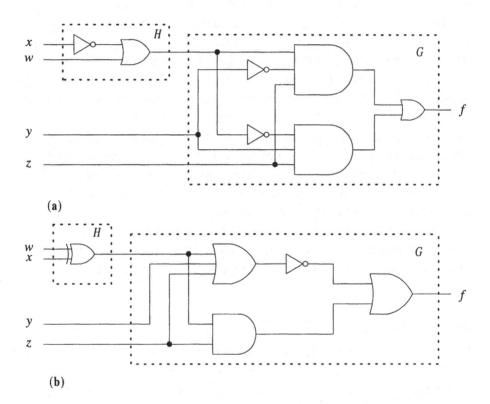

(a)

(b)

∎ **Figure 3.6** Logic circuit diagrams for Example 3.29

*3.9   IMPLEMENTATION OF SPECIAL BOOLEAN FUNCTIONS

As indicated in Section 3.8, various Boolean functions have special properties. Often, such functions can be implemented more easily than arbitrary Boolean functions and enable us to simplify the design of certain digital circuits. Here are some examples. A monotonic nondecreasing function is positive in all its variables. Since no variable is complemented, such a function can be implemented with AND and OR gates only. Symmetric functions often occur in practice. For example, the sum and carry and the difference and borrow functions in binary addition and subtraction, derived in Chapter 5, are symmetric functions. A circuit implementation of a totally symmetric function can have its inputs unlabeled since they are all interchangeable. The design of a large-scale system is often too complex but can be carried out more effectively if the system can be decomposed into a number of smaller-scale subsystems.

We are concerned in this book with the design of digital circuits implemented with two-state devices. There is, however, another type of logic device, called the *threshold gate*. A circuit implemented with threshold gates usually requires a fewer number of gates and interconnections than the implementation of the same circuit with conventional gates. While only a few implementations of threshold logic circuits are used now, the potential of this type of logic is very attractive and it may yet prove to be of practical importance.

Therefore, the purpose of this section is twofold: (1) to consider briefly some techniques for identifying and implementing special Boolean functions, and (2) to introduce the concept of threshold logic.

Symmetric Logic Circuits

If a function is totally symmetric, then by Definition 3.10 it is sufficient to specify only the number of variables that have to be 1 in order for the functional value to be 1. However, detecting total symmetry just by following the definition is not an easy task because there are $n!$ permutations of the n variables. Therefore, let us consider some simpler mechanisms for detecting symmetry.

---

* This section is optional.

□ **Theorem 3.14** An n-variable function f is symmetric (totally symmetric) if and only if it may be specified by a set of integers $A = \{a_0, a_1, \ldots, a_k\}$ (called **a-numbers**), where $0 \le a_j \le n$ ($j=0, 1, \ldots, k$, and $k = 0, 1, \ldots, n$), such that $f = 1$ when and only when a_j of the variables are 1. □

■ **Example 3.30** The set of a-numbers of the function $f(x, y, z) = xy + xz + yz$ is $A = \{2, 3\}$. ■

Since the allowable a-numbers are the $n+1$ integers $0, 1, \ldots, n$, a totally symmetric function may have any combination of these $n+1$ integers as a-numbers. Therefore, the number of totally symmetric functions is given by the combinatorial identity:

$$\sum_{j=0}^{n+1} \binom{n+1}{j} = 2^{n+1}$$

It can be similarly shown that there are $2^{(j+1)2^{n-j}}$ partially symmetric functions (Definition 3.11) of n variables that are symmetric in j ($j < n$) of the variables.

Let us denote a symmetric function by S_A^n, where A stands for the set of a-numbers and n indicates the number of variables. For example, the symmetric function $f(x, y) = x'y + xy'$ is designated by S_1^2, while the symmetric function given in Example 3.30 is designated by $S_{2,3}^3$.

Given two symmetric functions of the same variables, S_A^n and S_B^n, then $S_C^n = S_A^n + S_B^n$ and $S_D^n = S_A^n \cdot S_B^n$ are also symmetric functions. Their sets of a-numbers, C and D, are respectively equal to the union and intersection of the set of a-numbers A with the set of a-numbers B; that is, $C = A \cup B$ and $D = A \cap B$.. (These results follow directly from the definitions of the OR and AND operations.) Thus, for example, $S_{0,2,4}^4 + S_{3,4}^4 = S_{0,2,3,4}^4$ and $S_{0,2,4}^4 \cdot S_{3,4}^4 = S_4^4$.

Similarly, the complement of a totally symmetric function S_A^n, denoted by $\left(S_A^n\right)'$, is the totally symmetric function designated by S_{I-A}^n, where I is the set of integers $I = \{0, 1, \ldots, n\}$ and $I - A$ is the set of those integers included in I but not in A. For example, the complement of $S_{2,4}^4$ is the function $S_{0,1,3}^4$.

Consider the following two permutations:

$$\begin{pmatrix} x_1 & x_2 & \ldots & x_{n-2} & x_{n-1} & x_n \\ x_2 & x_3 & \ldots & x_{n-1} & x_1 & x_n \end{pmatrix} \qquad (3.16a)$$

and

$$\begin{pmatrix} x_1 & x_2 & \ldots & x_{n-2} & x_{n-1} & x_n \\ x_1 & x_2 & \ldots & x_{n-2} & x_n & x_{n-1} \end{pmatrix} \qquad (3.16b)$$

By repeatedly using these permutations, we can generate all possible permutations of the n variables. This fact leads to the following conclusion:

☐ **Theorem 3.15** A Boolean function $f(x_1, x_2, \ldots, x_n)$ is totally symmetric if and only if

$$f(x_1, x_2, \ldots, x_{n-1}, x_n) = f(x_2, x_3, \ldots, x_{n-1}, x_1, x_n)$$

and

$$f(x_1, x_2, \ldots, x_{n-1}, x_n) = f(x_1, x_2, \ldots, x_{n-2}, x_n, x_{n-1}) \qquad ☐$$

■ **Example 3.31** Consider the function

$$f(x_1, x_2, x_3, x_4, x_5) = x_1'x_2x_3x_4x_5 + x_1x_2'x_3x_4x_5 + x_1x_2x_3'x_4x_5$$
$$+ x_1x_2x_3x_4'x_5 + x_1'x_2'x_3'x_4'x_5' + x_1x_2x_3x_4x_5'$$

The function remains unchanged under the permutations (3.16a) and (3.16b) and is, therefore, a totally symmetric function. In fact, it is $S_{0,4}^5$. ■

A simple procedure to detect symmetry is first to represent the function in one of the two canonical forms. If we assume that the function is in canonical SOP form, then each a-number represents the number of 1's in the binary representation of the minterm. We refer to the a-number as the **weight** of the minterm. It follows then that, if all minterms of *identical* weights are contained in the representation of the function, the function is symmetric. The following examples illustrate this procedure by using a truth table. We can also use other forms of functional representation, such as the Karnaugh map which is introduced in Chapter 4.

■ **Example 3.32**   Consider the 4-variable function $f = \sum(0, 7, 11, 14)$. Its truth table is shown in Table 3.11, which lists the decimal and binary representations of the 16 4-literal minterms. The weight column lists the minterm weights, and the corresponding functional values of f are shown under the f column. $f = 1$ for minterm 0, whose weight is 0, and since no other minterm has a 0 weight, f includes all the minterms of 0 weight. $f = 1$ also for minterms 7, 11, and 14. The weight of each of these minterms is 3. However, because not all minterms of weight 3 are included in f (minterm 13 is not included), f is not totally symmetric.   ■

■ **Example 3.33** Let $f = \sum(0, 7, 11, 13, 14, 15)$. Since all minterms of weights 0, 3, and 4 are included in f, we conclude that f is the totally symmetric function $S_{0,3,4}^{4}$.   ■

Sometimes, a digital system can be characterized by a special class of symmetric functions, called **elementary symmetric functions**, that have only a *single a*-number. There are $n + 1$ elementary symmetric functions of n variables:

$$S_0^n = x_1' x_2' \cdots x_n'$$
$$S_1^n = x_1 x_2' \cdots x_n' + x_1' x_2 x_3' \cdots x_n' + \cdots + x_1' x_2' \cdots x_{n-1}' x_n$$
$$\vdots$$
$$S_n^n = x_1 x_2 \cdots x_n$$

Note that the weight of S_j^n ($j = 0, 1, \ldots, n$) is the binomial coefficient $\binom{n}{j}$. Also note that $S_i^n \cdot S_j^n = 0$ if $i \neq j$, and $\sum_{j=0}^{n+1} S_j^n = 1$.

■ **Table 3.11** Truth table for Example 3.32

Minterms		Weight	f
Decimal	Binary		
0	0000	0	1
1	0001	1	0
2	0010	1	0
3	0011	2	0
4	0100	1	0
5	0101	2	0
6	0110	2	0
7	0111	3	1
8	1000	1	0
9	1001	2	0
10	1010	2	0
11	1011	3	1
12	1100	2	0
13	1101	3	0
14	1110	3	1
15	1111	4	0

Decomposable Logic Circuits

There exist a variety of decomposition schemes, but only those digital systems represented by the class of *simple disjunctive decomposable functions* (Definitions 3.13 and 3.14) are considered in this section. We assume that $r = n - s$ and disregard the two trivial cases $s = 1$ or $s = n - 1$. To test for the nontrivial cases, $1 < s \leq n - 2$, we use a special array, called the **decomposition chart**, which requires an introduction.

Consider a truth table that represents the 2^n binary combinations of n variables. If "folded" about the middle, it can be represented in an array form, as shown in Figure 3.7(a) for $n = 4$. Each column is designated by one of the four binary combinations of the two variables (w and x) associated with the columns. Each row is similarly designated by one of the four binary combinations of the two variables (y and z) associated with the rows. The binary combination associated with each array *cell* is obtained by intersecting the appropriate row (column) with the appropriate column (row). For example, if w is the most significant variable and z is the least significant variable, then the cell marked 13 in Figure 3.7(a) corresponds to the binary combination $wxyz=1101$ obtained by intersecting the 11 column with the 01 row.

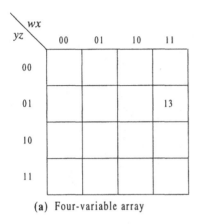

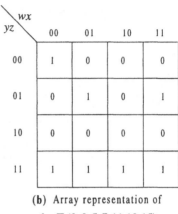

(a) Four-variable array

(b) Array representation of $f = \Sigma(0,3,5,7,11,13,15)$

■ **Figure 3.7** Array representation of Boolean functions

Because the array is just a rearrangement of the truth table, we can use it to represent Boolean functions. For example, Figure 3.7(b) shows the array representation of the 4-variable function $f(w,x,y,z) = \Sigma(0,3,5,7,11,13,15)$. As will be seen in Chapter 4, the Karnaugh map is just a variant of the array considered here.

Having represented a Boolean function in array form, let us observe the zero-one pattern in each column (row). The number of *distinct* columns (rows) in the array is called the **column (row) multiplicity**. For example, the column multiplicity of the array in Figure 3.7(b) is 3 because columns 01 and 11 have the same zero-one pattern, while columns 00 and 10 have different zero-one patterns. The row multiplicity of this array is 4, since all the rows are distinct. The reader is encouraged to verify that we can change either row or column multiplicity (or both) by rearranging the variables associated with the rows and columns.

Based on this introduction, we can now state a test for nontrivial disjunctive decomposition.

☐ **Theorem 3.16**   A Boolean function can be disjunctively decomposed into $f(X) = G[H(Y), Z]$ [Equation (3.15)] if and only if the column (row) multiplicity of the array having column headings $y_1, y_2, \ldots, y_s$ and row headings $z_1, z_2, \ldots, z_r$ is *at most* 2.   ☐

Let us apply this test to a specific example.

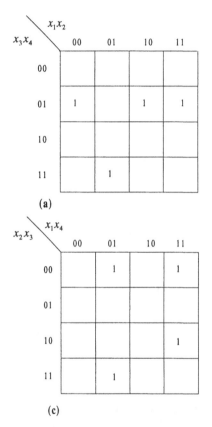

(a)

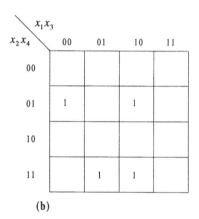

(b)

(c)

■ **Figure 3.8**   Array representations for Example 3.34

■ **Example 3.34**   Let $f(x_1, x_2, x_3, x_4) = \sum(1, 7, 9, 13)$.   Figure 3.8 shows three possible array representations for f obtained by rearranging the column and row headings. (We only show the 1 values of f in Figure

3.8; the empty cells correspond to the 0 values of f.) Table 3.12 lists the column and row multiplicities of the three arrays. As seen, only array (a) has column multiplicity that is not more than 2. Therefore, its column headings x_1 and x_2 correspond to y_1 and y_2, and its row headings x_3 and x_4 correspond to z_1 and z_2. Hence, f is decomposable into $f = G[H(x_1, x_2), x_3, x_4]$.

To find G and H explicitly, we have to manipulate f algebraically. We already did this for the same function in Example 3.29(a), where it was shown that $H(x_1, x_2) = x_1 + x_2'$ and $f = Hx_3', x_4 + H'x_3x_4$. ∎

The test is conceptually simple to apply but requires the generation of all nontrivial permutations of the n variables. This task is best accomplished using a computer program.

■ **Table 3.12** Column and row multiplicities for Example 3.34

	Column Multiplicity	Row Multiplicity
Array (a)	2	3
Array (b)	4	3
Array (c)	3	4

Threshold Logic Circuits

Threshold functions can be viewed as a generalization of the AND, OR, NAND, and NOR operations and often enable us to implement digital systems with fewer threshold gates than with conventional logic gates. Threshold functions are defined as follows.

☐ **Definition 3.15** A **threshold function** (also called a **linearly separable function**) of n binary variables is defined by the following inequalities:

$$f(x_1, x_2, \ldots, x_n) = 1 \text{ if and only if } \sum_{i=1}^{n} w_i x_i \geq T \qquad (3.17a)$$

$$= 0 \text{ if and only if } \sum_{i=1}^{n} w_i x_i < T \qquad (3.17b)$$

□

The values of the **threshold** T and the **weights** w_i (for $i = 1, 2, \ldots, n$) may be any real, finite, positive or negative numbers. The *sum* and *product* operations in Equation (3.17) are the *conventional arithmetic* operations (rather than Boolean operations). A schematic representation of a threshold gate is shown in Figure 3.9. A threshold gate is specified by its input variables and a **weight-threshold vector** $\{w_1, w_2, \ldots, w_n; T\}$.

Although designs for integrated-circuit threshold gates have been proposed, they have yet to become a commercial success. One problem associated with the production of reliable threshold gates is difficulty in controlling the parameters that determine the threshold and the weights.

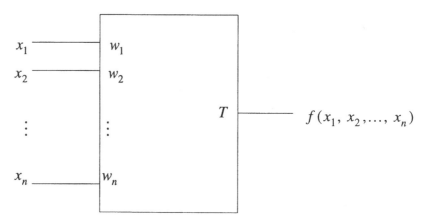

■ **Figure 3.9** Schematic representation of a threshold gate

■ **Example 3.35**   (a)   Let $f(x_1, x_2) = x_1 x_2$. Since $f = 1$ when $x_1 = x_2 = 1$, Equation (3.17a) yields $w_1 + w_2 \geq T$ as the first inequality that has to be satisfied. The other inequalities result from $f = 0$:

$$x_1 = 1, \; x_2 = 0 \quad \text{imply} \quad w_1 < T$$

$$x_1 = 0, \; x_2 = 1 \quad \text{imply} \quad w_2 < T$$
$$x_1 = 0, \; x_2 = 0 \quad \text{imply} \quad 0 < T$$

These four inequalities are satisfied, for example, by $T = 2$, $w_1 = 1$, and $w_2 = 1$. Hence, there exists a threshold gate implementing f that provides another realization of the AND operator.

(b) Let $w_1 = w_2 = 1$ and $T = 1$. Hence, by Equation (3.17a), $f = 1$ if and only if $x_1 + x_2 \geq 1$. Recall that the sum is the conventional arithmetic operation and, therefore, this inequality is satisfied if either x_1 or x_2 or both are equal to 1. Hence, this threshold gate implements the OR operation.

(c) Let $w_1 = w_2 = w_3 = 1$ and $T = 2$. Therefore, $f = 1$ if and only if $x_1 + x_2 + x_3 \geq 2$. To satisfy this inequality, the binary combination of the three variables must contain at least two 1's. Hence, $f = \sum(3, 5, 6, 7) = x_1 x_2 + x_1 x_3 + x_2 x_3$. Since $f = 1$ if at least two (a majority) of its variables are 1, this function is called a **majority function.** ∎

Example 3.35(a) indicates that the values of the threshold and weights need not be unique. In general, we have to solve the set of inequalities in order to obtain either a solution for the T and w_i's or a contradiction indicating that the function cannot be implemented by a single threshold gate. For an n-variable Boolean function, there are at most 2^n linear inequalities, some of which may be eliminated because they are implied by others. Example 3.35(c) shows that a *single* threshold gate can implement more complex functions than by using standard logic gates; to implement the 3-variable majority function, we require three AND gates and one OR gate.

To find the Boolean function implemented by a given threshold gate, we can use the truth table as illustrated in the following example. (Alternatively, we can use the Karnaugh map representation which is introduced in Chapter 4.)

■ **Example 3.36** A 4-variable function $f(x_1, x_2, x_3, x_4)$ implemented by a single threshold gate is specified by $w_1 = -1$, $w_2 = 1$, $w_3 = 2$, $w_4 = -3$, and $T = 2$. With x_1 the most significant bit, Table 3.13 shows the corresponding truth table. The rightmost column indicates whether or not Equation (3.17a) is satisfied. Since $f = 1$ only for those binary combinations that satisfy Equation (3.17a), then $f = \sum(2, 6, 14)$.   ∎

■ **Table 3.13** Truth table for Example 3.36

Minterms			
Decimal	Binary	$\sum_{i=1}^{4} w_i x_i$	$\sum_{i=1}^{4} w_i x_i \geq T$
0	0000	0	No
1	0001	-3	No
2	0010	2	Yes
3	0011	-1	No
4	0100	1	No
5	0101	-2	No
6	0110	3	Yes
7	0111	0	No
8	1000	-1	No
9	1001	-4	No
10	1010	1	No
11	1011	-2	No
12	1100	0	No
13	1101	-3	No
14	1110	2	Yes
15	1111	-1	No

Some Properties of Threshold Functions

□ **Property 3.1** Consider an n-variable function f that is implemented by a single threshold gate with a weight-threshold vector $\{w_1, w_2, \ldots, w_n; T\}$. If any one of the inputs, x_i say, is complemented, then f can be implemented by another single threshold gate with a weight-threshold vector $\{w_1, w_2, \ldots, -w_i, \ldots, w_n; T - w_i\}$ [i.e., with w_i replaced by $(-w_i)$ and a threshold equal to $T - w_i$].   □

☐ **Property 3.2**   If a function is implemented by a single threshold gate, then by an appropriate selection of complemented and uncomplemented input variables the function can be implemented by another single threshold gate whose weights have any desired sign distribution.   ☐

As a consequence of Property 3.2, the function can be implemented by a single threshold gate with only positive or negative weights.

☐ **Property 3.3**   Every threshold function is unate.   ☐

The converse of Property 3.3 is not true since there are unate functions that are not threshold functions, as illustrated by the following example.

■ **Example 3.37**   Consider the unate function $f(x_1, x_2, x_3, x_4) = x_1 x_2 + x_3 x_4$. This function takes the value 1 if

(a) $x_1 = x_2 = 1$ and $x_3 = x_4 = 0$
(b) $x_1 = x_2 = 0$ and $x_3 = x_4 = 1$

and the value 0 if

(c) $x_1 = x_3 = 0$ and $x_2 = x_4 = 1$
(d) $x_1 = x_3 = 1$ and $x_2 = x_4 = 0$

Conditions (a) to (d) imply the following inequalities:

$$w_1 + w_2 \geq T$$
$$w_3 + w_4 \geq T$$
$$w_2 + w_4 < T$$
$$w_1 + w_3 < T$$

Combining the first two inequalities, we get

$$w_1 + w_2 + w_3 + w_4 \geq 2T$$

and combining the last two inequalities yields

$$w_1 + w_2 + w_3 + w_4 < 2T$$

These two inequalities cannot hold simultaneously. Therefore, we have a contradiction and can conclude that the given unate function f

cannot be implemented by a single threshold gate.   ∎

☐ **Property 3.4**   If an n-variable function f can be implemented by a single threshold gate with a weight-threshold vector $\{w_1, w_2 \ldots, w_n; T\}$, then f' can be implemented by another single threshold gate whose weight-threshold vector is $\{-w_1, -w_2, \ldots, -w_n; -T\}$.   ☐

In the special case when all the weights are equal ($w_i = w \neq 0$, for $i = 1, 2, \ldots, n$), we get the following threshold function:

$$f(x_1, x_2, \ldots, x_n) = 1 \quad \text{if and only if } \sum_{i=1}^{n} x_i \geq \frac{T}{w}$$

$$= 0 \quad \text{if and only if } \sum_{i=1}^{n} x_i < \frac{T}{w}$$

Since the maximum value of $\sum_{i=1}^{n} x_i$ is n and the minimum value is 0, then f is identically equal to 1 if $T/w = 0$ and f is identically equal to 0 if $T/w > n$. For $0 < T/w \leq n$, $f = 1$ if k or more variables are equal to 1, and $f = 0$ if $k - 1$ or less variables are equal to 1, where $k - 1 < T/w \leq k$ with $k = 0, 1, 2, \ldots$. Since we do not have to distinguish between the variables (and it is only their total number that counts), this special case is, in fact, the class of totally symmetric functions $S_{k, k+1, \ldots, n}^{n}$. These functions are known as **symmetric threshold functions** (or **voting functions**). In addition, if n is odd and $k = (n+1)/2$, the function is a *majority function*, like the one already encountered in Example 3.35(c).

Problems

3.1   Prove Theorem 3.3.

3.2 Prove Theorem 3.4.

3.3 Prove Theorem 3.5.

3.4 Prove Theorem 3.6.

3.5 Prove Theorem 3.9.

3.6 The XOR operation is also called the symmetric difference (Δ) in Boolean algebra and is defined as $x\Delta y = xy' + x'y$. Prove that $x\Delta(x+y) = x'y$.

3.7 Write the duals of the following statements:

(a) $(x' + y')' = xy$.

(b) $xy' = 0$ if and only if $xy = x$.

(c) $x = 0$ if and only if $y = xy' + x'y$ for all y.

3.8 Prove the following identities:

(a) $(x \odot y \odot z)' = x \odot y \oplus z$.

(b) $(x \oplus y \oplus z)' = x \oplus y \odot z$.

3.9 Show that:

(a) If $f(x,y,z) = x \oplus y \oplus z$, then $f(x,y,z) = x \odot y \odot z$.

(b) If $g(x,y,z) = x \oplus y \odot z$, then $g(x,y,z) = x \odot y \oplus z$.

3.10 Using the rules of Boolean algebra, simplify the following Boolean expressions:

(a) $\left\{ \left[(xy)'x\right]'\left[(xy)'y\right]' \right\}'$

(b) $\left\{ \left[x'yw' + xwz\right]' + \left[(xwz)' + y'w'z' + ywz'\right]' \right\}'$

(c) $(x+y)'(x'+y')'$

(d) $y(wz' + wz) + xy$

(e) $xyz + x'y'z + x'yz + xyz' + x'y'z'$

3.11 Find the complements of the following Boolean expressions and reduce them to a minimum number of literals.

(a) $(xy' + w'z)(wx' + yz')$

(b) $wx' + y'z'$

(c) $x'z + w'xy' + wyz + w'xy$

3.12   Obtain the truth tables of the following functions.

   (a) $F_1(w,x,y,z) = xy + x'z$

   (b) $F_2(w,x,y,z) = wx' + yz + w'y'$

3.13   Obtain the truth tables of $F_1 + F_2$ and $F_1 F_2$, where F_1 and F_2 are given in Problem 3.12(a) and (b).

3.14   A *self-dual* Boolean function is a function whose truth table remains unchanged when all the 0's and 1's are interchanged. How many self-dual Boolean functions of n variables are there?

3.15   Establish, whether the following sets of operators are functionally complete.

   (a) $\{f,0\}$, where $f(x,y) = x' + y$

   (b) $\{f,1\}$, where $f(x,y,z) = x'y' + x'z' + y'z'$

3.16   How many Boolean functions of four variables take on the value 1 exactly twice? Three times?

3.17   Show that if $m_1 + \cdots + m_k$ is the canonical SOP form of $f(x_1,\ldots,x_n)$ then $m_1' \cdots m_k'$ is the canonical POS form of $f'(x_1,\ldots,x_n)$.

3.18   Find the canonical SOP form of f_1:

x	y	f_1
0	0	1
0	1	1
1	0	0
1	1	1

3.19   Find the canonical SOP form of each of the following functions:

x	y	z	f_2	f_3	f_4
0	0	0	1	1	1
0	0	1	1	1	0
0	1	0	0	1	0
0	1	1	0	0	0
1	0	0	1	0	1
1	0	1	0	1	0
1	1	0	1	1	0
1	1	1	1	1	1

3.20 Find the canonical POS form of the function f_2 in Problem 3.19.

3.21 Derive the Boolean expressions and truth tables for the logic circuits shown in Figures P3.21(a) and (b).

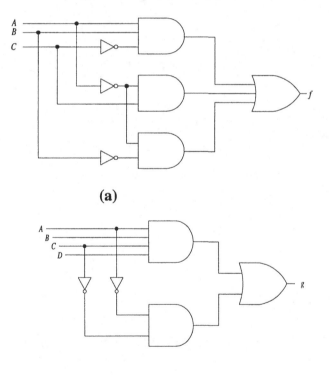

(a)

(b)

■ **Figure P3.21**

3.22 **(a)** Show that the circuits of Figures P3.22(a) and (b) are equivalent.

 (b) Show that the circuits of Figures P3.22(c) and (d) are equivalent.

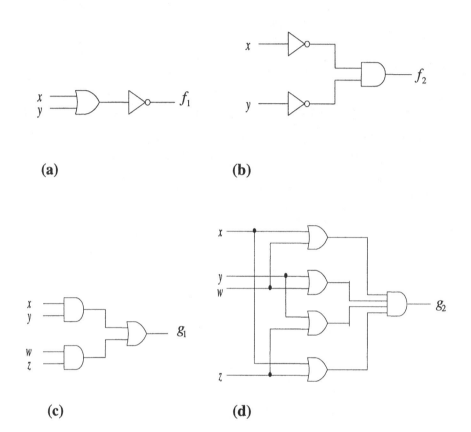

(a)   **(b)**

(c)   **(d)**

■ **Figure P3.22**

3.23 (a) Show that a positive-logic AND gate is a negative-logic OR gate, and vice versa.

 (b) Show that a positive-logic NAND gate is a negative-logic NOR gate, and vice versa.

 (*Note*: Positive logic and negative logic are defined in Section 1.2.)

3.24 Using multiple-input AND and OR gates, what is the maximum number of gates required to implement a Boolean function of n variables, $f(x_1, \ldots, x_n)$, in canonical SOP form?

3.25 Obtain the circuit logic diagram of a 3-input NAND gate using only 2-input NAND gates.

3.26 Explain why not all 4-input, single-output Boolean switching functions can be replaced by three 2-input, single-output Boolean switching functions as in Figure P3.26.

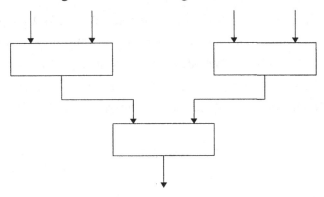

■ **Figure P3.26**

3.27 Express the following functions in canonical SOP and POS forms.

    (a) $F(w, x, y, z) = z(w' + x) + x'z$

    (b) $F(w, x, y, z) = w'x'z + y'z + wxz' + wxy'$

    (c) $F(x, y, z) = (xy + z)(y + xz)$

3.28 Convert the following to the other canonical form.

    (a) $F = \sum(0, 2, 6, 11, 13, 14)$ **(b)** $F = \prod(1, 2, 4, 6, 12)$

    (c) $F = \prod(0, 3, 6, 7)$ **(d)** $F = \sum(1, 3, 7)$

3.29 Design a circuit with three inputs and a single output so that the output is 1 precisely when two or three of the inputs have a value 1.

3.30 A *2's module* is a circuit that has two inputs, A and FLAGIN, and two outputs, B and FLAGOUT. If FLAGIN = 1, then $B = A'$ and

FLAGOUT = 1. If FLAGIN = 0 and $A = 1$, then FLAGOUT = 1. If FLAGIN = 0 and $A = 0$, then FLAGOUT = 0. If FLAGIN = 0, then $A = B$. Design a circuit to implement the 2's module.

3.31 Using 2's modules (Problem 3.30), design a circuit that computes the 2's complement $y_3 y_2 y_1$ of the 3-bit binary number $x_3 x_2 x_1$.

3.32 A 4-bit binary number is represented as $A_3 A_2 A_1 A_0$ (A_0 is the least significant bit). Design a logic circuit that will determine whenever the binary number is greater than 6.

3.33 Figure P3.33 shows a multiplier circuit that takes two 2-bit binary numbers $x_1 x_0$ and $y_1 y_0$ and produces an output binary number $z_3 z_2 z_1 z_0$ that is equal to the arithmetic product of the two input numbers. Assume that x_0, y_0, and z_0 are the least significant bits (*lsb*). Design a logic circuit for the multiplier.

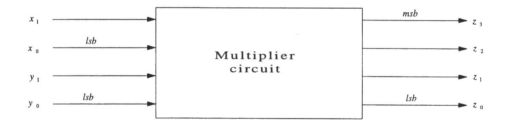

■ **Figure P3.33**

3.34 Let $x_1 x_0$ and $y_1 y_0$ represent two 2-bit binary numbers. Design a logic circuit, using x_1, x_0, y_1, and y_0 as inputs, whose output $z = 1$ when and only when the two numbers are equal.

3.35 Consider the block diagram shown in Figure P3.35. Given that $f_1 = ab + b'c' = x'$ and $f_2 = a'(bc' + b'c) = xy,'$ determine $x(a,b,c)$ and $y(a,b,c)$.

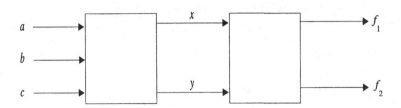

■ **Figure P3.35**

3.36 Design a logic circuit that accepts a 3-bit number and generates at its outputs a binary number equal to the square of the input number.

3.37 Four large tanks at a chemical plant contain different liquids being heated. Liquid-level sensors are used to detect whenever the level in tanks A and B rises above a predetermined level. Temperature sensors in tanks C and D detect when the temperature in these tanks drops below a prescribed temperature limit. Assume that the liquid-level sensor outputs (A and B) are LOW (0) when the level is satisfactory and HIGH (1) when the level is too high. Also, the temperature-sensor outputs (C and D) are LOW (0) when the temperature is satisfactory and HIGH (1) when the temperature is too low. Design a logic circuit that will detect whenever the level in tank A or tank B is too high at the same time that the temperature in either tank C or tank D is too low.

3.38 Figure P3.38 shows the intersection of two roads. Vehicle-detection sensors are placed along lanes A, B, C, and D. The sensor outputs are LOW (0) when no vehicle is present and HIGH (1) when a vehicle is present. The intersection traffic light is controlled as follows:

(1) The E-W traffic light will be green whenever both lanes C and D are occupied.

(2) The E-W light will also be green whenever either C or D is occupied but lanes A and B are not both occupied.

(3) The N-S light will be green whenever both lanes A and B are occupied but C and D are not both occupied.

(4)   The N-S light will also be green when either A or B is occupied while C and D are both vacant.

(5)   The E-W light will be green when no vehicles are present.

Design a logic circuit, whose inputs are the sensor outputs A, B, C, and D, to control the traffic light. There should be two outputs, N/S and E/W, which become 1 when the corresponding light is to be green.

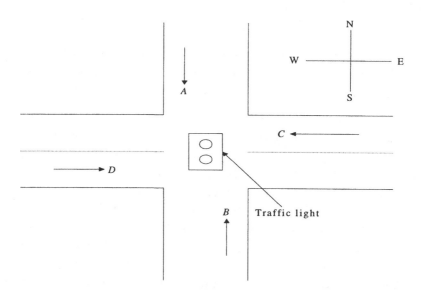

■   **Figure P3.38**

3.39   Design a logic circuit that tests the operation of a traffic light. If the control circuit of the traffic light malfunctions, it is possible that an invalid combination of signals will appear. The sole purpose of the test circuit is to detect any invalid combination and then generate an error signal that can be transmitted to the city maintenance area. The valid conditions are shown in Figure P3.39. Use E for the error signal and designate by $E = 1$ that the traffic light has malfunctioned.

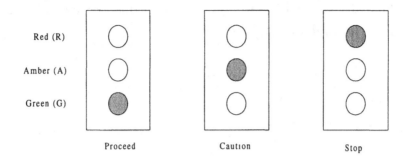

Proceed   Caution   Stop

■ **Figure P3.39**

3.40 Is the function $f(w,x,y,z) = xyz' + w'xyz$ unate? Justify your answer.

3.41 The truth table of a symmetric Boolean function remains unchanged for any permutation of the n columns of the variables. How many symmetric Boolean functions of n variables are there?

3.42 Express the following as symmetric functions:

(a) $A'S_{0,1,4}(B,C,D,E) + AS'_{0,3,4}(B,C,D,E)$

(b) $A'S_{0,1,4}(B,C,D,E) + AS_{0,3,4}(B,C,D,E)$

(c) $A'S_{0,1,4}(B,C,D,E) + AS_{0,3,4}(B',C',D',E')$

3.43 Let $f(w,x,y,z) = w'x'z' + wx'z + w'yz + wyz'$. Obtain a nontrivial decomposition of f.

3.44 Is the function $f(x_1,x_2,x_3,x_4) = x'_1x'_2x'_3x'_4 + x'_1x_2x_3x_4 + x_1x'_2x'_3x_4 + x_1x_2x_3x'_4$ decomposable?

3.45 Determine which of the following functions is a threshold function. For each threshold function, find a corresponding weight-threshold vector.

(a) $f_1(x_1,x_2,x_3) = \sum(1,2,3,7)$

(b) $f_2(x_1,x_2,x_3) = \sum(0,2,4,5,6)$

(c) $f_3(x_1,x_2,x_3) = \sum(0,3,5,6)$

3.46   Find the function $f(x_1, x_2, x_3, x_4)$ implemented by the threshold gate shown in Figure P3.46.

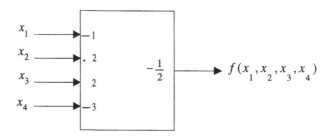

■   **Figure P3.46**

3.47   Determine the function $f(x_1, x_2, x_3, x_4)$ implemented by the threshold circuit shown in Figure P3.47.

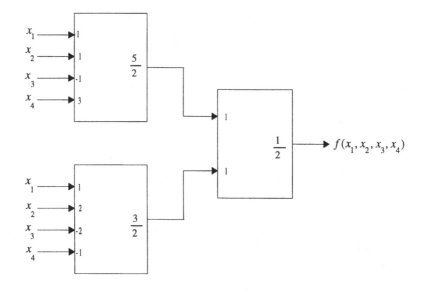

■   **Figure P3.47**

4 Combinational Logic Circuits

4.1 INTRODUCTION

Digital systems are classified either as **combinational systems** or **sequential systems**. The two types of systems have various common characteristics, but there are some features that distinguish between them. In a combinational system, shown in block diagram form in Figure 4.1, the output signals depend only on the *current* input signals. In other words, the output signals at any given instant of time are just the results of logical operations on the input variables at that instant. Therefore, the behavior of an n-input, m-output combinational circuit can be described with great simplicity: The 2^n possible values of the input variables can be listed, and the value of each of the m output functions can be determined and tabulated for each input value. Combinational circuits are discussed in this chapter and in Chapter 5.

By contrast, if information concerning *previous* inputs is also required to determine the present output signals, then *memory* is inherent within the system and we call it *sequential*. Actually, a sequential system consists of a combinational circuit with memory devices connected to form feedback paths, as shown in Figure 4.2. We will return to the concepts of feedback and memory in sequential systems in Chapters 6 and 7.

Much of our discussions in this chapter will be focused on providing tools to express Boolean switching functions in a *minimal* form so that *cost-effective* combinational circuits can be implemented. In essence, if we can reduce the number of gates and literals required to implement a given Boolean function, the resulting circuit would be implemented more economically. But why should we be concerned with minimization when integrated-circuit (IC) technology, the predominant technology used in digital systems design, provides us with multiple-gate chips and a large variety of other functions at low cost?

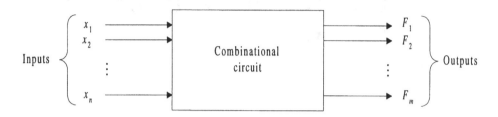

■ **Figure 4.1** A general combinational system

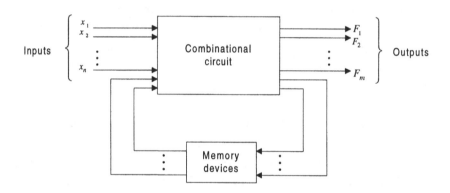

■ **Figure 4.2** A general sequential system

We can answer this question on two levels. First, the minimization of switching functions of a large number of variables is very important due to the extensive use of programmable arrays in the design of VLSI chips. (Programmable arrays will be discussed in Section 5.9.) The ideas presented in this chapter provide the conceptual framework that is prerequisite to understanding the minimization problem and process.

We can also answer the question by saying that a gate saved may enable us to reduce the number of chips required to implement the circuit and produce a more cost-effective design. Moreover, since the number of literals affects the *fan-in* of the gates (see Section 1.2), reducing the number of literals may enable us to utilize more effectively the various chips available for our design. Before illustrating these points, note first that one of the constraints in the design of IC chips is the number of input and output terminals (called *pins*) that a chip can have. If we assume that the number of pins is fixed at 21, we can design, for example, a chip containing two 2-input OR gates and five 2-input AND gates, a chip containing a single 4-input OR gate and four 3-input AND gates, or a chip having one 8-input OR gate and two 4-input AND gates. Call these chips C_1, C_2, and C_3 respectively, and consider now the implementation of a 4-input, 2-output circuit where the output functions are

$$F_1(a, b, c, d) = \Sigma(0, 2, 5, 7, 8, 10, 13, 15)$$

$$F_2(a, b, c, d) = \Sigma(4, 5, 6, 7, 8, 9, 10, 11)$$

For simplicity, assume that both the complemented and uncomplemented variables are available as inputs to the circuit.

In their canonical sum-of-products (SOP) form, each function consists of 32 literals and can be implemented with eight 4-input AND gates and one 8-input OR gate. Hence, we require two C_3 chips to implement the circuit. But, by algebraically manipulating F_1 and F_2, we find that they can be expressed as

$$F_1 = a'b'd' + ab'd' + a'bd + abd$$
$$F_2 = a'bc' + a'bc + ab'c' + ab'c$$

Each of these SOP expressions consists of 12 literals and can be implemented with four 3-input AND gates and a single 4-input OR gate. In this case, we must use two C_2 chips for the circuit. If we further simplify the functions, we can reduce them to their *minimal* SOP forms:

$$F_1 = b'd' + bd$$
$$F_2 = a'b + ab'$$

Each of these expressions contains only four literals and requires two 2-input AND gates and one 2-input OR gate. Hence, we now need only a single C_1 chip to implement the circuit.

We can safely assume that the three chips cost the same. Then, what our rather simplistic example demonstrates is that a considerable savings can be realized if the circuit functions are expressed in minimal form. Another important factor supporting this argument is presented in the following section where we introduce the minimization problem.

4.2   ANALYSIS AND DESIGN OF COMBINATIONAL CIRCUITS

A combinational circuit can be specified in several ways. The most common representations are (1) verbal descriptions of the behavior of the circuit, (2) circuit logic diagrams that show the interconnection of gates within the circuit, and (3) Boolean expressions (or truth tables) for each output function. From what we have learned in Chapter 3, we already know that circuit diagrams and Boolean expressions of outputs are essentially interchangeable.

Therefore, combinational circuit descriptions boil down to two: functional descriptions and structural descriptions. *Functional descriptions* are like the engineer's notion of a "black box": We know the inputs and the outputs but we do not know the exact details of the internal construction of the box. *Structural descriptions*, by contrast, expose the internal workings of the box, but what it actually does must be determined. In essence, the digital system designer is called on to translate from a functional description to a structural description of a circuit. Much of the remainder of this book is devoted to this basic translation process.

Throughout this text, we will emphasize time and again that the *design* of digital circuits is basically the reverse process of *analysis*. The design process starts from the specifications of a required function and terminates with a circuit logic diagram. The analysis process begins with a circuit logic diagram and terminates with a set of Boolean functions, a truth table, or a verbal description of the operation of the circuit.

We have already designed our first digital circuit in Section 3.7 and will continue in this vein in this and subsequent chapters. Design seems a more creative process than analysis. It is always fascinating to solve a problem, to turn an idea into a working circuit that does something useful. But, then, having designed a circuit, we usually invoke the analysis procedure to check and verify our design. We ask: Will the circuit work? Did we satisfy all the specifications? Was there something implied in the specifications that we did not take into account? But we may also be given a circuit designed by others and asked: How does it work? Does it work? Or, if it does not work, what can be done to make it work? So if we seem preoccupied with design, it is only because we rely on the duality between analysis and design.

Analysis Procedure

The conversion between a circuit logic diagram and the Boolean expressions of the outputs is accomplished by successively labeling each gate output with an expression derived from the gate inputs and the gate function. In general, proceed as follows:

1. In a combinational circuit there will always be at least one gate whose only inputs are the independent variables (the inputs to the circuit). Label the output of each such gate with a Boolean expression using the input symbols and the gate function.
2. Next there may be gates whose inputs come either from previously labeled gates, or from circuit inputs, or both. Find a Boolean expression for each of these gates.
3. Repeat step 2 until all the Boolean expressions for the outputs of the circuit are obtained.

We call this transformation process the **analysis procedure** and illustrate it in the following example.

■ **Example 4.1** The Boolean expression for the output of the combinational circuit diagram in Figure 4.3 is

$$x_1 x_3' + x_2(x_3 x_4 + x_1)$$ ■

- **Figure 4.3** Transforming a circuit diagram to a Boolean expression

A change in the input signals may cause a momentarily unstable response in the outputs of a combinational circuit. The reason for this is that, when an input signal is changed, the *propagation time delays* (see Section 1.2) within the circuit may cause some gates to receive one signal for a while and another signal a short time later. The outputs, therefore, may change more than once in response to a single change in the input signal. Each output will, nevertheless, stop changing eventually and remain **stable** for a given input. The Boolean expressions obtained in the analysis procedure are those giving the stable output values for each input combination. For this reason, these outputs are often referred to as **steady-state output functions**.

The reverse process, where we transform from any Boolean expression to a circuit diagram, is illustrated in the following example.

- **Example 4.2** The circuit diagram for the expression

$$F = x_1 x_2 (x_4 + x_5) + x_3 (x_4 + x_5') + x_1 x_2 x_6' x_7 + x_3 x_6 x_7$$

is shown in Figure 4.4   ■

Different connections of gates within the circuit may produce the same outputs for each input combination. This happens when the corresponding Boolean expressions for the different circuits generate the same truth table and are, therefore, *logically equivalent* (see Section 3.8). Hence, two different combinational circuits are equivalent if their output expressions are (logically) equivalent. Therefore, the set of combinational single-output circuits, just like the set of Boolean expressions, may be partitioned into *disjoint equivalence classes*, where each equivalence class is associated with a *unique* truth table, the one induced by each output expression for the circuits in that class.

On the other hand, for a given truth table of a Boolean function F, the canonical sum-of-products (product-of-sums) form of F is a particular member of the class of all Boolean expressions that induce F. Moreover, there is only one canonical sum-of-products (product-of-sums) expression

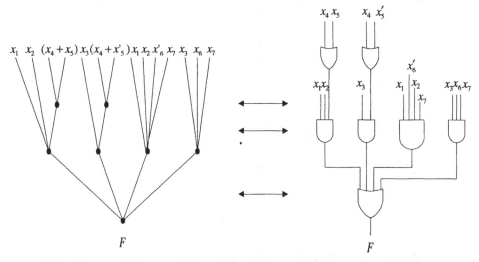

■   **Figure 4.4** Transforming a Boolean expression to a circuit diagram

in the class, so the canonical form is a representative, so to speak, for the whole class. There is, then, a one-to-one correspondence between equivalence classes of Boolean expressions and canonical forms, which, in turn, determines a one-to-one correspondence between canonical forms and truth tables. The diagram in Figure 4.5 illustrates in symbolic form what we have just said, and Examples 4.3 and 4.4 will refresh our memory of these correspondences.

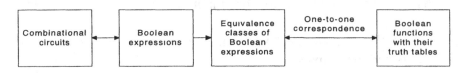

■   **Figure 4.5** Correspondences between combinational circuits, Boolean expressions, and truth tables

- **Example 4.3**   The Boolean expression $(x_1' x_2)'(x_1 + x_3)$ induces the truth table of Table 4.1.   ■

- **Table 4.1**   Truth table of $(x_1' x_2)'(x_1 + x_3)$

Decimal	x_1	x_2	x_3	F
0	0	0	0	0
1	0	0	1	1
2	0	1	0	0
3	0	1	1	0
4	1	0	0	1
5	1	0	1	1
6	1	1	0	1
7	1	1	1	1

- **Example 4.4** The Boolean expression

$$x_1' x_2' x_3' + x_1' x_2 x_3' + x_1 x_2' x_3' + x_1 x_2' x_3 + x_1 x_2 x_3$$
$$= m_0 + m_2 + m_4 + m_5 + m_7$$

is the canonical sum-of-products form for the function F whose truth table is listed in Table 4.2.   ■

Design Procedure

The design of combinational circuits starts with a word description of the problem and ends with a circuit logic diagram. In general, the **design procedure** involves the following steps:

1. State the problem.
2. Assign letter symbols to the input variables and the output functions.
3. Derive the truth tables that define the relationship between the inputs and the outputs.
4. Obtain Boolean expressions for each output.
5. Simplify (minimize) the Boolean expressions.
6. Draw the circuit logic diagram.

■ **Table 4.2** Truth table of $F = \Sigma(0, 2, 4, 5, 7)$

Decimal	x_1	x_2	x_3	F	
0	0	0	0	1	$m_0 = x_1' x_2' x_3'$
1	0	0	1	0	
2	0	1	0	1	$m_2 = x_1' x_2 x_3'$
3	0	1	1	0	
4	1	0	0	1	$m_4 = x_1 x_2' x_3'$
5	1	0	1	1	$m_5 = x_1 x_2' x_3$
6	1	1	0	0	
7	1	1	1	1	$m_7 = x_1 x_2 x_3$

The design process is summarized in the flow diagram shown in Figure 4.6. The *minimization problem* (step 5) will be considered later in this section, and the *minimization process* will be discussed in subsequent sections.

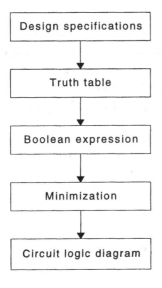

■ **Figure 4.6** Design flow diagram

Design Examples

For some problems it may be possible to go directly from a word description of the desired circuit to Boolean expressions for the output functions. In other cases, it may be better first to specify the truth table of the output functions and then derive Boolean expressions from it. The following example illustrates how to go directly from a word description to a Boolean expression.

■ **Example 4.5   Alarm circuit:** The alarm will ring if and only if the alarm switch is turned on and the door is not closed, or it is after 5 P.M. and the window is not closed.

The first step is to associate a variable with each phrase in the sentence. Let us, therefore, use the following assignments:

$$A = \text{alarm will ring}$$
$$S = \text{alarm switch is on}$$
$$D = \text{door is closed}$$
$$W = \text{window is closed}$$
$$T = \text{it is after 5 P.M.}$$

We assume that a variable has a value 1 when the phrase is true and 0 when it is false. Hence, this assignment implies that $A = 1$ if and only if the alarm will ring, $S = 1$ if the switch is turned on, and $T = 1$ if it is after 5 P.M. Also, since D represents "the door is closed," D' will represent "the door is not closed." Likewise, W' represents "the window is not closed."

Using this assignment of variables, the design specifications can be translated into the Boolean expression

$$A = SD' + TW'$$

This expression, in turn, produces the circuit diagram of Figure 4.7.

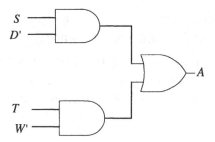

■ **Figure 4.7** Circuit diagram of the alarm system

Our choice of letters for the variables is, of course, completely arbitrary. We could just as easily let $x_1 = S$, $x_2 = D$, $x_3 = T$, and $x_4 = W$, so that $A = x_1 x_2' + x_3 x_4'$. ■

The following example illustrates how one might go from the design description to a truth table and from there to a circuit diagram.

■ **Example 4.6 Parity-bit generator and detector:** When binary code words are transmitted over a communication link, frequently an extra bit, called a *parity bit*, is added to detect transmission errors (see Section 2.5). The message, which *includes* the parity bit, is transmitted and then checked at the receiving end for errors, as shown in Figure 2.2. The circuit that generates the parity bit is called *parity generator*, and the circuit that checks the message in the receiver is a *parity detector*.

In Figure 4.8(a), we list the truth table for an *odd-parity* generator for 3-bit code words. The three bits (x_1, x_2, x_3) constitute the inputs to the parity generator circuit. The parity bit G is the circuit output and is generated so that the total number of 1's in x_1, x_2, x_3, and G is odd. Therefore, $G = 1$ if and only if the number of 1's among x_1, x_2, and x_3 is even. Figure 4.8(b) shows an implementation of G with AND and OR gates. We ask the reader to show that the circuit requires even fewer gates if XOR gates are used.

The parity detector output D is a function of four variables (x_1, x_2, x_3, G) and should be 1 whenever an error occurs in transmission, that is, when the number of 1's in the four inputs is even. Figure 4.8(c) gives the truth table for the odd-parity detector circuit. We leave it to the reader to implement the parity detector circuit with AND and OR gates. ■

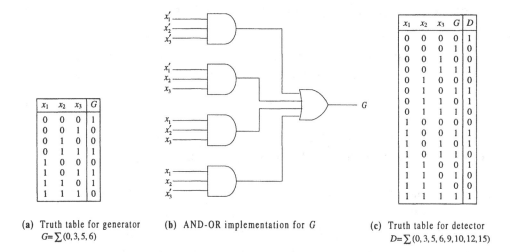

(a) Truth table for generator
$G = \sum(0, 3, 5, 6)$

(b) AND-OR implementation for G

(c) Truth table for detector
$D = \sum(0, 3, 5, 6, 9, 10, 12, 15)$

■ **Figure 4.8** Odd-parity generator and detector

The circuit design problem may call for more than one output function, as illustrated in the following examples.

■ **Example 4.7 Binary full-adder:** Since the addition of two numbers in any positional number system is performed column by column (see Section 2.4), only the two digits of a given column and a carry from the previous column need be considered at a time. Whenever the sum exceeds or equals the base, the carry operation adds one unit to the next higher column.

In base 2, the (binary) *full-adder* circuit consists of three inputs and two outputs. Two of the input variables (x_1 and x_2) represent the addend and augend bits, respectively, and the third variable (x_3) represents the *input carry* from the previous column. Since the sum $x_1 + x_2 + x_3$ is a 2-bit binary number, we let S represent its least significant bit and C the most significant bit. Therefore, C will be the *output carry* to the next column. For example, the sum of $x_1 = 1$, $x_2 = 1$, and $x_3 = 1$ is the 2-bit number 11, so that $C = 1$ and $S = 1$. But, on the other hand, if $x_1 = 1$, $x_2 = 0$, and $x_3 = 0$, then their sum is the 2-bit number 01, so that $C = 0$ and $S = 1$.

■ **Table 4.3** Truth table for a binary full-adder

Addend x_1	Augend x_2	Input Carry x_3	Output Carry C	Sum S
0	0	0	0	0
0	0	1	0	1
0	1	0	0	1
0	1	1	1	0
1	0	0	0	1
1	0	1	1	0
1	1	0	1	0
1	1	1	1	1

From the truth table for the full-adder, shown in Table 4.3, we can derive the following expressions for S and C:

$$S = \sum(1, 2, 4, 7) = x_1' x_2' x_3 + x_1' x_2 x_3' + x_1 x_2' x_3' + x_1 x_2 x_3$$

and

$$C = \sum(3, 5, 6, 7) = x_1' x_2 x_3 + x_1 x_2' x_3 + x_1 x_2 x_3' + x_1 x_2 x_3$$

Figure 4.9 shows an implementation of the full-adder circuit with AND and OR gates. The output S is implemented in its canonical sum-of-products form. We leave as an exercise for the reader to show that

S is also equivalent to $x_1 \oplus x_2 \oplus x_3$ and can, therefore, be implemented by a 3-input XOR gate. The output C is also be represented by the expression $x_1 x_2 + x_1 x_3 + x_2 x_3$. This expression can be derived using the Karnaugh map technique (as will be shown later), but we can also obtain this expression by applying the laws of Boolean algebra: Combine $x_1 x_2 x_3$ with each of the other three terms in C and apply the distributive law.   ■

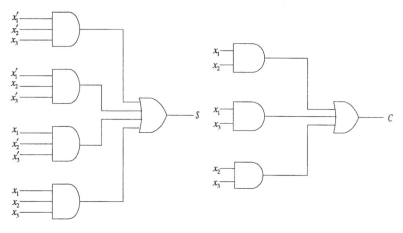

■   **Figure 4.9**   Binary full-adder circuit diagram

■   **Example 4.8**   **Monitoring device:** An electric motor is supplied by three generators. The operation of each generator is monitored by activating one or two warning devices as soon as the generator fails. The monitoring circuit activates the warning devices whenever the following conditions are satisfied: (1) A warning lamp lights up if one or two generators fail. (2) An acoustic alarm is activated if two or more generators fail.

Number the generators 1, 2, and 3 and let $x_1, x_2,$ and x_3 represent the inputs to the monitoring circuit from the respective generators. Let $x_i = 0$ mean that generator i is working and $x_i = 1$ designate that it has failed. The monitoring circuit has two output functions, $A(x_1, x_2, x_3)$ and $L(x_1, x_2, x_3)$. Assume that $A = 1$ if and only if the acoustic alarm sounds and $L = 1$ if and only if the warning lamp lights up.

From the design specifications, we conclude that the lamp will *not* light up only in two situations: (1) when all generators are operating, or (2) if all have failed. Hence,

$$L' = x_1' x_2' x_3' + x_1 x_2 x_3$$

and therefore

$$L = (x_1 + x_2 + x_3)(x_1' + x_2' + x_3')$$

Similar considerations yield

$$A = x_1 x_2 x_3' + x_1 x_2' x_3 + x_1' x_2 x_3 + x_1 x_2 x_3$$

Note that this expression can be simplified (as in Example 4.7) to obtain

$$A = x_1 x_2 + x_1 x_3 + x_2 x_3$$

The monitoring circuit diagram is shown in Figure 4.10. ∎

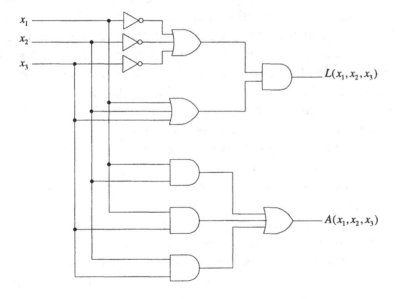

∎ **Figure 4.10** Circuit diagram for the monitoring device

The Minimization Problem

In general, any combinational design problem can be converted to a set of Boolean functions and, in turn, implemented by a combinational circuit. But two things impinge on the design of a circuit: the form of the Boolean expression obtained and the type of digital devices used to implement the expression. Moreover, different types of devices may need different forms of Boolean expression for convenience of implementation. Some devices, such as the multiplexers and read-only memories discussed in Chapter 5, require the canonical sum-of-products expression. But, if we are using gates to implement the circuit, then the canonical form may not be the simplest way to express a given Boolean function because it may contain **logical redundancies**. For example, the expressions $\sum(0, 2, 4, 5, 7)$ and $x_1'x_3' + x_1x_2' + x_1x_3$ are logically equivalent, but implementing a circuit for the former requires five 3-input AND gates and one 5-input OR gate, while the circuit for the latter requires three 2-input AND gates and one 3-input OR gate.

Logical redundancies simply mean that some terms carry information already contained in some other terms. Since the number of terms is directly proportional to the implementation complexity of the circuit, eliminating redundant information will invariably result in a *simpler* circuit. If a simpler circuit can be obtained, then the design will be more *cost effective*, as was already demonstrated in Section 4.1. Note, however, that eliminating redundancies may affect the *reliability* of the circuit. We will see in Section 4.8 that the absence of redundancies may cause the circuit to operate unreliably to the extent that *redundancies have to be reintroduced in order to restore reliable operation.*

Similarly, *physical space economy* must be considered during the design process. For example, integrated-circuit (IC) chips are mounted on printed-circuit (PC) boards so that, if the number of chips can be reduced, we would probably be able to mount more chips on a given PC board or, alternatively, implement the circuit on PC boards of smaller size.

Since our design will always be assessed by its cost effectiveness, we can state the **minimization problem** as follows: *For a given function F (or set of functions $F_1, F_2, ..., F_k$), design a combinational circuit that implements F (or $F_1, F_2, ..., F_k$) but that adheres to some minimality criterion.*

The remainder of this chapter is devoted to the discussion of some of the basic tools available to simplify digital circuits. The ideas presented here have been extended to develop *design automation* algorithms for use with digital computers to effectively minimize logic circuits containing a large number of Boolean variables.

Minimality Criteria

Various criteria are used to determine a simplest circuit; we list three common ones:

1. Minimal number of appearances of literals
2. Minimal number of literals in a sum-of-products (or product-of-sums) expression
3. Minimal number of terms in a sum-of-products (or product-of-sums) expression, provided there is no other such expression with the same number of terms and with fewer literals

For our purpose, we will adopt the third criterion as the standard by which we will measure minimality. We will restrict our attention to sum-of-products expressions, however, our discussion is equally applicable to product-of-sums expressions.

Let us clarify what our choice of minimality criterion means. Suppose that F is written as a sum-of-products (SOP) of literals. Let q_F denote the total number of literals in F, and let p_F denote the number of product terms in F. Then, for two SOP expressions F and G, G is *simpler* than F if $q_G \leq q_F$ and $p_G \leq p_F$ and one of these inequalities is strict. Then, G is **minimal** (or a **minimal sum**) if there is no simpler SOP expression equivalent to F. In other words, we are looking for the "shortest" sum-of-products expression with the least number of literals that is equivalent to F.

■ **Example 4.9** If $F(x_1, x_2, x_3, x_4) = \Sigma(0, 5, 7, 8, 15)$, then $q_F = 20$ and $p_F = 5$. However, F is logically equivalent to $G = x_1' x_3' x_4' + x_1' x_2 x_4 + x_2 x_3 x_4$, for which $q_G = 9$ and $p_G = 3$. Hence, G is simpler than F. ■

Two-Level Circuits

Later, we will be able to show that G in Example 4.9 is, in fact, a minimal SOP expression according to criterion 3. But this expression can also be written as $x_2 x_4 (x_1' + x_3) + x_1' x_3' x_4'$, requiring a fewer number of literals; however, it is no longer in a sum-of-products form. Rather, it is in a **mixed form**.

The circuit minimization tools to be presented in the following sections always result in minimal sum-of-products (or product-of-sums) expressions that satisfy criterion 3. Sometimes these expressions can be manipulated algebraically into a mixed form to further reduce the number of terms and/or literals. Mixed forms differ from sum-of-products (or product-of-sums) forms in the number of *levels* required to implement the circuit. If we assume that each input variable is available both complemented and uncomplemented, then the number of levels is decided according to the following definition.

□   **Definition 4.1**   The number of **levels** of a combinational circuit is the maximum number of gates from any input line to any output line of the circuit.

                                                     □

Examples of 2-, 3-, and 4-level circuits are shown in Figures 4.8(b), 4.4, and 4.3, respectively. In general, a sum-of-products (or a product-of-sums) expression is *always* implemented as a 2-level circuit as shown in Figures 4.11. The 2-level expressions are the appropriate form for several implementation methodologies, in particular for the programmable array implementations discussed in Section 5.9. In contrast, mixed-form expressions always result in a circuit implementation with more than two levels. As we will see in Section 5.3, there are trade-offs between 2-level and mixed-form implementations. For example, a 2-level circuit has the least amount of *propagation time delay* (see Section 1.2) between its inputs and outputs, but a multilevel circuit is obtained when we impose maximal *fan-in* constraints on the gates used to implement it.

Obtaining a Minimal Sum: Use of Boolean Algebra Theorems

Given our definition of a minimal sum, how can we obtain such an expression? We can attempt to simplify a Boolean expression by using the

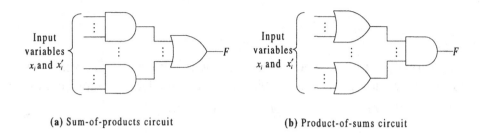

(a) Sum-of-products circuit   (b) Product-of-sums circuit

■ **Figure 4.11** General description of 2-level circuits

laws of Boolean algebra, as we did in Section 3.2. In general, a minimal sum can always be derived from the canonical sum-of-products form by applying the logical adjacency theorem $(xy + xy' = x)$, the idempotent law $(x + x = x)$, and the consensus theorem (see Table 3.1). This is illustrated in the following examples.

■ **Example 4.10** Consider the Boolean expression

$$F(x_1, x_2, x_3) = \Sigma(0, 2, 3, 4, 5, 7)$$
$$= x_1'x_2'x_3' + x_1'x_2x_3' + x_1'x_2x_3 + x_1x_2'x_3' + x_1x_2'x_3 + x_1x_2x_3$$

First, we add $m_0 + m_7$ to obtain the expression

$$F = x_1'x_2'x_3' + x_1'x_2x_3' + x_1'x_2x_3 + x_1x_2'x_3' + x_1'x_2x_3$$
$$+ x_1x_2x_3 + x_1x_2'x_3 + x_1x_2x_3$$

Then, combining m_0 and m_2, m_0 and m_4, m_3 and m_7, and m_5 and m_7, we obtain

$$F = x_1'x_3'(x_2' + x_2) + x_2x_3'(x_1' + x_1) + x_2x_3(x_1' + x_1) + x_1x_3(x_2' + x_2)$$
$$= x_1'x_3' + x_2x_3' + x_2x_3 + x_1x_3$$

But if, instead, we combine m_0 and m_2, m_4 and m_5, and m_3 and m_7, then we obtain

$$F = x_1'x_3' + x_1x_2' + x_2x_3$$

Likewise, by combining m_2 and m_3, m_0 and m_4, and m_5 and m_7, we obtain an equivalent expression:

$$F = x_1' x_2 + x_2' x_3' + x_1 x_3$$

The Karnaugh map developed in the next section will reveal the reasons why adding the two minterms m_0 and m_7 led to simplification and why only the latter two expressions are minimal. ∎

- **Example 4.11** Repeated applications of logical adjacency may be necessary to simplify a Boolean expression:

$$
\begin{aligned}
F(x_1, x_2, x_3, x_4) &= \sum (12, 13, 14, 15) \\
&= x_1 x_2 x_3' x_4' + x_1 x_2 x_3' x_4 + x_1 x_2 x_3 x_4' + x_1 x_2 x_3 x_4 \\
&= x_1 x_2 x_3' (x_4 + x_4') + x_1 x_2 x_3 (x_4 + x_4') \\
&= x_1 x_2 x_3' + x_1 x_2 x_3 = x_1 x_2 (x_3 + x_3') = x_1 x_2
\end{aligned}
$$
∎

The characteristic property of four minterms that can be simplified as in Example 4.11 is that all but two of the variables are the same (either primed or unprimed) in all four terms. We can also observe this from their corresponding binary information. Minterms $m_{12} = x_1 x_2 x_3' x_4'$ and $m_{13} = x_1 x_2 x_3' x_4$ are logically adjacent because their corresponding binary representations, 1100 and 1101, respectively, are identical in all but one bit position. Likewise, the four minterms m_{12}, m_{13}, m_{14}, and m_{15} can be simplified because 1100, 1101, 1110, and 1111 are identical in all but two bit positions. For such a group of four minterms, the simplified product term is derived by eliminating the variables that correspond to the changing bit positions. In other words, because the last two bit positions changed in each of the four binary numbers, x_3 and x_4 are eliminated.

Eight minterms can also be simplified if all but three of the variables are identical in all eight terms. The three variables that can be eliminated occur as primed or unprimed in all eight possible ways. Thus,

$$G(x_1, x_2, x_3, x_4) = \sum (1, 3, 5, 7, 9, 11, 13, 15)$$

simplifies to x_4 because it can be written as

$$G = [x_1'x_2'x_3' + x_1'x_2'x_3 + x_1'x_2x_3' + x_1'x_2x_3 + x_1x_2'x_3' + x_1x_2'x_3$$
$$+ x_1x_2x_3' + x_1x_2x_3]x_4$$

and only the fourth bit position is identical in all eight corresponding binary numbers. Likewise

$$H(x_1, x_2, x_3, x_4, x_5) = \sum (9, 11, 13, 15, 25, 27, 29, 31)$$

simplifies to x_2x_5 because the eight binary numbers differ in all but the second and fifth bit positions.

The general rule is that *a sum of 2^m minterms in n variables, where all but m of the variables are identical, can be simplified to a single product term of (n–m) literals where the m variables that change are eliminated.* The elimination of literals depends on identities like the following:

$$x_i + x_i' = 1$$
$$x_i'x_j' + x_i'x_j + x_ix_j' + x_ix_j = 1$$
$$x_i'x_j'x_k' + x_i'x_j'x_k + x_i'x_jx_k' + x_i'x_jx_k + x_ix_j'x_k' + x_ix_j'x_k$$
$$+ x_ix_jx_k' + x_ix_jx_k = 1$$

and so on. As you might expect, these identities are nothing more than a result of repeated applications of logical adjacency.

While it is true that any Boolean function can be simplified by applying Boolean algebra theorems, three basic problems arise nevertheless:

1. Using the theorems requires ingenuity, and applying them becomes increasingly difficult as the number of variables increases.
2. The algebraic procedure lacks specific rules to predict each step in the process.
3. It is not always apparent when we have arrived at a minimal sum.

The Karnaugh map minimization method studied in Section 4.4 and the Quine-McCluskey procedure considered in Section 4.9 overcome these difficulties by providing systematic methods for minimizing Boolean expressions. But we have already discovered the clue to understanding each procedure: It is the simple observation that *logical adjacency can be described by binary information.* The Karnaugh map and Quine-McCluskey procedures simply follow this clue to its logical conclusion.

4.3 KARNAUGH MAPS

A **Karnaugh map** (or **K-map** for short) is a graphical method for representing the truth table of a Boolean function. K-maps may be used to represent functions of any number of variables, but are most often used for functions of at most six variables.

If n is the number of variables, then the K-map has 2^n **cells**, one for each of the 2^n minterms (maxterms) and, therefore, one for each row of the truth table. Moreover, these cells are arranged such that the minterms (maxterms) are displayed in a geometric pattern so that logically adjacent terms are visibly apparent. Minimization, therefore, can be accomplished by recognizing basic patterns on the K-map.

Each cell of the K-map is assigned a coordinate that forms an n-bit binary number corresponding to a minterm (maxterm) of n literals. Two cells are **adjacent** if their corresponding binary combinations differ by only one bit. If a function F is expressed in canonical sum-of-products (SOP) form, then we complete the K-map of F by placing a 1 in those cells whose coordinates correspond to minterms that occur in the canonical form, and placing a 0 in each of the remaining cells. If F is in canonical product-of-sums (POS) form, we place a 0 in those cells that correspond to maxterms of the canonical form and a 1 in all the others.

If the functional value 1 occurs in a cell of the K-map of a function F, then the cell is called a *1-cell* of F; if a 0 value occurs, the cell is called a *0-cell*. Hence, two adjacent 1-cells (0-cells) immediately identify the presence of a redundant variable in the corresponding minterms (maxterms); four adjacent 1-cells (0-cells) imply the existence of two redundant variables; and so on.

We introduce the notion of K-maps assuming that F is in SOP form. We will show in Section 4.7 how POS forms can be handled similarly through the principle of duality or by using DeMorgan's laws. In this section, we consider K-maps of up to four variables; extensions to five and six variables will be introduced in Section 4.5.

One-Variable Maps

Consider a function F of only one variable (x). Since x can take on only two values, the K-map needs only two cells, as shown in Figure 4.12(a). The region where $x=0$ is the left half of the map, and the value $F(0)$ should be entered in that cell. The right half of the K-map corresponds to $x=1$, and $F(1)$ is entered in that cell. Likewise, in Figure 4.12(b), the locations of minterms are indicated so that a 1 is entered in the left cell if the minterm m_0 appears in the canonical form of F, and a 1 is entered in the right cell if the minterms m_1 appears.

Higher-order K-maps can be obtained by successively *doubling* the size of the 1-variable K-map. This can be done, for example, by simply "folding out" the larger map from the smaller, as will be shown in the following sections.

(a)   (b)

■ **Figure 4.12** One-variable K-maps

Two-Variable Maps

A 2-variable K-map corresponding to $F(x_1,x_2)$ can be obtained by either of the fold-out procedures in Figure 4.13(a) or (b). (The hinges are drawn to indicate the possibility of folding.)   The cells correspond to the minterms $m_0=x_1'x_2'$, $m_1=x_1'x_2$, $m_2=x_1x_2'$, and $m_3=x_1x_2$. Note that each cell in the new row represents a minterm whose subscript is 2 greater than the subscript of the minterm of the cell from which it is unfolded. This particular arrangement forces minterms differing in only one variable to be adjacent. Figures 4.13(a), (b), and (c) depict general forms of 2-variable K-maps, and Figure 4.13(d) shows the K-map for $F(x_1,x_2)=\sum(2,3)$.

Either representation of $F(x_1,x_2)$ in Figure 4.13(a), (b), or (c) is useful; each conveys the information that certain minterms are logically adjacent to others. Figure 4.13(c) only indicates that, if the coordinates of the cells are reversed with respect to Figure 4.13(a), then the minterms are located differently.

The adjacency information is contained in Table 4.4. That m_2 is adjacent to m_0 is obvious if we consider their binary values, but may not be immediately apparent from the K-map in Figure 4.13(b). To understand this, you must think of the map as bent into a cylinder, as in Figure 4.14, where the right edge of the cell for m_0 is identified with the left edge of the cell for m_2. The idea of the folding out also suggests these adjacencies.

Note an interesting feature of the 2-variable K-map: Every minterm is adjacent to two other minterms. As we will see, every minterm in a

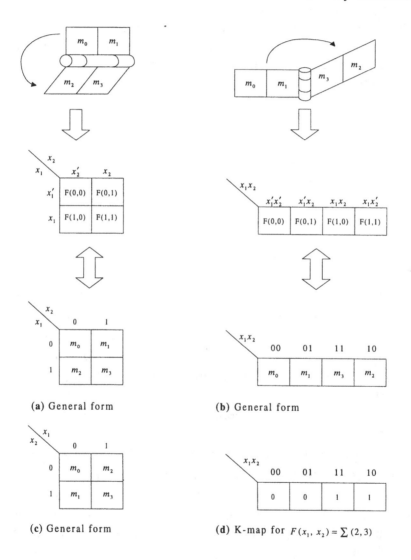

Figure 4.13 Two-variable K-maps

■ **Table 4.4** Adjacency in a 2-variable K-map

Minterm:	Is adjacent to:
m_0	m_1, m_2
m_1	m_0, m_3
m_2	m_0, m_3
m_3	m_1, m_2

3-variable K-map is adjacent to three other minterms; every minterm in a 4-variable K-map is adjacent to four other minterms; and so on. In general, *every minterm in an n-variable K-map is adjacent to n other minterms*. Furthermore, the headings of the columns in Figure 4.13(b) are particularly important because they follow the *cyclic* code sequence {00, 01, 11, 10}. This idea will be repeated over and over again for K-maps of higher orders.

Three-Variable Maps

The 3-variable K-map of Figure 4.15(b) is produced by a fold out from the 2-variable map, as shown in Figure 4.15(a). Note that each cell in the new row has a minterm number that is 2^2 greater than the cell from which it is unfolded.

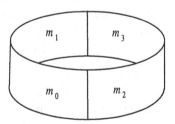

■ **Figure 4.14** Three-dimensional configuration of a 2-variable K-map

In Figure 4.15(b), the cell assigned to $m_5 = x_1 x_2' x_3$ corresponds to row 1 and column 01 because binary 101 is equal to decimal 5. In other words, m_5 appears at the *intersection* of the row marked x_1 and column belonging to $x_2' x_3$ (column 01). Note that there are four cells where each variable

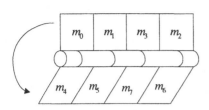

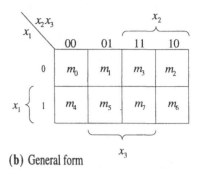

(a) K-map obtained from a foldout of a
2-variable map

(b) General form

(c) K-map for $F(x_1, x_2, x_3) = \sum(0, 3, 5, 7)$

■ **Figure 4.15** Three-variable K-maps

takes on the value 1 and four cells where each takes the value 0. In the
first case the variable is uncomplemented, but it is complemented in the
second. For convenience, we have written in Figure 4.15(b) the letter
symbol for the variable with braces to indicate the four cells where it is
uncomplemented. Figure 4.15(c) shows the K-map for $F = \sum(0, 3, 5, 7)$.

The adjacency information is listed in Table 4.5. Every minterm is
adjacent to three other minterms because the column headings in Figure
4.15(b) follow the cyclic code sequence {00, 01, 11, 10}. Observe that
minterms m_0 and m_2, and m_4 and m_6, are adjacent. We can visualize this
by bending the K-map into a cylinder, where the right edge of the K-map
is identified with the left edge.

Four-Variable Maps

Four-variable K-maps can be obtained, as indicated in Figure 4.16(a),
from the 3-variable map by folding out the map and again doubling its

■ **Table 4.5** Adjacency in a 3-variable K-map

Minterm:	Is adjacent to:
m_0	m_1, m_2, m_4
m_1	m_0, m_3, m_5
m_2	m_0, m_3, m_6
m_3	m_1, m_2, m_7
m_4	m_0, m_5, m_6
m_5	m_1, m_4, m_7
m_6	m_2, m_4, m_7
m_7	m_3, m_5, m_6

size. The new cells represent minterms with subscripts that are 2^3 greater than their unfolded counterparts. Again, we have indicated in Figure 4.16(b) the regions where the variables take on the value 1. To verify the location of minterms, let us consider $m_{13} = x_1 x_2 x_3' x_4$. Since decimal 13 is equivalent to binary 1101, m_{13} is located at the intersection of the row with heading 11 and the column with heading 01. Figure 4.16(c) shows the K-map for $F = \sum(1, 6, 7, 9, 12, 15)$.

Every minterm in the 4-variable K-map is adjacent to four other minterms because both the column and row headings follow the cyclic code sequence {00, 01, 11, 10}. Minterms m_0 and m_2, m_4 and m_6, m_{12} and m_{14}, and m_8 and m_{10} are adjacent if we view the right edge of the K-map as identified with the left edge. Likewise, minterms m_0 and m_8, m_1 and m_9, m_3 and m_{11}, and m_2 and m_{10} are adjacent by identifying the top edge of the K-map with the bottom edge. You might visualize this by thinking of the 4-variable K-map as a doughnut-shaped object with the cells labeled on the surface of the doughnut.

4.4 PRIME IMPLICANTS AND MINIMIZATION

Once a Boolean function F has been plotted on a K-map, the identification of minterms that can be combined by logical adjacency is facilitated by the

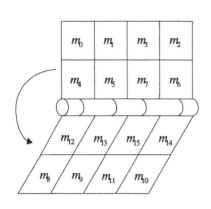

(a) K-map obtained from a foldout of a
   3-variable map

(b) General form

(c) K-map for $F(x_1, x_2, x_3, x_4) = \sum(1, 6, 7, 9, 12, 15)$

■ **Figure 4.16** Four-variable K-maps

geometrical arrangements on the K-map. The simplest case is the
combination of two adjacent cells. But, in general, if m is a nonnegative
integer, then a collection of 2^m 1-cells forms what we call a **cluster** *if each*
1-cell in the cluster is adjacent to m other 1-cells in the cluster. We say
that the cluster **covers** the 1-cells in the collection.

Hence, as we have already observed in Section 4.2, each cluster can be represented by a product of $(n-m)$ literals, where n is the number of variables of the function. The m literals *not* contained in the product term have been eliminated by repeated applications of logical adjacency (see the discussion following Examples 4.10 and 4.11). Clusters are, therefore, nothing more than a way of visualizing this simplification.

Figure 4.17 illustrates a variety of clusters of one, two, four, or eight 1-cells on a 4-variable K-map. (The 0's are not entered for convenience.) These clusters correspond to $m = 0, 1, 2,$ and 3. If $m = 4$ (this case is not shown in Figure 4.17), then the cluster will contain $2^4 = 16$ 1-cells, indicating that the entire 4-variable map consists of 1's and, therefore, the function is identically equal to 1. Each cluster in Figure 4.17 covers all the 1-cells of the function, and each 1-cell is adjacent to m other 1-cells in the cluster. In each cluster, some of the coordinates remain fixed, while the remaining coordinates take on *all* possible combinations of values. Hence, each cluster can be represented by a product term that is formed with the variables whose coordinates remain unchanged. The number of literals in each product is $(n-m)$, where $n = 4$ in our case. As we can see, the clusters and their corresponding product terms are easily determined from the K-maps by inspection. We will demonstrate this feature of the K-map repeatedly in subsequent examples.

But now comes a crucial observation: Clusters on the K-map of F produce **implicants** of F (see Definition 3.3), and clusters *not contained* in any larger cluster produce **prime implicants** of F (see Definition 3.4). To illustrate this, let us reconsider the K-map for $E(x_1, x_2, x_3, x_4)$ in Figure 4.17. The 2-cell cluster $m_4 + m_5 = x_1'x_2x_3'$ is an implicant of E. However, it is not a prime implicant of E because we can delete a literal (x_1' in this case) from it and obtain a new product term, x_2x_3', which also implies E. We can reach the same conclusion by inspecting the K-map: The 2-cell cluster of m_4 and m_5 is contained in the (larger) 2^2-cell cluster of m_4, m_5, m_{12}, and m_{13}. In fact, any single-cell or 2-cell cluster formed with the 1-cells of E is an implicant of E, but is not a prime implicant of E. Only the cluster indicated for E in Figure 4.17 is a prime implicant of E. (In fact, all the clusters shown in Figure 4.17 are prime implicants of their respective functions.) Once we identify a prime implicant, the resulting product term will consist of a *minimal* number of literals.

To highlight the importance of prime implicants, let us consider the

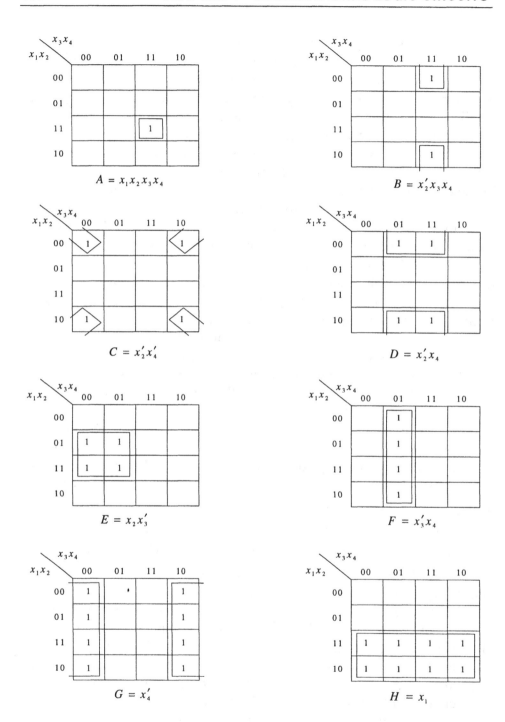

■ **Figure 4.17** Typical clusters on a 4-variable K-map

following theorem and its corollary.

☐ **Theorem 4.1** Any minimal sum-of-products (SOP) expression of a Boolean function F is equivalent to a sum of prime implicants of F.

Proof Suppose that $F = P + R$, where P is a product of literals that is not a prime implicant of F, and R is the sum of the remaining product terms of F. Then P implies F ($P \Rightarrow F$); but, since P is not a prime implicant of F, it must be possible to remove a literal from P to obtain a new product term Q such that $Q \Rightarrow F$. Since $Q = 1$ whenever $P = 1$, then $F = Q + R$. This last SOP expression contains the same number of terms but one fewer literal than originally. Thus, the original expression was not a minimal sum expression for F. ·

☐

In fact, the preceding proof can be modified to show that if the *cost* of implementing the circuit increases as the number of literals increases, then any Boolean expression for F that minimizes this cost must be equivalent to a sum of prime implicants. In short, the theorem applies in more general settings than the definition of minimality that we have adopted.

☐ **Corollary 4.1** A Boolean expression F in n variables is equivalent to the sum of *all* prime implicants of F.   ☐

Proof Let G be a minimal SOP expression of F and assume that H is the sum of all the remaining prime implicants of F not contained in G. Since $H \Rightarrow F$, then $F = G + H$ and, therefore, $G + H$ is the sum of all prime implicants of F.   ☐

If we call the sum of all prime implicants of F the *complete sum* of F, then, just as there is only one canonical SOP form, there is only one complete sum for each Boolean function. However, while Corollary 4.1 tells us that any Boolean expression is equivalent to its complete sum, Theorem 4.1 tells us that a minimal sum is always a sum of *some* of the prime implicants. As it turns out, for most functions, a minimal sum and the complete sum are *not identical*. Based on these observations, therefore, we can formulate a plan for obtaining a minimal sum by dividing the task into two parts:

1. Determine all prime implicants of F.
2. Select those prime implicants whose sum will be a minimal sum expression for F.

Let us remember this plan of attack, for it is the guiding principle for all that we do in the remainder of this chapter. In the following two sections we show how to determine all prime implicants from the K-map and select those that make up the minimal sum. Another procedure will be introduced in Section 4.9.

K-Map Minimization Procedure

The K-map minimization method consists of the following steps:

1. Plot the functional values of F in the cell locations on the K-map.
2. Circle all prime-implicant clusters (those clusters that are not contained in larger clusters).
3. Find a minimum set of prime-implicant clusters covering all the 1-cells of the K-map. (If there is more than one such set, choose a set with the minimum number of literals.)
4. Determine the product terms corresponding to the selected clusters to form a minimal sum.

We apply this procedure first to K-maps of up to four variables and defer its application to 5- and 6-variable K-maps to Section 4.5.

- **Example 4.12**   Apply the K-map procedure to minimize $F(x_1, x_2)$ whose truth table is given in Figure 4.18(a).

Step 1: The canonical SOP expression determines the location of the functional values on the K-map, as illustrated by Figure 4.18(b).

Step 2: Encircle clusters of 1-cells, as shown in Figure 4.18(c).

Steps 3 and 4: The two clusters are summed to form a minimal sum

$$F(x_1, x_2) = x_1 + x_2$$ ∎

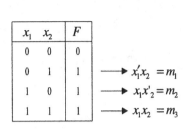

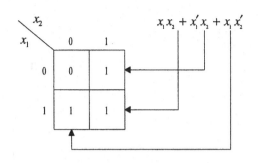

(a) Truth table of $F(x_1, x_2)$

(b) K-map of $F(x_1, x_2)$

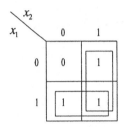

(c) K-map with clusters exhibited

■ **Figure 4.18** The steps of the K-map minimization procedure

■ **Example 4.13** Simplify the Boolean function $F(x_1, x_2, x_3) = \Sigma(2, 3, 4, 5)$ whose corresponding K-map is shown in Figure 4.19. The next step is to cluster certain 1-cells so that logical adjacency applies. Hence, the upper right cluster represents

$$m_2 + m_3 = x_1'x_2x_3' + x_1'x_2x_3 = x_1'x_2(x_3 + x_3') = x_1'x_2$$

And the lower left cluster represents

$$m_4 + m_5 = x_1x_2'x_3' + x_1x_2'x_3 = x_1x_2'(x_3' + x_3) = x_1x_2'$$

Thus, F simplifies to

$$F = x_1'x_2 + x_1x_2'$$

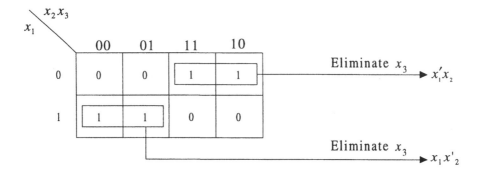

■ **Figure 4.19** K-map for F = Σ(2, 3, 4, 5) of Example 4.13

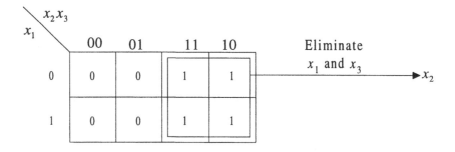

■ **Figure 4.20** K-map for F = Σ(2, 3, 6, 7) of Example 4.14

■ **Example 4.14** The K-map for $F(x_1, x_2, x_3) = \Sigma(2, 3, 6, 7)$ is shown in Figure 4.20. The cluster of four adjacent 1-cells corresponds to the repeated applications of logical adjacency:

$$m_2 + m_3 = x_1'x_2x_3' + x_1'x_2x_3 = x_1'x_2$$
$$m_6 + m_7 = x_1x_2x_3' + x_1x_2x_3 = x_1x_2$$

But $x_1'x_2 + x_1x_2 = (x_1' + x_1)x_2 = x_2$, so $F = x_2$.

■

■ **Example 4.15** the K-map for $F(x_1, x_2, x_3) = \Sigma(0, 2, 4, 5, 6)$ in Figure 4.21 shows that

$$F = x_1 x_2' + x_3'$$

The four cells in the columns headed by 00 and 10 indicate that the sum of these four minterms reduces to x_3'. The other cluster of two minterms indicates a reduction to $x_1 x_2'$.   ■

■ **Example 4.16** The cluster shown in Figure 4.22 for $F(x_1, x_2, x_3, x_4) = \Sigma(12, 13, 14, 15)$ results in

$$F = x_1 x_2$$

(Refer to Example 4.11 to see how this comes from repeated applications of logical adjacency.)   ■

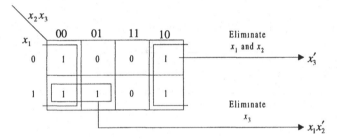

■ **Figure 4.21** K-map for $F = \Sigma(0, 2, 4, 5, 6)$ of Example 4.15

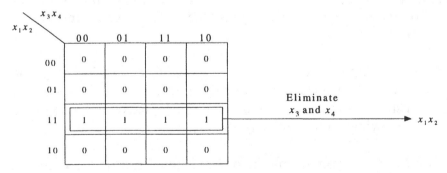

■ **Figure 4.22** K-map for $F = \Sigma(12, 13, 14, 15)$ of Example 4.16

■ **Example 4.17** The K-map minimization procedure applied to $F(x_1, x_2,$ $x_3, x_4) = \Sigma(1, 3, 8, 10, 12, 13, 14, 15)$ in Figure 4.23 gives the minimal sum expression

$$F = x_1' x_2' x_4 + x_1 x_2 + x_1 x_4' \qquad\qquad ■$$

Minimal Sums and Essential Prime Implicants

Prime-implicant clusters contained in the *union* of other clusters give *redundant* product terms by the consensus theorem. There are, therefore, prime implicants that are *not* included in a minimal sum expression.

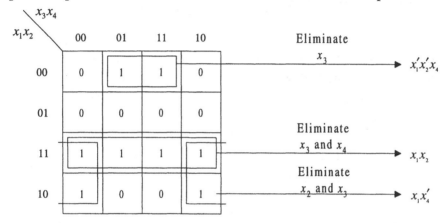

■ **Figure 4.23** K-map for $\Sigma(1, 3, 8, 10, 12, 13, 14, 15)$ of example 4.17

■ **Example 4.18** The dashed clusters on the K-maps of Figures 4.24(a) and (b) give redundant product terms since all their cells are covered by other clusters.   ■

Example 4.18 points out the need for a new concept. If some prime implicants are not contained in a minimal sum expression, then how can we distinguish between them and those that are contained in the minimal sum? The following definition points the way.

□ **Definition 4.2** A prime implicant of a function F is called an **essential prime implicant** if it covers at least one minterm of F that is not covered by any other prime implicant. Such a minterm is called a *distinguished minterm*, and the corresponding cell is called a *distinguished cell.*   □

In Figure 4.24(a), $x_1'x_3$ and x_1x_2 are essential prime implicants of G because $x_1'x_3$ is the only prime implicant of G that covers m_1 and x_1x_2 is the only prime implicant that covers m_6. In other words, the clusters $\Sigma(1, 3)$ and $\Sigma(6, 7)$ determine essential prime implicants because the distinguished 1-cells for m_1 and m_6 are each covered by only one prime-implicant cluster. These distinguished 1-cells are indicated by an asterisk on the K-map in Figure 4.24(a). Thus, G has one nonessential and two essential prime implicants. The prime implicant x_2x_3 is nonessential because each 1-cell of the cluster $\Sigma(3, 7)$ is covered by two or more prime implicants. Likewise, $x_1'x_3'x_4$, $x_1'x_2x_3$, $x_1x_3x_4$, and $x_1x_2x_3'$ are the essential prime implicants of the function H of Figure 4.24(b), and minterms m_1, m_6, m_{11}, and m_{12} are distinguished.

Since every minterm of a function F must be covered by some prime implicant, we conclude that *every essential prime implicant must appear in a minimal sum of F.*

Now comes a crucial point: In each example that we have presented so far, the minimal sum has been exactly the sum of all essential prime implicants. But is this always the case? Unfortunately, no; the following Example 4.19 shows that there are functions F for which the sum of all essential prime implicants does not cover all the minterms of F.

■ **Example 4.19** The K-map in Figure 4.25 for $F(x_1, x_2, x_3\ x_4) = \Sigma(2, 5, 6, 7, 13)$ determines two different minimal sums. This is because the two essential prime implicants $x_2x_3'x_4$ and $x_1'x_3x_4'$ leave one 1-cell uncovered and that cell can be covered by either of two different prime implicants, $x_1'x_2x_4$ or $x_1'x_2x_3$. Thus, there are two minimal sums and we write:

$$F = x_2x_3'x_4 + x_1'x_3x_4' + \begin{cases} x_1'x_2x_4 \\ \text{or} \\ x_1'x_2x_3 \end{cases}$$

■

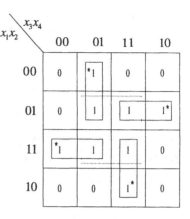

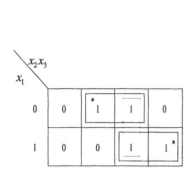

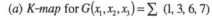

(a) K-map for $G(x_1, x_2, x_3) = \sum (1, 3, 6, 7)$ **(b)** K-map for $H(x_1, x_2, x_3, x_4)$

$\sum (1, 5, 6, 7, 11, 12, 13, 15)$

■ **Figure 4.24** Essential prime implicants on K-maps

In fact, as illustrated in the following example, a function need not have any essential prime implicants at all.

■ **Example 4.20** The function $F(x_1, x_2, x_3) = \Sigma(0, 2, 3, 4, 5, 7)$, first discussed in Example 4.10, has no essential prime implicants at all.

The K-map for such a function is called a **cyclic** prime-implicant map since no prime implicant is essential (or, equivalently, since every 1-cell is covered by at least two prime implicants).

The different ways of combining the minterms of F correspond to different ways of circling prime-implicant clusters of 1-cells on the K-maps of Figure 4.26. The clustering produces four product terms in Figure 4.26(a), but only three product terms in Figures 4.26(b) and (c). This is because the clusters of Figure 4.26(a) share common 1-cells. ■

Therefore, we ask, besides containing the essential prime implicants, what are some other characteristics of a minimal sum? For one thing, a minimal sum must be irredundant in the sense of the following definition.

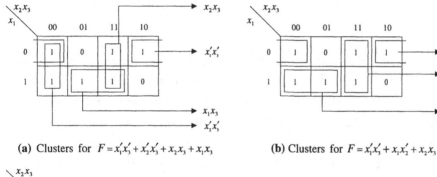

■ **Figure 4.25** A K-map that determines two minimal sums (Example 4.19)

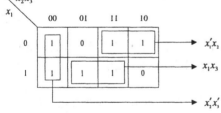

(a) Clusters for $F = x_1'x_3' + x_2'x_3' + x_2x_3 + x_1x_3$

(b) Clusters for $F = x_1'x_3' + x_1x_2' + x_2x_3$

(c) Clusters for $F = x_1'x_2 + x_2'x_3' + x_1x_3$

■ **Figure 4.26** A cyclic prime-implicant K-map (Example 4.20)

□ **Definition 4.3** A sum-of-products (SOP) expression for a function F is **irredundant** if and only if deleting any product term and any literal renders the resulting sum *not equivalent* to F. □

Thus, each product term of an irredundant sum for F *must* be a prime implicant of F. Moreover, all essential prime implicants of F *must* appear in any irredundant sum of F. In general, one finds an irredundant sum by (1) choosing enough prime-implicant clusters to cover every 1-cell of the function, and (2) choosing clusters so that no cluster is contained in the union of other clusters.

We might hope that irredundancy completely characterizes a minimal sum, but this hope is short-lived, because the following example shows an irredundant sum that is *not* minimal.

■   **Example 4.21**   The function $F(x_1, x_2, x_3, x_4) = \Sigma\ (1, 3, 4, 5, 7, 10, 11, 12, 14, 15)$ has two irredundant sums that can be derived from the K-maps of Figures 4.27(a) and (b). By contrast, the third sum, derived from Figure 4.27(c), has a redundant term. These three sums are

$$F = x_1'x_4 + x_1x_3 + x_1'x_2x_3' + x_1x_2x_4' = x_1'x_4 + x_1x_3 + x_2x_3'x_4'$$
$$= x_1'x_4 + x_1x_3 + x_2x_3'x_4' + x_1'x_2x_3'$$

Since none of the clusters in Figures 4.27(a) and (b) are contained within a union of other clusters, or within a larger cluster, the first two sums are irredundant. The term $x_1'x_2x_3'$ is redundant in the third sum because its cluster is contained in the union of the clusters for $x_1'x_4$ and $x_2x_3'x_4'$. Only the expression for Figure 4.27(b) is minimal because it has three product terms, whereas the other irredundant sum [Figure 4.27(a)] has four product terms.   ■

Note that a similar situation was shown in Example 4.20. All three sums are irredundant, but only those given in Figures 4.26(b) and (c) are minimal.

This then is the crux of the problem: There may be some nonessential prime implicants that must be selected and included in a minimal sum. Whereas the selection of nonessential prime implicants is not always straightforward, the difficulty is somewhat reduced using the following procedure: *To determine a minimal sum, first determine all essential prime implicants. Then determine all irredundant sums, and, among them, find those "shortest" sums with the smallest number of prime implicants. Finally, among the shortest irredundant sums, find one with the least number of literals.*

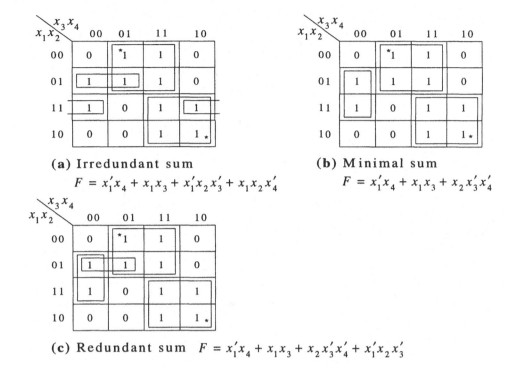

(a) Irredundant sum
$$F = x_1'x_4 + x_1x_3 + x_1'x_2x_3' + x_1x_2x_4'$$

(b) Minimal sum
$$F = x_1'x_4 + x_1x_3 + x_2x_3'x_4'$$

(c) Redundant sum $F = x_1'x_4 + x_1x_3 + x_2x_3'x_4' + x_1'x_2x_3'$

■ **Figure 4.27** Redundant and irredundant sums for F of Example 4.21

4.5 FIVE- AND SIX-VARIABLE K-MAPS

A 5-variable K-map is formed by using two 4-variable K-maps (Figure 4.28, and a 6-variable K-map consists of four 4-variable K-maps (Figure 4.29). We draw the individual 4-variable maps on different levels to enhance our ability to visualize the adjacencies between them.

For the 5-variable K-map, each 4-variable map is for the variables x_2, x_3, x_4, and x_5. The top level represents all rows of the truth table for which $x_1 = 0$, and the bottom level represents all rows for which $x_1 = 1$. Similarly, for the 6-variable K-map, the first-level 4-variable map represents all rows of the truth table for which $x_1 = 0$ and $x_2 = 0$; the second-level map represents all rows for which $x_1 = 0$ and $x_2 = 1$; the third-level map corresponds to rows for which $x_1 = 1$ and $x_2 = 1$; and the fourth-level map represents all rows for which $x_1 = 1$ and $x_2 = 0$.

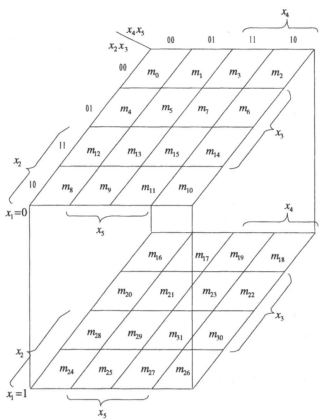

■ **Figure 4.28** Five-variable K-map

The basic rule for forming clusters on 5- and 6-variable K-maps is the same as for 4-variable maps: Combine any set of 1-cells for which some of the coordinates remain fixed, while the remaining coordinates take on all possible combinations of values. For cells on the *same* 4-variable map level, the patterns for clusters are the same patterns discussed in connection with 4-variable maps. Cells or clusters on *different* 4-variable map levels can be combined in a cluster only if they occupy the same relative positions on their respective 4-variable maps. For example, in Figure 4.28, the cells m_5 and m_{21} can be combined, but the cells m_5 and m_{20} cannot be combined.

For the 6-variable map of Figure 4.29, cells from two *different* 4-variable map levels can be combined only if the respective 4-variable maps are *vertically adjacent* (i.e., determine adjacent floors). As seen in

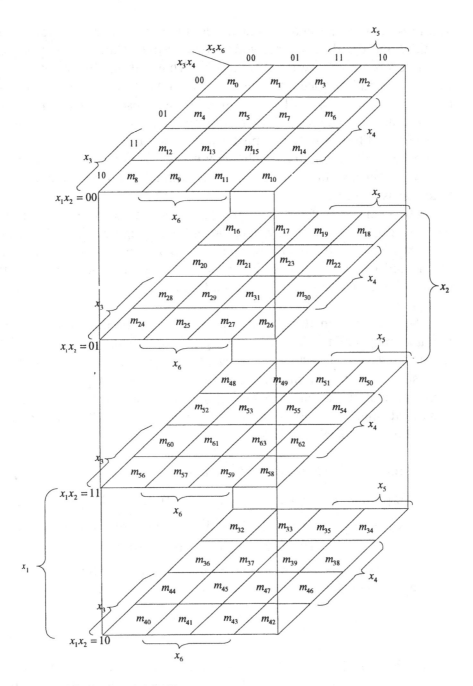

■ **Figure 4.29** Six variable K-map

Figure 4.29, the first and second levels, the second and third levels, and the third and fourth levels are vertically adjacent. But, then, the first and fourth levels are also vertically adjacent. Thus, informally speaking, each floor is vertically adjacent to a floor above and below it. For example, the cells for m_{13} and m_{29} can be combined, as can those for m_{13} and m_{45}, m_{23} and m_{55}, and m_{55} and m_{39}. They can be combined because their respective coordinates differ in only one bit position. But a cell from the first level, where $x_1x_2 = 00$, cannot be combined with one from the third level, where $x_1x_2 = 11$, because the two cells already differ in two coordinates, rather than one as required.

Adjacencies are determined the same as before on each level *plus* vertical adjacencies. For example, the cell m_7 in Figure 4.29 is adjacent to four cells (m_3, m_5, m_6, and m_{15}) on level 1, to one cell (m_{23}) on level 2, and to one cell (m_{39}) on level 4. Four cells, like m_7, m_{23}, m_{55}, and m_{39}, that occupy the same relative position on all four levels can be combined into a cluster.

The first step in the K-map minimization procedure for 5- and 6-variable maps is to determine the prime implicants for each of the individual 4-variable map levels *separately*. The second step is to compare each such cluster with clusters of vertically adjacent maps. If there is an identical cluster in an adjacent map, the two separate clusters are combined into one prime-implicant cluster. The following example illustrates this procedure.

- **Example 4.22** Consider the K-map for

$$F(x_1, x_2, x_3, x_4, x_5) = \Sigma(0, 2, 4, 6, 7, 8, 10, 12, 14, 15, 22, 23, 30, 31)$$

shown in Figure 4.30. The set A on level 1 determines the prime implicant $x_2'x_5'$. The sets B and C form prime implicants in their respective levels, but because they occupy the same relative positions, they combine to form a new prime implicant x_3x_4. Thus,

$$F = x_1'x_5' + x_3x_4 \qquad\blacksquare$$

The K-maps in Figure 4.31 illustrate two other points. First, note that on one 4-variable map in Figure 4.31(a) there is a prime implicant A that is identical with a subset C of a prime implicant B in an adjacent map. Therefore, a new prime implicant can be formed from the original prime

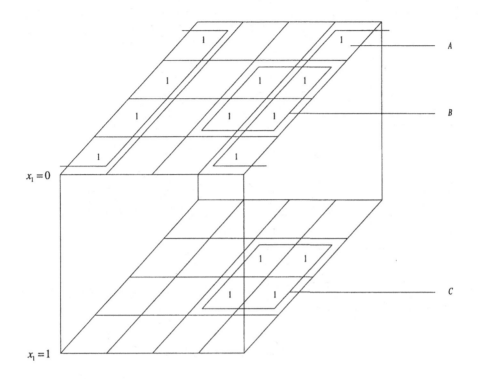

■ **Figure 4.30** K-map for F of Example 4.22

implicant A and the subset C. In such a case, the prime implicant A of one level is no longer a prime implicant for the 5-variable map since it is contained in the larger prime implicant $x_2x_3x_5$ determined by the union of A and C. Nevertheless, the set B still determines a prime implicant $(x_1'x_3x_5)$. Thus, minterms $\{m_{13}, m_{15}, m_{29}, m_{31}\}$ determine the prime implicant $x_2x_3x_5$, while the minterms $\{m_5, m_7, m_{13}, m_{15}\}$ determine the prime implicant $x_1'x_3x_5$. Hence,

$$F(x_1, x_2, x_3, x_4, x_5) = x_2x_3x_5 + x_1'x_3x_5$$

A second situation is illustrated in Figure 4.31(b). There may be two prime-implicant clusters (like D and E) on two adjacent 4-variable maps

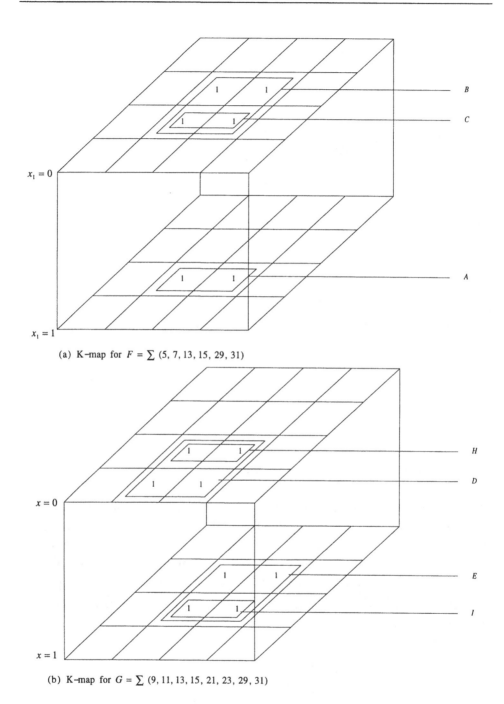

(a) K-map for $F = \sum (5, 7, 13, 15, 29, 31)$

(b) K-map for $G = \sum (9, 11, 13, 15, 21, 23, 29, 31)$

■ **Figure 4.31** Prime implicants on a 5-variable K-map

that do not occupy identical relative positions. Neither is identical to a subset of the other; however, both have identical subsets (H and I). The two identical subsets, determined by $\{m_{13}, m_{15}\}$ and $\{m_{29}, m_{31}\}$, combine to form a new prime implicant $x_2x_3x_5$. Both original prime implicants, $x_1'x_2x_5$ and $x_1x_3x_5$ (determined by $\{m_9, m_{11}, m_{13}, m_{15}\}$ and $\{m_{21}, m_{23}, m_{29}, m_{31}\}$, respectively), remain as prime implicants for the 5-variable map. Note that the prime implicant $x_2x_3x_5$ is redundant (and, therefore, is a nonessential prime implicant) since it is covered by the sum of the other two prime implicants. Thus,

$$G(x_1, x_2, x_3, x_4, x_5) = x_1'x_2x_5 + x_1x_3x_5$$

is the minimal sum.

- **Example 4.23** Simplify the function

$$F = \Sigma(0, 1, 4, 5, 13, 15, 20, 21, 22, 23, 24, 26, 28, 30, 31)$$

The 5-variable K-map for F is shown in Figure 4.32. The prime implicants determined by $A = \Sigma\,(0, 1, 4, 5)$ and $E = \Sigma\,(24, 26, 28, 30)$ are the only essential prime implicants. Since all the other 1-cells are covered by at least two different prime implicants, we proceed by trial and error. After a few tries, we see that we can cover all 1-cells with three more prime implicants. If we choose the prime implicants determined by $B = \Sigma\,(13, 15)$ and $D = \Sigma\,(22, 23, 30, 31)$, then the two remaining 1-cells can be covered by two different clusters of four 1-cells, $C = \Sigma\,(20, 21, 22, 23)$ or the union of $H = \Sigma\,(4, 5)$ and $G = \Sigma\,(20, 21)$. Thus, the resulting minimal sum is

$$F = x_1'x_2'x_4' + x_1x_2x_5' + x_1'x_2x_3x_5 + x_1x_3x_4 + \begin{cases} x_2'x_3x_4' \\ \text{or} \\ x_1x_2'x_3 \end{cases}$$

- **Example 4.24** The K-map of the 6-variable function

$$F = \Sigma\,(0,1,4,5,16,17,20,21,37,42,43,46,47,53,58,59,62,63)$$

appears in Figure 4.33. Let

$A \ = \Sigma\,(0, 1, 4, 5)$
$B = \Sigma\,(16, 17, 20, 21)$
$C = \Sigma\,(58, 59, 62, 63)$
$D = \Sigma\,(42, 43, 46, 47)$
$E \ = m_5$
$G = m_{21}$
$H = m_{53}$
$I \ = m_{37}$

Then the three unions $\{A$ and $B\}$, $\{C$ and $D\}$, and $\{E, G, H,$ and $I\}$ form three essential prime implicants, so the minimal sum is

$$F(x_1, x_2, x_3, x_4, x_5, x_6) = x_1' x_3' x_5' + x_1 x_3 x_5 + x_3' x_4 x_5' x_6 \qquad \blacksquare$$

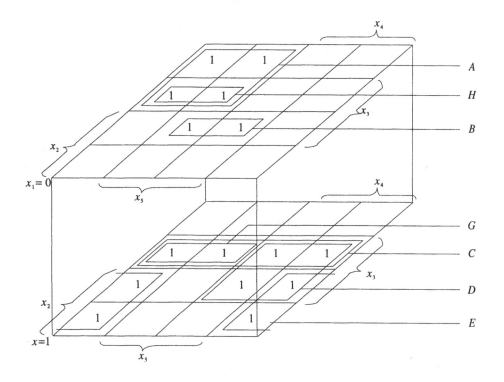

■ **Figure 4.32**   K-map for F of Example 4.23

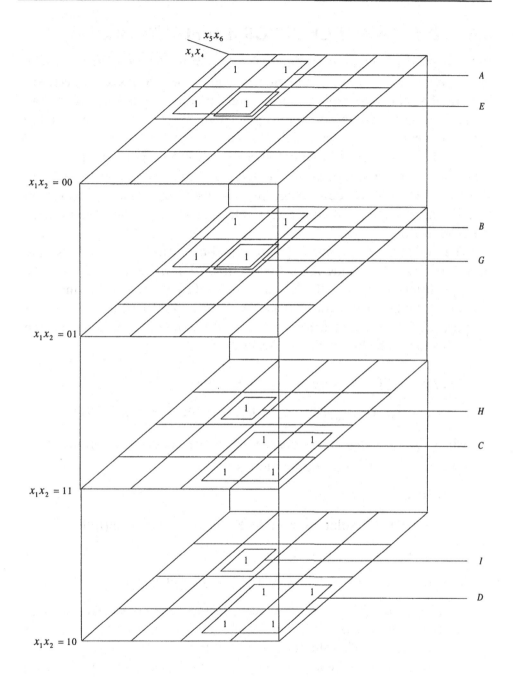

- **Figure 4.33** K map for *F* of Example 4.24

4.6 INCOMPLETELY SPECIFIED FUNCTIONS

When the functional value is known for each of the 2^n input combinations of n variables, the Boolean function is said to be **completely specified.** Nevertheless, in many applications there are some input conditions that, for one reason or another, do not occur, cannot occur, or are not allowed to occur. (One such case, the design of a seven-segment display circuit, was discussed in Section 3.7) We call these **don't-care** input conditions, and functions that have don't-care inputs are said to be **incompletely specified.** Don't-care conditions can be used to great advantage in simplifying the circuit required to implement an incompletely specified Boolean function.

The outputs corresponding to don't-care input conditions are designated by an $\times$ on the K-map (sometimes d or $\emptyset$ are used). During the K-map clustering process, the $\times$'s are assigned as either 0 or 1, whichever gives the simpler expression. In each case, the choice depends *only* on the simplification that can be achieved. The following example illustrates the minimization procedure with don't-cares.

■   **Example 4.25**   Simplify the function given by

$$F(x_1, \ x_2, \ x_4) = \sum (1, 3, 7, 8, 9, 10, 12, 13, 14, 15) + \sum_d (4, 5, 11)$$

where $\sum_d (4, 5, 11)$ represents the don't-care conditions. From the K-map of Figure 4.34(a), the solution is

$$F = x_1 + x_2' x_4 + x_3 x_4$$

Alternatively, the solution from the K-map of Figure 4.34(b) is

$$F = x_1 + x_4 \qquad\qquad ■$$

The points to notice in Example 4.25 are the following:

1. In Figure 4.34(a), the don't care in the cell m_{11} is taken as 1 to enable us to make a cluster of eight 1-cells and two clusters of four 1-cells. The other don't-cares are set to 0.
2. The result obtained from Figure 4.34(a) is correct but is not a minimal solution. A minimal expression is obtained by taking the don't-cares at both m_{11} and m_5 to be 1 as shown in Figure 4.34(b).

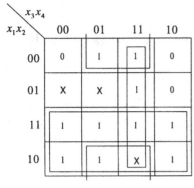

x_1x_2 \ x_3x_4	00	01	11	10
00	0	1	1	0
01	X	X	1	0
11	1	1	1	1
10	1	1	X	1

(a) One choice of don't-cares

x_1x_2 \ x_3x_4	00	01	11	10
00	0	1	1	0
01	X	X	1	0
11	1	1	1	1
10	1	1	X	1

(b) A second choice of don't-cares

■ **Figure 4.34** Simplifying with don't-care conditions

We leave as an exercise for the reader to redesign the seven-segment display circuit of Section 3.7. Obtain the minimal expressions for each of the seven segments by utilizing the don't-care conditions.

4.7 PRODUCT-OF-SUMS SIMPLIFICATION

The minimized functions derived thus far have been expressed as minimal sums, but it is also possible to express a function as a minimal product. The definition of a minimal product is analogous to that of a minimal sum: It is the product-of-sums (POS) form that contains the fewest (sum) terms and the fewest literals.

The K-map procedure for obtaining a minimal product for a function F follows the same basic pattern for minimal sums except that we take clusters of 0-cells rather than 1-cells. Alternatively, we can find the minimal sum for the complementary function F' and then use DeMorgan's theorem (10b) (see Table 3.1) to obtain the minimal product for F. The clusters of 0-cells are expressed by sums of literals. If a cluster is not contained in a larger cluster of 0-cells, we call the corresponding sum of literals a **prime implicate** of F (see Definition 3.4).

■ **Example 4.26** Find the minimal sum and the minimal product for

$$F(x_1, x_2, x_3, x_4) = \sum(5, 6, 7, 9, 10, 11, 13, 14, 15)$$

whose K-map is shown in Figure 4.35(a).

The minimal sum for F is obtained from the four clusters of 1-cells:

$$F = x_2 x_4 + x_2 x_3 + x_1 x_4 + x_1 x_3$$

and is implemented in Figure 4.35(b). Likewise, the two clusters of 0-cells [shown dashed in Figure 4.35(a)] give

$$F = (x_1 + x_2)(x_3 + x_4)$$

and this minimal POS expression is implemented in Figure 4.35(c). As can be seen, the minimal product form is the more economical circuit in this case.

We can also find the minimal POS for F by using the two clusters of 0-cells to obtain the minimal sum for F' (simply treat the 0-cells as though they were 1-cells):

$$F' = x_1' x_2' + x_3' x_4'$$

and then applying DeMorgan's theorem (10b) of Table 3.1 to derive F. ∎

For some functions, the minimal sum leads to a more economical circuit, but for other functions the minimal product is more economical. There is no known method for determining which form will lead to a more economical circuit so, in general, the designer should examine both forms and then select the one that is cheapest to implement.

4.8 HAZARDS IN COMBINATIONAL CIRCUITS

When an input to a combinational circuit changes, an erratic response may occur at the output because different signal paths through the circuit have different *propagation time delays* (see Section 1.2). For example, consider the circuit of Figure 4.36(a). When all the inputs are 1, the output of gate 1 is 1, the output of gate 2 is 0, and the output of gate 3 is 1. Moreover, if

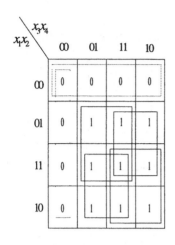

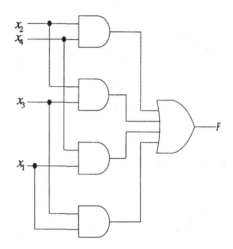

(a) K-map for $F=\Sigma(5, 6, 7, 9, 10, 11, 13, 14, 15)$ (b) AND-OR implementation

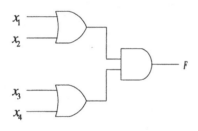

(c) OR-AND implementation

■ **Figure 4.35** Obtaining a minimal product

$x_1 = x_3 = 1$ and x_2 changes from 1 to 0, the output of gate 3 *should* remain a constant 1. But then the output of the circuit can *momentarily* go to 0 if the propagation time delay of gate 1 is *smaller* than the combined delays of the inverter and gate 2. In this case, when x_2 changes to 0, the output of gate 1 will change to 0, while gate 2 will not show a change until some time later. Hence, the output of gate 2 will remain at 0 temporarily, and the output of gate 3 will be 0 for this period of time. After a delay, the output of gate 2 will change to 1 and cause the output of gate 3 to change back to 1.

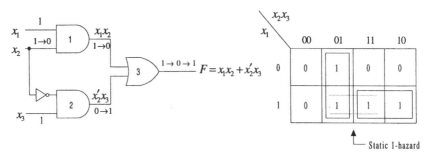

(a) Circuit with a static 1-hazard

(b) K-map indicating the static 1-hazard

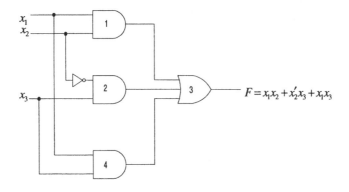

(c) Circuit with a static 1-hazard removed

■ **Figure 4.36** Detecting and eliminating hazards

In general, if in response to an input change, a circuit output momentarily goes to 0 when it should remain a constant 1, we say that the circuit has a **static 1-hazard**. On the other hand, if the output momentarily changes to 1 when it should remain a constant 0, we say that the circuit has a **static 0-hazard**. Another type of hazard, called a **dynamic hazard**, occurs when the output is supposed to change from 0 to 1 (or from 1 to 0) but in fact changes three or more *odd* times. The three types of hazard are illustrated in Figure 4.37. In each case, the steady-state output of the circuit is correct, but an unwarranted transient change appears at the output when an input is changed.

We can relate the definitions to the circuit of Figure 4.36(a) and conclude that it contains a static 1-hazard. In fact, we can detect the hazard

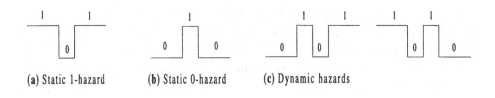

(a) Static 1-hazard   (b) Static 0-hazard   (c) Dynamic hazards

■ **Figure 4.37**   Types of hazards

on the K-map of the output shown in Figure 4.36(b). Observe that, if two adjacent 1-cells are covered by the same cluster, changing the input to the circuit between the two corresponding input states *cannot* cause a hazard. For example, changing the input state from $x_1 x_2 x_3 = 001$ to 101 cannot cause a static 1-hazard because the corresponding 1's on the map are both covered by the term $x_2' x_3$. Hence, when x_1 is changing, the term $x_2' x_3$ remains a constant 1 and the output will remain unchanged. In contrast, if two adjacent 1-cells on the map are not covered by the same cluster, then a static 1-hazard is present. For example, a change from $x_1 x_2 x_3 = 101$ to 111 will cause a static 1-hazard because the corresponding 1-cells are covered by different clusters.

We can eliminate the static 1-hazard by including the implicant $x_1 x_3$ shown by the dashed cluster in Figure 4.36(b). Hence, the circuit of Figure 4.36(c) with gate 4 added will be *hazard free*. Note that $x_1 x_3$ (or gate 4) is *redundant* as far as the minimal sum representation of the function is concerned. Nevertheless, we must introduce a redundant term (gate) in order to ensure a *reliable* operation of the circuit.

By contrast, a product-of-sums (POS) implementation of the same circuit could have static 0-hazards. Thus, from the K-map of Figure 4.36(b), $F = (x_2 + x_3)(x_1 + x_2')$, and what we just said about detecting static 1-hazards can be applied now to static 0-hazards. Since the 0-cells for m_0 and m_2 are adjacent but are not covered by the same cluster, a static 0-hazard is present in the POS implementation of F. Therefore, appending the redundant term $(x_1 + x_3)$ to F will eliminate the 0-hazard so that the POS implementation $F = (x_2 + x_3)(x_1 + x_2')(x_1 + x_3)$ will be hazard free.

A circuit may have a dynamic hazard even if it is free of static hazards. Since a dynamic hazard involves at least three changes in the output, the effect of an input change must reach the output at three different times; i.e., the changing input variable must have three different paths to the output. The procedure used to detect static hazards can be extended to dynamic hazards, but this is beyond the scope of this text.

To this point, we have emphasized minimization as a design criterion for digital circuits. But now we see that **reliability** is also an important design objective. Unfortunately, these two design objectives may work against each other because minimization is attained by eliminating redundant terms, while reliability, generally speaking, is enhanced by increasing the number of redundant terms. Therefore, the designer may be forced to compromise on one objective to attain the other.

4.9   THE QUINE-McCLUSKEY TABULATION TECHNIQUE

The K-map minimization technique has two major weaknesses: (1) It relies on visualization of geometric patterns, and (2) its effectiveness is limited in practice to six variables or less. The **Quine-McCluskey method** (also called the **tabulation method**), on the other hand, does not suffer from these same weaknesses, but instead offers an algorithm that can be applied to functions of any number of variables. The great advantage of this procedure is that it can be programmed on a digital computer.

The tabulation method follows the same general plan used for the K-map minimization procedure: We determine all prime implicants and then select those that will appear in a minimal sum. Let us describe, first, how the Quine-McCluskey method is used to determine all prime implicants.

Determining All Prime Implicants

Recall that the essence of logical adjacency can be described by binary information. Each product term in n variables is represented in the Quine-McCluskey procedure by an n-bit string of 0's, 1's, and dashes (−); a 0 or 1 in the ith position indicates that the variable x_i is primed or unprimed, respectively, and a dash (−) indicates that x_i has been eliminated altogether. For example, with $n = 5$, the minterms $m_5 = x_1' x_2' x_3 x_4' x_5$ and

$m_{21} = x_1 x_2' x_3 x_4' x_5$ can be described by the bit strings 00101 and 10101, respectively, and the fact that $m_5 + m_{21}$ simplifies to $x_2' x_3 x_4' x_5$ can be described by the string -0101 because x_1 has been eliminated by the logical adjacency theorem. Likewise, minterms $m_7 = x_1' x_2' x_3 x_4 x_5$ and $m_{23} = x_1 x_2' x_3 x_4 x_5$ can be represented by 00111 and 10111, and the simplified sum $m_7 + m_{23} = x_2' x_3 x_4 x_5$ by -0111. But then $\sum(5, 7, 21, 23)$ simplifies to $x_2' x_3 x_5$ because the strings -0101 and -0111 can be combined to form the string $-01-1$, where the two dashes indicate that both variables x_1 and x_4 have been eliminated.

In general, two binary strings p and q can be combined if they differ in *exactly* one bit position. In fact, if

$$p = k_1 k_2 \ldots k_{i-1} k_i k_{i+1} \ldots k_n$$

and

$$q = k_1 k_2 \ldots k_{i-1} k_i' k_{i+1} \ldots k_n$$

where k_j, for $j \neq i$, is taken from $\{0, 1, -\}$ but k_i is from $\{0, 1\}$. Then the two strings combine to

$$p + q = k_1 k_2 \ldots k_{i-1} (-) k_{i+1} \ldots k_n$$

where a dash has been inserted in the ith position.

With this in mind, let us describe the basic steps of the Quine-McCluskey procedure for determining all the prime implicants as we work out the details of an example.

■ **Example 4.27** Simplify the function

$$F = (x_1, x_2, x_3, x_4) = \sum(0, 1, 2, 3, 5, 6, 8, 11)$$

Step 1: Convert all minterms to $\{0, 1\}$ bit strings. Partition the strings into groups according to the *number* of 1's in each string. Tabulate the strings by the number of 1's in ascending order. Label each minterm string (row in the table) with its decimal equivalent; this will record the minterms covered by the product string.

Applying step 1 to F, we obtain the following table (for now, disregard the check marks (✓) which result from Step 2 below):

Number of 1's	Decimal Label	Product String $(x_1 x_2 x_3 x_4)$	
0	0	0000	✓
1	1	0001	✓
	2	0010	✓
	8	1000	✓
2	3	0011	✓
	5	0101	✓
	6	0110	✓
3	11	1011	✓

Step 2: Form a new table similar to the one in step 1. Start at the top of the table in step 1 and compare each string with r 1's with each string in the neighboring group with $(r+1)$ 1's to determine if the two strings can be combined. When two strings p and q can be combined, place a check (✓) in the previous table beside each of p and q, and place the combined string in the new table, labeling it with a combination of the decimal labels of both p and q.

The table resulting from step 2 is shown next.

Number of 1's	Decimal Label	Product String $(x_1 x_2 x_3 x_4)$	
0	0, 1	000 –	✓
	0, 2	00 – 0	✓
	0, 8	– 000	
1	1, 3	00 – 1	✓
	1, 5	0 – 01	
	2, 3	001 –	✓
	2, 6	0 – 01	
2	3, 11	– 011	

For example, since m_0 and m_1 can be combined, rows 0 and 1 are checked in the table of step 1, and the combined string 000- is labeled with the pair of decimal numbers $\{0, 1\}$ and is placed in the new table. The checks beside rows 0 and 1 indicate that the two product terms m_0 and m_1 have been combined and, hence, neither m_0 nor m_1 can be a prime implicant of F. (When we finish, the unchecked strings will determine all the prime implicants.) In fact, m_0 can be combined with m_2 and m_8, so each of these rows must be checked also. (There is no need to place more than one check beside m_0.)

Step 3: Repeat step 2 to successively generate tables of combined strings. Observe that *strings are combined only if they differ in one bit position and the dashes in both strings coincide.* The process terminates when no further combinations of strings are possible. Then *the unchecked strings (**in all the tables**) constitute the set of prime implicants of the function.*

The table resulting from step 3 follows:

Number of 1's	Decimal Label	Product String $(x_1x_2x_3x_4)$	
0	0, 1, 2, 3	00 - -	
	0, 2, 1, 3	00 - -	Duplicate

In the table of step 2, the combination of the row labeled $\{0, 1\}$ with the row labeled $\{2, 3\}$ yields the same information as the combination of rows $\{0, 2\}$ and $\{1, 3\}$ (as shown in the table of Step 3). We eliminate the duplication by having a single row labeled $\{0, 1, 2, 3\}$, but checking all four rows in the table of step 2.

Since we cannot apply step 2 to the last table, the process terminates. *All* the unchecked strings in *all* the tables determine the prime implicants of F, as follows:

Decimal Label	String	Prime Implicant
0, 1, 2, 3	00--	$x_1' x_2'$
0, 8	-000	$x_2' x_3' x_4'$
1, 5	0-01	$x_1' x_3' x_4$
2, 6	0-10	$x_1' x_3 x_4'$
3, 11	-011	$x_2' x_3 x_4$

Note that, since F involves only four variable, we could have obtained the same information from its K-map in Figure 4.38.

■

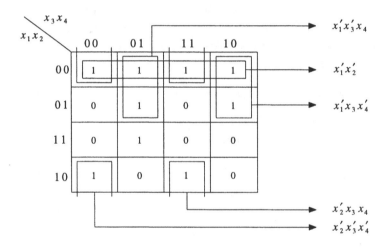

■ **Figure 4.38** K-map for the function of Example 4.27

Selecting Prime Implicants for a Minimal Sum

The Quine-McCluskey procedure also provides the means for selecting those prime implicants that appear in a minimal sum by constructing the **prime-implicant table**. Each row of this table corresponds to a prime implicant, and each column corresponds to a minterm for which the function has a value 1. It is good practice to list the prime implicants in descending order according to the number of minterms they cover. Thus, prime implicants located higher in the prime-implicant table are expressed with fewer literals than those located lower in the table. It is also good practice to list the minterms in ascending order according to their decimal equivalents.

We place a cross (X) in the prime-implicant table at the intersection of a row and a column if the prime implicant *covers* the corresponding minterm. Note that the decimal labels placed beside each prime implicant in steps 1, 2, and 3 above show where to place the crosses in the prime-implicant table; these decimal labels are nothing but the subscripts of the minterms covered by the prime implicants. As an example, Table 4.6 shows the prime-implicant table for the function $F = \sum(0, 1, 2, 3, 5, 6, 8, 11)$ of Example 4.27.

■ **Table 4.6** Prime-implicant table for $F = \sum(0, 1, 2, 3, 5, 6, 8, 11)$

Prime Implicants		Minterms							
Decimal Label	String	0	1	2	3	⑤	⑥	⑧	⑪
0, 1, 2, 3	00--	X	X	X	X				
*0, 8	-000	X						X	
*1, 5	0-01		X			X			
*2, 6	0-10			X			X		
*3, 11	-011				X				X

Having constructed the prime-implicant table, we then determine the essential prime implicants (if they exist). Essential prime implicants are found by searching for columns with only a *single* cross. For any such column, the single cross means a distinguished minterm, and the row where this cross occurs corresponds, therefore, to an essential prime implicant. We mark the rows corresponding to essential prime implicants with an asterisk and circle the labels of the distinguished minterms, as shown in Table 4.6.

With the essential prime implicants found, we can now delete from the table each essential row and each column containing a cross in the essential row. This leaves only those minterms not covered by the essential prime implicants. For Example 4.27, this "striking out" process would leave a table with all columns removed, because the essential prime implicants in this case completely determine the minimal sum:

$$F = x_2'x_3'x_4' + x_1'x_3'x_4 + x_1'x_3x_4' + x_2'x_3x_4$$

But this function represents a special case; for most functions, the essential prime implicants do not cover all minterms. In general, after the essential prime implicants have been found and the corresponding rows and columns removed, we obtain a reduced prime-implicant table. We then continue the process until all the minterms are covered. The following example emphasizes the main steps of the procedure.

■ **Example 4.28** Simplify the function

$$F(x_1, x_2, x_3, x_4) = \Sigma(0,1,3,6,7,14,15)$$

We combine all the steps of the Quine-McCluskey procedure in Table 4.7   In parts (a), (b), and (c) of the table we generate the prime implicants, and in part (d) we show the prime-implicant table. The

■ **Table 4.7** Combined steps of the Quine-McCluskey procedure

(a) Step 1

Number of 1's	Decimal Label	String	
0	0	0000	√
1	1	0001	√
2	3	0011	√
	6	0110	√
3	7	0111	√
	14	0110	√
4	15	1111	√

(b) Step 2

Number of 1's	Decimal Label	String	
0	0, 1	000 –	
1	1, 3	00 –1	
2	3, 7	0 –11	
	6, 7	011 –	√
	6, 14	– 110	√
3	7, 15	– 111	√
	14, 15	111–	√

(c) Step 3

Number of 1's	Decimal Label	String
2	6, 7, 14, 15	–11 –

(d) Prime-implicant table

Decimal		Minterms						
Label	String	⓪	1	3	⑥	7	⑭	⑮
*6, 7, 14,15	-11-				X	X	X	X
*0, 1	000-	X	X					
1, 3	00-1		X	X				
3, 7	0-11			X		X		

first two rows in Table 4.7(d) correspond to the essential prime implicants x_2x_3 and $x_1'x_2'x_3'$.   After striking out these rows and the columns that contain a cross in either one of them, we obtain a reduced prime-implicant table with only the column for minterm m_3 remaining.   This minterm may be covered by either prime implicant $x_1'x_2'x_4$ (the third row) or $x_1'x_3x_4$ (the fourth row).   Therefore, both expressions

$$F = x_2x_3 + x_1'x_2'x_3' + \begin{cases} x_1'x_2'x_4 \\ \text{or} \\ x_1'x_3x_4 \end{cases}$$

are minimal sums.   ∎

There are two other ways to reduce the prime-implicant table: removal of dominated rows and removal of dominating columns.   We consider them in the following two sections.

Dominated Row Removal

A row (prime implicant) P_i in the prime-implicant table **dominates** a row P_j if P_i covers all minterms covered by P_j (i.e., P_i has a cross in each column where P_j has a cross) and if, in addition, the number of literals in P_i is less than or equal to the number of literals in P_j.   If row P_i dominates row P_j, then row P_j can be eliminated from the table.   We illustrate this procedure in the following example.

■ **Example 4.29**   Consider the function

$$F(x_1, x_2, x_3, x_4, x_5) = \sum(1,3,4,5,6,7,10,11,12,13,14,15,18,19,20,21,23,25,26,27)$$

We leave the details of determining the prime implicants to the reader and list only the prime-implicant table in Table 4.8(a). The four essential prime implicant rows are indicated by an asterisk, and the distinguished minterm numbers are encircled.

The reduced prime-implicant table obtained by removing the essential rows and the columns covered by the essential prime implicants is shown in Table 4.8(b). Some of the rows in this table dominate other rows and may be deleted. For example, row 7 (prime implicant $x_3' x_4 x_5$) has a cross in the columns for minterms m_{11} and m_{19}, while row 8 only has a cross in the column for m_{19}. Therefore, row 7 dominates row 8 and we remove row 8 from the reduced table. Likewise, row 9 can be removed because row 7 dominates row 9; and rows 4 and 6 can be removed because they are dominated by rows 3 and 5, respectively.

■ **Table 4.8**   Minimization of F of Example 4.29

(a) Prime-implicant table

Prime Implicants	Minterms																		
	①	3	4	5	6	7	10	11	⑫	⑬	14	15	18	19	⑳	㉑	22	23	㉕	26	27
*(1) $x_2' x_3$		X	X	X	X										X	X	X	X			
*(2) $x_1' x_3$		X	X	X	X				X	X	X	X									
(3) $x_1 x_3' x_4$											X	X							X	X	
(4) $x_1 x_2' x_4$											X	X			X	X					
(5) $x_2 x_3' x_4$					X	X													X	X	
(6) $x_1' x_2 x_4$					X	X					X	X									
(7) $x_3' x_4 x_5$	X							X						X							
(8) $x_2' x_4 x_5$	X				X									X				X			
(9) $x_1' x_4 x_5$	X				X	X							X								
*(10) $x_1' x_2' x_5$	X	X		X	X																
*(11) $x_1 x_2 x_3' x_5$																				X		X

(b) Reduced prime-implicant table

Prime Implicants	Minterms				
	10	11	18	19	26
(3) $x_1 x_3' x_4$			X	X	X
(4) $x_1 x_2' x_4$			X	X	
(5) $x_2 x_3' x_4$	X	X			X
(6) $x_1' x_2 x_4$	X	X			
(7) $x_3' x_4 x_5$		X		X	
(8) $x_2' x_4 x_5$				X	
(9) $x_1' x_4 x_5$		X			

(c) Final reduced table

Prime Implicants	Minterms				
	⑩	11	⑱	19	26
**(3) $x_1 x_3' x_4$			X	X	X
**(5) $x_2 x_3' x_4$	X	X			X
(7) $x_3' x_4 x_5$		X		X	

The final reduced table, shown in Table 4.8(c), contains two columns with a single cross. Therefore, the corresponding rows *must* be contained in some minimal sum. These rows, marked with a double asterisk, are called *secondary essential rows*. Note that, while essential prime implicants occur in every minimal sum, **secondary essential prime implicants** may not occur in some of the minimal sums.

After removing from Table 4.8(c) those columns that have a cross in the rows corresponding to the secondary essential prime implicants, we obtain an empty table. Thus, a minimal sum for F is the sum of the essential prime implicants from rows 1, 2, 10, and 11 of Table 4.8(a) and the secondary essential prime implicants from rows 3 and 5 of Table 4.8(c):

$$F = \underbrace{x_2' x_3 + x_1' x_3 + x_1' x_2' x_5 + x_1 x_2 x_3' x_5}_{\text{essential}} + \underbrace{x_1 x_3' x_4 + x_2 x_3' x_4}_{\text{secondary essential}}$$ ∎

Dominating Column Removal

Prime-implicant tables can also be reduced by deleting certain columns. A column C_i **dominates** another column C_j if C_i has a cross in every row in which C_j has a cross. If column C_i dominates column C_j, then column C_i can be deleted from the prime-implicant table without affecting the search for a minimal sum. For example, in Table 4.8(b), column 11 dominates column 10 and column 19 dominates column 18.

Note that for rows the *dominated* row is removed, but for columns the *dominating* column is removed. We may remove dominated rows and dominating columns and, in turn, create new dominated rows that can be removed, and so on.

Cyclic Prime-Implicant Tables

The process of removing essential rows, dominated rows, and dominating columns simplifies the search of a minimal sum. It does not, however, always yield *all* minimal sums, nor does it always lead to a clear-cut choice of prime implicants for a minimal sum. It may be that a prime-implicant table is such that each column contains at least two crosses (so that there are no essential prime implicants). Such tables are called **cyclic prime-implicant tables**.

With a cyclic prime-implicant table, we must resort to even more drastic measures: We select a row with the fewest number of literals for possible inclusion in the minimal sum, remove that row from the table, and then apply the reduction techniques to the resulting table. This entire process must then be repeated for each row that could replace the original selected row. The final minimal sum is obtained by choosing the simplest expression that resulted from all the selected rows.

A cyclic prime-implicant table comes into the picture in the following example.

■ **Example 4.30** Simplify the function

$$F(x_1, x_2, x_3, x_4) = \Sigma(0, 2, 5, 6, 7, 8, 9, 12, 13, 15)$$

The entire procedure is summarized in Table 4.9. In the reduced table [Table 4.9(e)], row domination gives one minimal sum, but two additional ones can be found by other observations.

Nevertheless, there is something we can do to find a minimal sum. Since all the rows involve the same number of literals, we can remove any one arbitrarily. Observe that, since each minterm must be covered by some prime implicant, one additional row selection is all that is required. By inspecting Table 4.9(e) we see that, if we select either rows 3 and 5, rows 3 and 6, or rows 4 and 5, then all columns are covered. Moreover, all these selections involve the same number of literals, so one selection is no better than another.

Therefore, adding these selections to the essential prime implicants, we obtain, correspondingly, three minimal sums:

$$F = x_1 x_3' + x_2 x_4 + \begin{cases} x_1' x_2' x_4' + x_1' x_3 x_4' \\ \text{or} \\ x_1' x_2' x_4' + x_1' x_2 x_3 \\ \text{or} \\ x_2' x_3' x_4' + x_1' x_3 x_4' \end{cases}$$

■

As an alternative, we may use the **branching method** to handle a cyclic prime-implicant table. In this method, we take a column with a minimal number of crosses. Since in Table 4.9(e) there are two crosses in each column, we may choose any column, say the one for minterm m_6. To ensure that m_6 is covered by some prime implicant, we must select either the prime implicant of row 5 or row 6. Hence, we obtain the two parts of Table 4.10 as follows: In Table 4.10(a), we select row 6 and eliminate that row and every column that has a cross in it; in Table 4.10(b) we do the same for row 5. (In general, if there are n crosses in the selected column, there would be n tables.)

The reduced Table 4.10(a) is also a cyclic table and we could apply the branching method again. But, in this simple case, it is obvious that selecting row 3 will yield a minimal sum. In Table 4.10(b), we can choose either row 3 or 4. These selections give the same three minimal sums listed in Example 4.30.

■ **Table 4.9** Simplifying a cyclic prime-implicant table

(a) Step 1

Number of 1's	Decimal Label	String	
0	0	0000	√
1	2	0010	√
	8	1000	√
2	5	0101	√
	6	0110	√
	9	1001	√
	12	1100	√
3	7	0111	√
	13	1101	√
4	15	1111	√

(b) Step 2

Number of 1's	Decimal Label	String	
0	0, 2	00-0	
	0, 8	-000	
1	2 ,6	0-10	
	8 ,9	100-	√
	8 ,12	1-00	√
2	5, 7	01-1	√
	5, 13	-101	√
	6, 7	011-	
	9, 13	1-01	√
	12, 13	110-	√
3	7, 15	-111	√
	13, 15	11-1	√

(c) Step 3

Number of 1's	Decimal Label	String
1	8, 9, 12, 13	1-0-
2	5, 7, 13, 15	-1-1

(d) Prime-implicant table

Decimal Label		String	Minterms									
			0	2	⑤	6	7	8	⑨	⑫	13	⑮
*(1)	8, 9, 12,13	1-0-						X	X	X	X	
*(2)	5, 7, 13,15	-1-1			X		X				X	X
(3)	0, 2	00-0	X	X								
(4)	0, 8	-000	X					X				
(5)	2, 6	0-10		X		X						
(6)	6, 7	011-				X	X					

(e) Reduced prime-implicant table

String	Minterms		
	0	2	6
(3) 00-0	X	X	
(4) -000	X		
(5) 0-10		X	X
(6) 011-			X

To summarize, the basic steps of the Quine-McCluskey procedure are:

1. Determine the prime implicants according to steps 1, 2, and 3 in Example 4.27.

2. Construct the prime-implicant table.
3. Determine the essential prime implicants and include them in the minimal sum. Remove the essential rows and corresponding columns from the table.
4. If all the columns are covered, *stop*; if not, remove dominated rows and dominating columns, and find the secondary essential prime implicants.
5. Repeat steps 3 and 4 as many times as applicable.
6. If a cyclic table occurs, apply either of the cyclic table reduction techniques, and return to step 3.

■ **Table 4.10**   Tables illustrating the branching method

(a) Table 4.9(e) with
      row 6 removed

(b) Table 4.9(e) with
      row 5 removed

		Minterms	
	String	0	2
(3)	00-0	X	X
(4)	-000	X	
(5)	0-10		X

		Minterms
	String	0
(3)	00-0	X
(4)	-000	X
(6)	011-	

Don't-Care Conditions

A function with don't-care conditions can be simplified by the Quine-McCluskey procedure after a slight modification. The don't-care minterms are *included* in the list of minterms of the function when the prime implicants are determined in steps 1, 2, and 3 of Example 4.27. This allows for the derivation of prime implicants with the least number of literals, but it also has the effect of increasing the number of possible prime implicants. On the other hand, when the prime-implicant table is set up, the don't-care minterms are *not included* in the list of minterms because they do not have to be covered by the selected prime implicants. If we did list them, this would amount to choosing all the don't-care values as 1, which might not be the best choice. By not listing them, we are leaving the choice open.

■ **Example 4.31** Simplify the function with don't-care conditions

$$F = (x_1, x_2, x_3, x_4) = \Sigma(2, 6, 7, 8, 13) + \Sigma_d(0, 5, 9, 12, 15)$$

We leave to the reader to verify in steps 1, 2, and 3 (or by K-maps) that the prime implicants of

$$G(x_1, x_2, x_3, x_4) = \Sigma(0, 2, 5, 6, 7, 8, 9, 12, 13, 15)$$

are $x_2 x_4$, $x_1 x_3'$, $x_1' x_2 x_3$, $x_1' x_3 x_4'$, $x_2' x_3' x_4'$, and $x_1' x_2' x_4'$.

The prime-implicant table is shown in Table 4.11(a). Note that, if the don't-care conditions were included in the prime-implicant table, then we would have had two essential prime implicants, $x_1 x_3'$ and $x_2 x_4$, with distinguished minterms m_5, m_9, m_{12}, and m_{15}. But, because these are all don't-care minterms, we have no essential prime implicants in Table 4.11(a) and every column has two crosses.

In Table 4.11(a), note that row 2 (for the prime implicant $x_1 x_3'$) dominates row 5 (for $x_2' x_3' x_4'$). Hence, we eliminate row 5 and obtain the reduced Table 4.11(b). Row 2 of this table is a secondary essential prime implicant since column 8 has only one cross. If we delete row 2 and columns 8 and 13, we obtain the cyclic prime-implicant table of Table 4.11 (c). We could apply the branching method to this table, but it is clear that the first and third rows yield the only minimal sum. Hence the unique minimal sum is

$$F = x_1 x_3' + x_2 x_4 + x_1' x_3 x_4' \qquad\qquad ■$$

*4.10 MULTIPLE-OUTPUT MINIMIZATION

So far, we have discussed methods for minimizing combinational circuits with a single-output function. But many circuits are designed to implement several output functions, requiring, therefore, that several expression be minimized *collectively*. To do that, we can extend the K-map and the Quine-McCluskey methods to the multiple-output case with a few modifications.

---

* This section is optional

■ **Table 4.11** Tabulation method with don't-care conditions

(**a**) Prime-implicant table

	Decimal Label	Prime Implicant	2	6	7	8	13
(1)	5, 7, 13, 15	x_2x_4			X		X
(2)	8, 9, 12, 13	x_1x_3'				X	X
(3)	6, 7	$x_1'x_2x_3$		X	X		
(4)	2, 6	$x_1'x_3x_4'$	X	X			
(5)	0, 8	$x_2'x_3'x_4'$				X	
(6)	0, 2	$x_1'x_2'x_4'$	X				

(**b**) Reduced prime-implicant table (**c**) Cyclic prime-implicant table

Prime Implicant	2	6	7	(8)	13
(1) x_2x_4			X		X
**(2) x_1x_3'				X	X
(3) $x_1'x_2x_3$		X	X		
(4) $x_1'\ x_3x_4'$	X	X			
(6) $x_1'x_2'x_4'$	X				

Prime Implicant	2	6	7
(1) x_2x_4			X
(3) $x_1'x_2x_3$		X	X
(4) $x_1'x_3x_4'$	X	X	
(6) $x_1'x_2'x_4'$	X		

For the multiple-output case, we could minimize each output function *separately* and then implement it in a separate circuit. While this greedy approach has the advantage of simplicity, in general, it does not lead to the most economical multiple-output circuit. In other words, minimizing separate components of the system may not lead to an overall minimized system. The following examples will illustrate this fact.

■ **Example 4.32** Consider the multiple-output combinational system block diagram of Figure 4.39(a) with three input variables (x_1, x_2, x_3) and two output functions:

$$F_1(x_1, x_2, x_3) = \sum(0, 1, 5)$$

$$F_2(x_1, x_2, x_3) = \sum(5, 6)$$

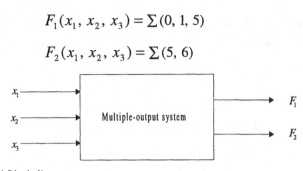

(a) Block diagram

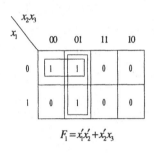

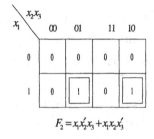

(b) K-maps for $F_1 = \sum(0,1,5)$ and $F_2 = \sum(5,6)$

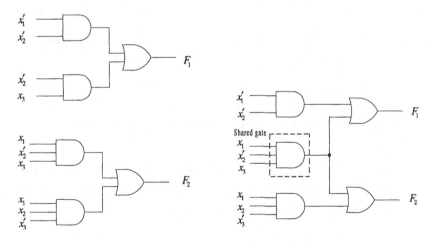

(c) Circuit obtained from separate minimization   (d) Minimal circuit for the entire system

■ **Figure 4.39**   Different approaches to minimizing a multiple-output system

If the functions are minimized separately, then using the K-maps of Figure 4.39(b) we obtain

$$F_1 = x_1'x_2' + x_2'x_3$$

$$F_2 = x_1x_2'x_3 + x_1x_2x_3'$$

and can implement the circuit as shown in Figure 4.39(c).

But inspecting the K-maps, we recognize that the prime implicant $x_2'x_3$ of F_1 can be replaced by the minterm $m_5 = x_1x_2'x_3$. Considering F_1 alone, this would be a poor choice because it would increase the number of literals in the expression for F_1. Nevertheless, the implicant $x_1x_2'x_3$ is an essential prime implicant for F_2 and can be *shared* with F_1 at no additional cost. This more economical circuit is shown in Figure 4.39(d) with the shared gate indicated.   ∎

∎ **Example 4.33**   Consider a multiple-output system where the output functions are

$$F_1(x_1, x_2, x_3) = \Sigma(1, 3, 7)$$

$$F_2(x_1, x_2, x_3) = \Sigma(2, 6, 7)$$

The K-maps are shown in Figure 4.40. Separate minimization gives

$$F_1 = x_1'x_3 + x_2x_3$$

$$F_2 = x_1x_2 + x_2x_3'$$

and implementing these functions requires four AND gates and two OR gates.

Nevertheless, the minimal circuit for *both* functions contains a shared gate for the minterm $m_7 = x_1x_2x_3$. Therefore, by sharing this one gate, the circuit implementing the expressions

$$F_1 = x_1'x_3 + x_1x_2x_3$$

and

$$F_2 = x_1x_2x_3 + x_2x_3'$$

requires only three AND gates and two OR gates.   ■

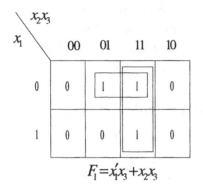
$$F_1 = x_1'x_3 + x_2x_3$$

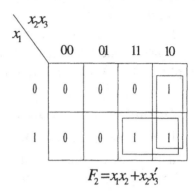
$$F_2 = x_1x_2 + x_2x_3'$$

■ **Figure 4.40** K-maps for Example 4.33

In general, for a single-output circuit, only prime implicants need be considered in determining a minimal two-level circuit. But Example 4.33 shows that it is not sufficient to consider only the prime implicants of the individual output functions in designing multiple-output systems (because m_7 is not a prime implicant for either of the output functions).

Basically, the task of minimizing a multiple-output system is one of identifying the prime implicants that cover each minterm of each output function and then carrying out a search for *shared* terms. But how can we determine those terms that can be shared in order to reduce the overall cost of the entire system? To find the answer, let us reconsider the shared terms in Examples 4.32 and 4.33. In each example, the shared terms turn out to be prime implicants for the product (AND) of the functions F_1 and F_2. This then is the clue to minimizing a multiple-output system.

Given two functions $F_1(x_1, x_2, \ldots, x_n)$ and $F_2(x_1, x_2, \ldots, x_n)$, the **product function**, F_1F_2, is defined as follows:

$$(F_1F_2)(x_1, x_2, \ldots, x_n) = \begin{cases} 0 & \text{if either } F_1(x_1,x_2,\ldots,x_n)=0 \text{ or } F_2(x_1,x_2,\ldots,x_n)=0 \\ 1 & \text{if both } F_1(x_1,x_2,\ldots,x_n) = 1 \text{ and } F_2(x_1,x_2,\ldots,x_n)=1 \\ \times & \text{if both } F_1(x_1,x_2,\ldots,x_n) \text{ and } F_2(x_1,x_2,\ldots,x_n) \text{ are} \\ & \text{don't-cares, or if one is a don't-care and the value of} \\ & \text{the other is 1} \end{cases}$$

While we have defined the product in terms of only two functions, the same rules apply to the product of any number of functions. Note that, if $n \leq 4$, the product functions can be easily determined by using K-maps.

Since prime implicants of product functions come into play in multiple-output minimization, we need to generalize the concept of prime implicant to that of a multiple-output prime implicant.

□ **Definition 4.4**   A **multiple-output prime implicant** of a set of functions, $\{F_1(x_1, x_2, \ldots, x_n),\ F_2(x_1, x_2, \ldots, x_n), \ldots,\ F_m(x_1, x_2, \ldots, x_n)\}$, is a product of literals that is either a prime implicant of one of the functions, or a prime implicant of any one of the product functions.

□

For each multiple-output system there exists a minimal sum-of-products (SOP) implementation for the *entire* system in which each product is a multiple-output prime implicant. Moreover, the implementation consists of two levels, where each multiple-output prime implicant is obtained as the output of an AND gate in the first level, and each output function is obtained as the output of an OR gate in the second level. The number of OR gates is equal to the number of output functions. Besides that, the SOP implementation is minimal with respect to the number of AND gates and the *total* number of gate inputs (counting the inputs to both AND gates and OR gates).

To find a minimal multiple-output SOP implementation, we follow the same procedure established earlier for single-output functions. First we generate the set of multiple-output prime implicants (using the K-map or the tabulation method); then we select a *minimal* cost subset (using the prime-implicant table) to implement the multiple-output system. Since all output functions and all possible products of the output functions must be considered, the multiple-output minimization problem can be very involved, especially when there are more than four output functions. If the number of functions is large, the process is often too cumbersome even with the aid of a computer.

To illustrate the details involved in executing the multiple-output minimization procedure, let us consider an example of a multiple-output combinational system, where the output functions are

$$F_1(x_1, x_2, x_3, x_4) = \Sigma(2, 3, 5, 7, 8, 9, 10, 11, 13, 15)$$

$$F_2(x_1, x_2, x_3, x_4) = \Sigma(2, 3, 5, 6, 7, 10, 11, 14, 15)$$

$$F_3(x_1, x_2, x_3, x_4) = \Sigma(6, 7, 8, 9, 13, 14, 15)$$

If we minimize each of these functions *separately*, we obtain

$$F_1 = x_2 x_4 + x_2' x_3 + x_1 x_2'$$

$$F_2 = x_1' x_2 x_4 + x_3$$

$$F_3 = x_2 x_3 + x_1 x_2' x_3' + \begin{cases} x_1 x_2 x_4 \\ \text{or} \\ x_1 x_3' x_4 \end{cases}$$

Implementing these expressions requires seven AND gates and three OR gates with a total of 25 gate inputs.

To minimize the given functions *collectively*, we can use the K-maps of Figure 4.41 to find the multiple-output prime implicants. We start with the K-map of the product $F_1 F_2 F_3$; then we consider the K-maps for the product functions $F_1 F_2$, $F_1 F_3$, and $F_2 F_3$; finally, we consider the K-maps for the individual functions F_1, F_2, and F_3. Prime implicants that are identified on the K-map for $F_1 F_2 F_3$ need not be shown on any other maps; similarly, prime implicants that are identified on the K-maps for $F_1 F_2$, $F_1 F_3$, and $F_2 F_3$ need not be show on the maps for F_1, F_2, and F_3. Only *new* prime implicants formed by new clusters need be marked in each case. This process continues until *all* the multiple-output prime implicants on *all* the maps have been discovered.

Note that no prime implicant of F_3 is identified on the K-map because all have been previously isolated as prime implicants of $F_1 F_3$ or $F_2 F_3$. In the K-map of F_2, only the prime implicant $x_1' x_2 x_4$ has been isolated earlier as a prime implicant of $F_1 F_2$. Similarly, in the K-map of F_1, the prime implicants $x_2' x_3$ and $x_3 x_4$ have been identified previously.

For functions of more than four variables, the Quine-McCluskey

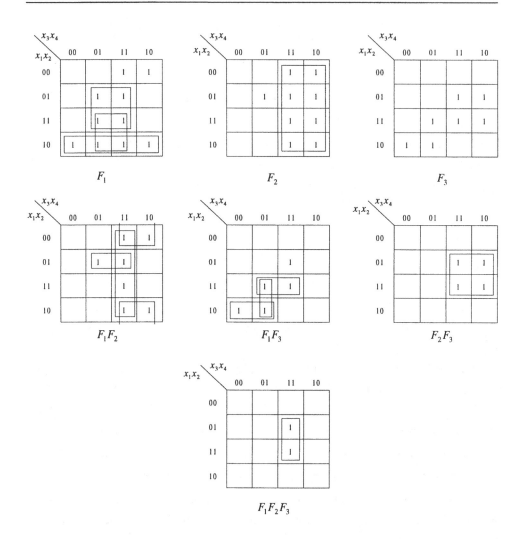

■ **Figure 4.41** K-maps for multiple-output minimization

procedure usually is the better one to use to generate the multiple-output prime implicants. When applying this procedure, first generate *all* the prime implicants of the individual functions and of the product functions and then eliminate those that are duplicated.

From the K-maps of Figure 4.41, we can tabulate the multiple-output prime implicants:

Prime Implicant	Minterms Covered	Map
$x_2 x_3 x_4$	7, 15	$F_1 F_2 F_3$
$x_2 x_3$	6, 7, 14, 15	$F_2 F_3$
$x_1 x_2 x_4$	13, 15	$F_1 F_3$
$x_1 x_2' x_3'$	8, 9	$F_1 F_3$
$x_1 x_3' x_4$	9, 13	$F_1 F_3$
$x_2' x_3$	2, 3, 10, 11	$F_1 F_2$
$x_3 x_4$	3, 7, 11, 15	$F_1 F_2$
$x_1' x_2 x_4$	5, 7	$F_1 F_2$
x_3	2, 3, 6, 7, 10, 11, 14, 15	F_2
$x_1 x_2'$	8, 9, 10, 11	F_1
$x_2 x_4$	5, 7, 13, 15	F_1
$x_1 x_4$	9, 11, 13, 15	F_1

The prime-implicant table for the multiple-output system, shown in Table 4.12, is constructed in a manner similar to the single-output case except that, now, the table shows sets of minterm columns corresponding to each of the output functions. Each row of the table corresponds to a multiple-output prime implicant. If a minterm occurs in more than one function, then that minterm is represented (by an X) on more than one column of the table. For example, minterms m_{13} and m_{15} occur in both F_1 and F_3 and are covered by the multiple-output prime implicant $x_1 x_2 x_4$ of $F_1 F_3$. Therefore, crosses are placed in these two columns in *both* F_1 and F_3.

Just as in the single-output case, enough rows must be selected so that there is at least one cross in each column. A column that contains only one cross determines a distinguished minterm for the function containing this column. Note, however, that this same minterm may not be distinguished for another function. By inspection, we can identify five distinguished minterms in Table 4.12: those whose labels are circled. The corresponding rows determine *essential multiple-output prime implicants* (designated

■ **Table 4.12** Multiple-output prime-implicant table

Multiple-Output		F₁										F₂									F₃						
Prime Implicant	Map	②	3	5	7	8	9	10	11	13	15	2	3	⑤	6	7	10	11	14	15	⑥	7	⑧	9	13	⑭	15
(1) x_3	F_2											X	X		X	X	X	X	X	X							
(2) $x_2 x_4$	F_1			X	X					X	X																
(3) $x_1 x_4$	F_1						X		X	X	X																
(4) $x_1 x_2'$	F_1					X	X	X	X																		
*(5) $x_2 x_3$	$F_2 F_3$														X	X			X	X	X	X				X	X
*(6) $x_2' x_3$	$F_1 F_2$	X	X					X	X			X	X				X	X									
(7) $x_3 x_4$	$F_1 F_2$		X		X				X		X		X			X		X		X							
*(8) $x_1' x_2 x_4$	$F_1 F_2$			X	X									X		X											
(9) $x_1 x_2 x_4$	$F_1 F_3$									X	X														X		X
(10) $x_1 x_3' x_4$	$F_1 F_3$						X			X														X	X		
*(11) $x_1 x_2' x_3'$	$F_1 F_3$					X	X																X	X			
(12) $x_2 x_3 x_4$	$F_1 F_2 F_3$				X						X					X				X		X					X

(designated with an asterisk), but these are essential *only* for the respective functions. Thus, minterm m_2 is distinguished for F_1, minterm m_5 is distinguished for F_2, and minterms m_6, m_8, and m_{14} are distinguished for F_3. Therefore, $x_2'x_3$ is an essential prime implicant of F_1, $x_1'x_2x_4$ is an essential prime implicant of F_2, and x_2x_3 and $x_1x_2'x_3'$ are essential prime implicants of F_3.

We may reduce the multiple-output prime-implicant table just as we did for a single-output system, but now we can only reduce those portions of the table for which the prime implicants are essential. In other words, we can remove the columns of the individual function that have a cross in common with its essential row, but we *cannot* remove this same row from other portions of the table for which the row is not essential. For example, row 6 is essential for F_1 so that columns 2, 3, 10, and 11 of F_1 can be removed from the table; but since row 6 also covers some minterms of F_2 and it is not essential for F_2, it must appear in the reduced table. On the other hand, if an essential row covers some minterms of only one function, then the corresponding columns *and* the essential row can be removed from the table. Since each of the essential rows in Table 4.12 is common to more than one function, only columns can be removed, resulting in Table 4.13.

We can apply the same dominance rules, as in the single-output case, to remove columns and rows from Table 4.13. We must, however, apply these rules with some care. A dominating column can be removed only if it and the dominated column belong to the same function. Row dominance must be applied across the table. For example, row 2 in Table 4.13 dominates row 12 for F_1, but row 12 cannot be removed because it also covers minterm 15 of F_2. On the other hand, row 12 can be removed because it is dominated by row 7 across the entire table. In considering row dominance, we must also take into account the "cost" involved in removing a dominated row. For example, row 2 in Table 4.13 dominates row 8. But row 8 is an essential prime implicant of F_2, so if we remove it now from the table, we would lose the possibility of including it as a multiple-output prime implicant for F_1 and *sharing* it with F_2. In addition, if we do remove row 8, then some other row(s) will have to be used to cover F_1 at the expense of additional gates and gate inputs.

■ **Table 4.13** Reduced multiple-output prime-implicant table

Multiple-output Prime implicant		Map	F_1						F_2							F_3
			5	7	8	9	13	15	2	3	6	10	11	14	15	13
(1)	x_3	F_2							X	X	X	X	X	X	X	
(2)	x_2x_4	F_1	X	X			X	X								
(3)	x_1x_4	F_1				X	X	X								
(4)	x_1x_2'	F_1			X	X										
(5)	x_2x_3	F_2F_3										X		X	X	
(6)	$x_2'x_3$	F_1F_2							X	X	X	X				
(7)	x_3x_4	F_1F_2		X				X	X			X			X	
(8)	$x_1'x_2x_4$	F_1F_2	X	X												
(9)	$x_1x_2x_4$	F_1F_3				X	X									X
(10)	$x_1x_3'x_4$	F_1F_3			X	X										X
(11)	$x_1x_2'x_3'$	F_1F_3			X	X										
(12)	$x_2x_3x_4$	$F_1F_2F_3$	X				X								X	

Taking these observations into account, we see that in Table 4.13, row 1 dominates rows 5 and 6, and row 7 dominates row 12. But only row 12 can be removed from the table. Row 11 is equal to row 4, but since row 11 is an essential prime implicant of F_3, we choose to remove row 4. For F_1, column 7 dominates column 5 and column 9 dominates column 8, so columns 7 and 9 can be removed. For F_2, columns 3 and 11 dominate column 2, and column 15 dominates column 6. Hence, columns 3, 11, and 15 can be removed. Moreover, column 10 is equal to column 2, and column 6 is equal to column 14, so columns 10 and 14 can also be removed.

The removal of dominating columns and dominated rows from Table 4.13 results in a table in which row domination allows the removal of rows 3, 7, and 10 to give Table 4.14. We now identify minterm 8 of F_1 and minterm 13 of F_3 as (secondary) distinguished minterms and designate accordingly the *secondary essential prime implicants* with a double asterisk. To cover the remaining minterms, we choose the essential rows 5 and 6 to cover minterms 2 and 6 of F_2, and essential row 8 to cover minterm 5 of F_1.

The multiple-output minimal expressions resulting from Tables 4.12 and 4.14 are

$$F_1 = x_2'x_3 + x_1x_2'x_3' + x_1'x_2x_4 + x_1x_2x_4$$

$$F_2 = x_1'x_2x_4 + x_2'x_3 + x_2x_3$$

$$F_3 = x_2x_3 + x_1x_2'x_3' + x_1x_2x_4$$

As we can see, to implement these expressions we now require five AND gates and three OR gates with a total of 23 gate inputs. This is in contrast to the seven AND gates, three OR gates, and the 25 gate inputs required when the functions are minimized separately.

■ **Table 4.14** Reduced multiple-output prime-implicant table with dominating columns and dominated rows removed.

Multiple-Output prime implicant	Map	F_1				F_2		F_3
		5	⑧	13	15	2	6	⑬
(1) x_3	F_2					X	X	
(2) x_2x_4	F_1	X		X	X			
*(5) x_2x_3	F_2F_3							X
*(6) $x_2'x_3$	F_1F_2					X		
*(8) $x_1'x_2x_4$	F_1F_2	X						
**(9) $x_1x_2x_4$	F_1F_3			X	X			X
**(11) $x_1x_2'x_3'$	F_1F_3		X					

PROBLEMS

4.1 Determine a Boolean expression for the output of each of the circuits in Figure P4.1.

4.2 Draw circuit diagrams for the following Boolean expressions using

AND and OR gates and inverters.
    (a) $(x_1' + x_2)(x_1 + x_3)$ **(b)** $(x_1 + x_2)' + x_1'x_3$
    (c) $x_1 x_2' x_3 + x_1' x_2 x_3 + x_1 x_2 x_3$

4.3   Find the truth tables for the following Boolean expressions:
    (a) $x_1 x_2' + x_3$ **(b)** $x_1(x_2 + x_3)' + x_2$

4.4   Find the canonical sum-of-products and product-of-sums forms for the truth tables in Table P4.4.

4.5   **(a)**   Draw a two-level AND-OR circuit for the function F_3 in Problem 4.4(c).
    (b)   Use properties of Boolean algebra to reduce the expression in Problem 4.4(c) to an equivalent expression whose circuit diagram requires only four logic gates (counting inverters).
    (c)   Draw the simplified circuit diagram.

4.6   Give the truth table for each of the following Boolean functions:
    (a) $F(x_1, x_2) = \Sigma(0, 1, 3)$
    (b) $F(x_1, x_2, x_3) = \Sigma(2, 5, 6, 7)$
    (c) $F(x_1, x_2, x_3, x_4) = \Sigma(0,1,4,6,9,11,13,14)$
    (d) $F(x_1, x_2, x_3, x_4, x_5) = \Sigma(0,2,9,14,15,17,28,30,31)$

4.7   **(a)**   Use properties of Boolean algebra to reduce the Boolean function $F(x_1, x_2, x_3) = \Sigma(0,1,4,5,7)$ to $x_1 x_3 + x_2'$.
    (b)   Show that the two Boolean expressions $x_1 x_3 + x_1' x_2$ and $(x_1 + x_2)(x_1' + x_3)(x_2 + x_3)$ are equivalent by determining the truth table for each.
    (c)   Write the canonical sum-of-products form equivalent to the two expressions in part (b).
    (d)   Use properties of Boolean algebra to convert one of the expressions of part (b) to the other.

4.8   A combinational circuit has four inputs and one output F. The output F is 1 if and only if three or more of the inputs are 1.
    (a)   Obtain the truth table for F.
    (b)   Find the canonical sum-of-products form for F.
    (c)   Find the canonical product-of-sums form for F.
    (d)   Draw a two-level AND-OR circuit for F.

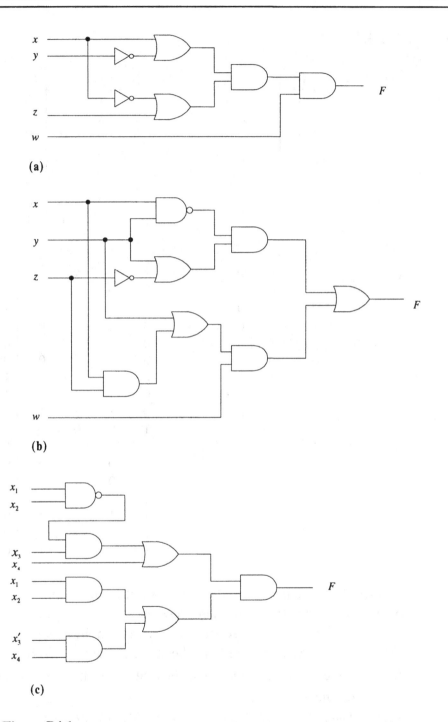

(a)

(b)

(c)

■   **Figure P4.1**

■ **Table P4.4**

(a)

x_1	x_2	F_1
0	0	0
0	1	1
1	0	0
1	1	1

(b)

x_1	x_2	x_3	F_2
0	0	0	0
0	0	1	1
0	1	0	0
0	1	1	1
1	0	0	0
1	0	1	0
1	1	0	0
1	1	1	1

(c)

x_1	x_2	x_3	F_3
0	0	0	1
0	0	1	0
0	1	0	0
0	1	1	0
1	0	0	1
1	0	1	1
1	1	0	0
1	1	1	0

(d)

x_1	x_2	x_3	x_4	F_4
0	0	0	0	0
0	0	0	1	1
0	0	1	0	0
0	0	1	1	0
0	1	0	0	0
0	1	0	1	1
0	1	1	0	1
0	1	1	1	1
1	0	0	0	0
1	0	0	1	0
1	0	1	0	0
1	0	1	1	1
1	1	0	0	1
1	1	0	1	0
1	1	1	0	0
1	1	1	1	0

4.9   In a special test car for safe driving, the following conditions must be met before the driver may start the car:

(1)   The key must be inserted (denote this condition by K).

(2)   The driver's seat belt must be attached (denote this by B_1).

(3)   If there is a passenger (P), the seat belt of his/her seat must be attached (B_2).

(4)   The driver must have passed a written test (T).

(5)   All doors of the car must be closed (D).

Use these conditions to write a sum-of-products expression that allows the car to be started only after all the conditions are met.

4.10   In a production process there are three motors operating, but only two are allowed to operate at the same time. Design a combinational circuit that prevents more than two motors from being switched on simultaneously.

4.11   **(a)**   A light (L) is controlled by three switches (x_1, x_2, x_3). The light is turned on precisely when a majority of the switches are on. Design a combinational circuit for controlling the operation of the light.

     (b)   A *majority* circuit is a combinational circuit with an odd number of inputs such that its output is 1 if and only if a majority of the inputs are 1 (see Chapter 3). Thus, part (a) calls for a majority circuit with three inputs. Design a 5-input majority circuit.

     (c)   A *minority* circuit is a circuit with an odd number of inputs such that its output is 1 if and only if a minority of its inputs are 1. Design a 3-input minority circuit.

4.12   Design a combinational circuit with four inputs (x_1, x_2, x_3, x_4) so that the output F is 0 if the input combination is a valid excess 3 coded decimal digit and F is 1 if any other combination of inputs occurs.

4.13   An oil pipeline is fed by three pipelines x_1, x_2, and x_3. Design a combinational circuit that switches off the pipeline at three points S_1, S_2, and S_3 such that oil runs only in the following situations: S_1 and S_3 are both open or both closed but S_2 is open; S_1 is open and S_2 and S_3 are closed.

4.14   A light in a room is to be controlled independently by three wall switches located at three entrances to the room. Flipping the switches changes the state of the light (off to on, or on to off). Derive a Boolean expression for the control circuit.

4.15   Design an even-parity generator for 3-bit messages.

4.16   An automatic control device supervises packaging of orders received by a mail-order wholesale firm. Toothpaste, hair spray, lipstick, and nail polish can be ordered. As a bonus item, shampoo is included with any order that includes hair spray or with any order that includes toothpaste, lipstick, and nail polish.

    (a)   Give a Boolean expression for a circuit that outputs 1 when shampoo is packaged with an order.

    (b)   Obtain the truth table for the output function of part (a).

    (c)   Derive the canonical sum-of-products form of the output function.

4.17   Derive a Boolean expression for a "keep awake" circuit that turns on a fan F in a computer terminal room if and only if the room temperature is over 80° Fahrenheit (designate this condition by T_0), or at least 10 people (P_1) have been in the room for at least 20 minutes (L) and the room temperature is at least 74° (T_1), or there are at least 15 users in the room (P_2), or there are at least 5 users (P_3) and the time (N) is between 12:00 A.M. and 6:00 P.M.

4.18   Design a division detector device that receives a 4-bit binary number $x_4x_3x_2x_1$ and outputs a 1 whenever the number is divisible by either 4 or 5.

    (a)   Obtain the truth table for this device.

    (b)   Find the canonical sum-of-products form for the output function F.

    (c)   Use K-map to obtain a minimal sum for F.

    (d)   Cluster the 0's to obtain a minimal sum for F'.

    (e)   Derive a minimal product-of-sums form for F.

    (f)   Compute the *complexity* of the circuits that implement the two minimal expressions for F. (We define the complexity of a circuit to be the sum of the number of gates and the number of input lines.)

4.19   Design a combinational circuit that activates a printer that prints a red dot on student application forms submitted by students that are male and over 21 years of age or female and under 21 years. (The printer does not care about female student applicants that are over 21 years.)

4.20 Plot each of the following functions on a K-map.

(a) $F_1 = \Sigma(1, 4, 7)$ (b) $F_2 = \Sigma(0, 2)$

(c) $F_3 = \Sigma(0, 2, 5, 6, 9, 11, 12, 15)$ (d) $F_4 = \Pi(4, 6, 10, 11, 12, 14, 15)$

4.21 Plot each of the following partially simplified functions on a K-map.

(a) $F = x_1'x_3x_4 + x_2x_3'x_4' + x_2x_4 + x_1'x_4'$

(b) $F = x_1x_3x_4 + x_1'x_2 + x_4'$

(c) $F = (x_1 + x_2' + x_3)(x_1' + x_3 + x_4')(x_3' + x_4')(x_2 + x_3')$

4.22 Obtain a minimal sum for each of the following:

(a) $F = \Sigma(0,2,3)$ (b) $F = \Sigma(2,3,6,7)$

(c) $F = \Sigma(4,5,6,7)$ (d) $F = \Sigma(1,2,3,6)$

(e) $F = \Sigma(0,1,2,3)$ (f) $F = \Pi(0,1,4,5)$

(g) $F = x_1'x_2' + x_2x_3 + x_1'x_2x_3'$ (h) $F = \Sigma(0,1,2,5,6,7)$

4.23 Obtain a minimal sum for each of the following functions:

(a) $F = \Sigma(7,13,14,15)$

(b) $F = \Sigma(2,3,4,5,6,7,10,11)$

(c) $F = \Sigma(1,3,4,5,10,12,13)$

(d) $F = \Sigma(0,2,3,5,6,7,8,10,11,14,15)$

(e) $F = \Sigma(0,2,4,5,6,7,8,10,13,15)$

(f) $F = \Pi(1,3,4,6,9,11,14,15)$

(g) $F = \Pi(4,5,6,7,12,13,14,15)$

(h) $F = \Sigma(0,1,2,3,4,6,8,9,10,11)$

(i) $F = \Sigma(0,2,4,5,6,8,10,12)$

(j) $F = \Sigma(0,2,6,8,10,12,14)$

(k) $F = \Sigma(1,3,4,5,9,10,11,12,13)$

(l) $F = \Sigma(0,1,5,7,8,10,14,15)$

4.24 For each of the following functions, show a K-map, find all prime implicants, indicate which prime implicants are essential, and determine a minimal sum and whether or not this minimal sum is unique.

(a) $F = \Sigma(0,2,4,5,6)$

(b) $F = \Sigma(3,4,6,7)$

(c) $F = \sum(0,1,2,6,8,9,10)$

(d) $F = \sum(0,1,2,3,4,6,7,8,9,11,15)$

(e) $F = \sum(1,3,4,5,7,8,9,11,15)$

(f) $F = \sum(0,1,2,4,5,6,8,9,12,13,14)$

(g) $F = \sum(2,3,4,6,7,8,12,13,15)$

(h) $F = \sum(1,2,4,5,6,11,12,13,14,15)$

(i) $F = \sum(4,5,8,9,12,13,14,15,16,17,20,21,22,23,24,25,26,28,29,30,31)$

4.25   Using K-maps, find a minimal sum for each of the following functions:

(a) $F = \sum(0,2,4,6,9,11,13,15,17,21,25,27,29,31)$

(b) $F = \sum(0,1,4,5,16,17,21,25,29)$

(c) $F = \sum(0,2,3,4,5,6,7,11,15,16,18,19,23,27,31)$

(d) $F = \sum(0,1,4,5,6,10,14,16,20,21,22,26,30)$

(e) $F = \sum(0,4,8,11,12,15,16,20,24,27)$

(f) $F = \sum(5,7,13,15,29,31)$

(g) $F = \sum(9,11,13,15,21,23,24,31)$

(h) $F = \sum(1,2,6,7,9,13,14,15,17,22,23,25,29,30,31)$

(i) $F = \sum(5,13,21,29,33,37,41,45,53,55,61,63)$

(j) $F = \sum(0,1,4,5,16,17,20,21,32,33,36,37,52,53)$

4.26   Simplify the incompletely specified Boolean functions using the don't-care conditions. Write your answers in both sum-of-products and product-of-sums forms.

(a) $F = \sum(1,3,7,11,15) + \sum_d(0,2,5)$

(b) $F = \sum(3,12,13) + \sum_d(5,6,7,15)$

(c) $F = \sum(0,2,4,6,8) + \sum_d(10,11,12,13,14,15)$

(d) $F = \sum(0,2,3,6,7) + \sum_d(5,8,10,11,15)$

(e) $F = \sum(0,1,2,3,7,8,10) + \sum_d(5,6,11,15)$

(f) $F = \sum(3,4,13,15) + \sum_d(1,2,5,6,8,10,12,14)$

(g) $F = \sum(0,1,5,6,9) + \sum_d(2,8,10,13,15)$

(h) $F = \sum(0,2,4,9,12,15) + \sum_d(1,5,7,10)$

(i) $F = \sum(0,3,4,7,8) + \sum_d(10,11,12,13,14,15)$

(j) $F = \sum(5,6,7,8,9) + \sum_d(10,11,12,13,14,15)$

(k) $F = \sum(0,3,5,7,10,15,16,18,24,29,31) + \sum_d(2,8,13,21,23,26)$

(l) $F = \sum(0,2,6,8,9,11,15,16,18,22,24) +$
$$\sum_d(4,10,12,13,14,20,26,28,30)$$

(m) $F = \sum(3,5,7,11,12,29,31) + \sum_d(1,2,6,10,28)$

(n) $F = \sum(0,2,5,7,8,10,13,15) + \sum_d(1,3)$

4.27   Find the minimal sum-of-products and minimal product-of-sums for the following functions:

(a) $F = \prod(0,1,4,5)$

(b) $F = \sum(1,3,4,6)$

(c) $F = x_1'x_3' + x_2'x_3' + x_2x_3' + x_1x_2x_3$

(d) $F = \prod(1,4,5,6,11,12,13,15)$

(e) $F = \prod(0,1,2,3,4,10,11)$

(f) $F = \sum(0,1,2,5,8,9,10)$

(g) $F = \prod(1,3,5,7,13,15)$

(h) $F = (x_1 + x_2' + x_4)(x_1' + x_2 + x_4)(x_3 + x_4)(x_3' + x_4')$

(i) $F = (x_1' + x_2' + x_4')(x_1 + x_2' + x_3')(x_1' + x_2 + x_4')(x_2 + x_3' + x_4')$

(j) $F = x_2'x_4x_5' + x_1x_2'x_3 + x_1x_2'x_5' + x_1'x_2x_5 + x_1'x_2'x_4'x_5'$

4.28   Analyze the circuit that implements

$$F = x_2'x_4' + x_1x_3 + x_1'x_3' \ (x_2 + x_4)$$

for (static) hazards. Redesign the circuit so that it becomes hazard free.

4.29   Find all static hazards in the circuit diagram of Figure P4.29. For each hazard, specify the values of the variables that are constant and the values of those changing when the hazard occurs. Also specify the order in which the gate outputs must change. Indicate how the hazards could be eliminated by adding gates to the existing circuit.

4.30   Find if there are static 1-hazards in the following functions; indicate where they occur and remove them.

(a) $F(x_1, x_2, x_3) = x_1'x_3' + x_2x_3 + x_1x_2'$

(b) $F(x_1, x_2, x_3, x_4) = x_1'x_2' + x_2x_3x_4 + x_1x_2x_3'$

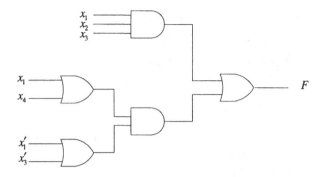

■ **Figure P4.29**

4.31   Show how a static 1-hazard can occur in the circuit of FigureP4.31.

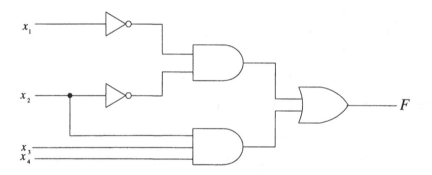

■ **Figure P4.31**

4.32   Use the Quine-McCluskey method to obtain the minimal sum for the following functions:
   (a) $F = \Sigma(2,6,7,8,10)$
   (b) $F = \Sigma(1,4,6,7,8,9,10,11,15)$
   (c) $F = \Sigma(0,1,2,3,5,8,11)$
   (d) $F = \Sigma(1,3,6,7,11,12,13,15)$
   (e) $F = \Sigma(0,4,5,6,7,8,9,10,11,14,15)$
   (f) $F = \Sigma(0,1,2,5,6,7,8,9,10,14)$
   (g) $F = \Sigma(20,28,52,60)$
   (h) $F = \Sigma(0,2,3,5,7,9,11,13,14,16,18,24,26,28,30)$

(i) $F = \Sigma(20,28,38,39,52,60,102,103,127)$

(j) $F = \Sigma(6,9,13,18,19,25,27,29,41,45,57,61)$

(k) $F = \Sigma(2,3,7,9,11,13) + \Sigma_d(1,10,15)$

(l) $F = x_1'x_2 + x_1'x_2'x_4 + x_1x_2'x_4' + x_1x_2'x_4$

(m) $F = \Sigma(0,3,6,9) + \Sigma_d(10,11,12,13,14,15)$

4.33 Design a combinational circuit that detects an error in the representation of a decimal digit in BCD code (i.e., the 8421 code).

4.34 Design a combinational code converter circuit that converts BCD messages into excess 3 code. Minimize each output function separately. [See Problem 4.26(c), (i), and (j).]

4.35 Design a combinational code converter circuit that converts:
(a) From $84(-2)(-1)$ code to BCD
(b) From 2421 code to $84(-2)(-1)$ code.
Minimize each output function separately.

4.36 Obtain the minimized output functions for a circuit that converts a 4-bit binary number to a decimal number in BCD code. (Note that two decimal digits are needed since the binary numbers range from 0 to 15.)

4.37 Design a combinational circuit that accepts a 3-bit binary number and generates a 6-bit binary number equal to the square of the input number. Minimize each output function separately. (See Problem 3.36.)

4.38 Design combinational circuits, each with four input lines representing a decimal digit in BCD code, that generate:
(a) The 9's complement of the input digit.
(b) The 2's complement of the input digit.
Minimize the output functions of each of the circuits separately.

4.39 An elevator services three floors of a hotel. On each floor there is a call button to call the elevator. We assume that at the moment of call the elevator cabin is stationary at one of the three floors. Thus, the calls (c_1, c_2, c_3) can come from any one of the floors (f_1, f_2, f_3), and the cabin can be located at any one of them. Using these six input variables, design a control circuit that moves the elevator motor in the right direction. Let $c_i = 0$ (or 1) mean that no call (or a call) comes from the ith floor. Let $f_i = 1$ mean that the elevator cabin is at the ith floor. Let M_1 and M_2 be output functions determining the direction of the motor. Thus, $M_1 = 1$ means the motor is moving the elevator cabin upward and $M_2 = 1$ means the motor is moving the cabin downward. Minimize each output function separately.

4.40   Let $T(x_1, x_2, x_3) = \Sigma(3,5,6)$ define a gate, called the T-gate.

     (a)   Prove that the T-gate and the constant 1 form a functionally complete set of operations (see Section 3.5).

     (b)   Implement $F(x_1, x_2, x_3, x_4) = \Sigma(0,1,2,4,7,8,9,10,12,15)$ by two

4.41   A certain 4-input gate, called L-gate, implements the function $L(x_1, x_2, x_3, x_4) = x_2 x_3 (x_1 + x_4)$. Implement the function $F = \Sigma(0,1, 6,9,10,11,14,15)$ with only three L-gates and one OR gate. (*Hint*: Draw the K-map for the L-gate and utilize possible "patches" on the K-map for F.)

4.42   A 3-input E-gate, where $E(x_1, x_2, x_3) = \Sigma(0,2,3,4)$, has been mass produced by an unfortunate company. Experimental evidence shows that the input combinations 101 and 010 cause the gate to "explode." Your task is to determine whether the gate is completely useless or can be externally modified so that it may be efficiently used to implement any Boolean function without causing explosions.

4.43   Let $g(x, y, z) = xy + y'z'$ and $h(x, y, z) = x'y'z + x'yz'$. Given that $g = f'$ and $h = fs'$, determine minimal sum expressions for both f and s.

4.44   Let $F_1 = \Sigma(1, 5, 9, 10, 15) + \Sigma_d(4, 6, 8)$ and $F_2 = \Sigma(0,2,3,4,7,15) + \Sigma_d(9,14)$. Find the minimal sum expressions for $F_3 = F_1 F_2$ and $F_4 = F_1 + F_2$.

4.45   Let $F = \Sigma(5,6,13)$ and $F_1 = \Sigma(0,1,2,3,5,6,8,9,10,11,13)$. Find F_2 such that $F = F_1 F_2'$. Discuss the uniqueness of the result for F_2 and obtain the simplest function F_2.

4.46 Design a minimal-cost, 2-level (AND-OR), multiple-output circuit for the following output functions:
$$F_1 = \Sigma(0,2,5,6,13)$$
$$F_2 = \Sigma(0,5,11,13,15)$$
$$F_3 = \Sigma(0,8,12)$$

4.47 Repeat Problem 4.46 for the following output functions:
$$F_1 = \Sigma(0,2,5,6,13)$$
$$F_2 = \Sigma(0,5,11,13,15)$$
$$F_3 = \Sigma(0,4)$$

4.48 Repeat Problem 4.46 for each of the following sets of output functions:

(a) $F_1 = \Sigma(0,4,6,7)$
$$F_2 = \Sigma(0,3,5,7)$$
$$F_3 = \Sigma(0,3,4,6)$$

(b) $F_1 = \Sigma(2,3,7,10,11,14) + \Sigma_d(1,5,15)$
$$F_2 = \Sigma(0,1,4,7,13,14) + \Sigma_d(5,8,15)$$

5 Combinational Circuit Implementation

5.1 INTRODUCTION

In Chapter 4 we presented procedures for obtaining minimal-cost, 2-level combinational circuits. Basically, our emphasis was on minimizing Boolean switching functions; now we turn our attention to implementation.

The basic building blocks for most integrated-circuit (IC) families are the NAND or NOR gates. Since each of these operators is *functionally complete* (see Section 3.5), most circuits are implemented with NAND or NOR gates. Often, however, we design a circuit with other types of operators (typically AND and OR gates), so we need to convert it to a circuit using only NAND or NOR gates. Therefore, in the following two sections we discuss first the conversion from 2-level circuits using AND and OR gates to 2-level circuits implemented with NAND or NOR gates; then we consider situations where a 2-level design is not appropriate and a multilevel implementation is required.

In the remainder of this chapter, we present a variety of MSI and LSI combinational circuits and illustrate their applications to the design of logic circuits through numerous examples. If an IC chip cannot be found to implement exactly the functions needed, the designer may be able to incorporate other MSI components in the circuit. This **modular** design approach is very important when ICs are used to implement the circuit. In this case, the cost of implementation is not determined by the number of gates used but by the number of IC chips and interconnecting wires required to implement the given function.

5.2 NAND AND NOR IMPLEMENTATIONS

A 2-level AND-OR circuit easily converts to a circuit composed of either NAND gates or NOR gates. We just use the fact that $F = (F')'$ for any function F, and then apply DeMorgan's laws (see Table 3.1)

$$(F_1 + F_2 + \cdots + F_n)' = F_1'F_2'\cdots F_n' \tag{5.1}$$

$$(F_1F_2\cdots F_n)' = F_1' + F_2' + \cdots + F_n' \tag{5.2}$$

In general, a Boolean function F can be written as a sum of literals $L_1, L_2, \cdots, L_m$ and product terms $P_1, P_2, \cdots, P_k$, where each P_i involves two or more literals:

$$F = L_1 + L_2 + \cdots + L_m + P_1 + P_2 + \cdots + P_k \tag{5.3}$$

Then, by DeMorgan's law, we get

$$F = (L_1'L_2'\cdots L_m'P_1'P_2'\cdots P_k')' \tag{5.4}$$

NAND Circuits

Consider the AND-OR implementation of Equation (5.3) shown in Figure 5.1(a). We can use Equation (5.4) to transform it to a NAND implementation of F as follows: First, we replace the second-level OR gate with a NAND gate with inputs $L_1', L_2' \ldots, L_m', P_1', P_2', \ldots, P_k'$. Then, since the first-level product terms $P_1, \ldots, P_k$ in Figure 5.1(a) are each realized with an AND gate, the terms $P_1', \ldots, P_k'$ can each be realized with a NAND gate in the first level of the transformed circuit. The resulting circuit is shown in Figure 5.1(b). Note that the inputs to all the first-level NAND gates are the same as those to the first-level AND gates in Figure 5.1(a).

Since the literals $L_1, \ldots, L_m$ must be inverted for the transformed circuit, we have two choices: (1) We can input the inverted literals directly into the second-level NAND gate as in Figure 5.1(b), or (2) for each literal at the first level, we can insert a NAND gate connected as an inverter and then input $L_1', \ldots, L_m'$ into the second-level NAND gate. This second

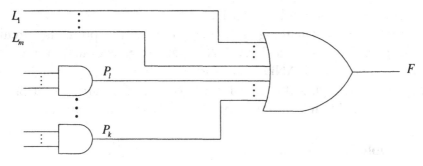

(a) Before conversion

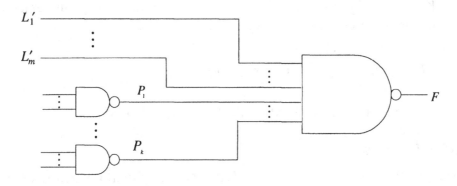

(b) After conversion, with inverted literals

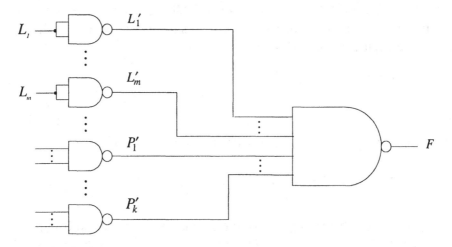

(c) After conversion, with inverting NAND gates

■ **Figure 5.1** AND-OR to NAND-NAND circuit conversion

approach is illustrated in Figure 5.1(c).

Hence, simply changing all gates in a sum-of-products circuit to NAND gates results in a 2-level NAND implementation, provided we insert an inverting NAND gate for each term with a single literal. Similarly, by applying this transformation to a *minimal* sum-of-products for F, we can obtain a *minimal-cost* 2-level NAND implementation.

■ **Example 5.1** The function $F(x_1, x_2, x_3, x_4) = \Sigma(1, 3, 5, 7, 9, 11, 13,$ 14, 15) has the minimal sum-of-products $x_1 x_2 x_3 + x_4$ derived from the K-map of Figure 5.2(a). Therefore, this function can be implemented by the AND-OR circuit of Figure 5.2(b). But, since

$$F = ((x_1 x_2 x_3 + x_4)')' = ((x_1 x_2 x_3)'(x_4)')'$$

it can also be realized by the NAND circuit of Figure 5.2(c). ■

NOR Circuits

The NOR operator is the dual of the NAND function; therefore, the procedures for a NOR implementation are the dual of the corresponding procedures for a NAND implementation. Consequently, the implementation of a Boolean function F with NOR gates requires that F be simplified in a product-of-sums form, resulting in the OR-AND circuit shown in Figure 5.3(a). Thus, if

$$F = (L_1 L_2 \cdots L_m)(S_1 S_2 \cdots S_k) \tag{5.5}$$

where $L_1, L_2, \ldots, L_m$ are literals and $S_1, S_2, \ldots, S_k$ are sum terms involving at least two literals, then, by using DeMorgan's law, we get

$$F = (L_1' + L_2' + \cdots + L_m' + S_1' + S_2' + \cdots + S_k')' \tag{5.6}$$

Using Equation (5.6), we get the transformation indicated in Figure 5.3(b). Alternatively, if an inverting NOR gate is inserted for each term involving a single literal, then we obtain the circuit shown in Figure 5.3(c) in which all the inputs to the first level are not changed.

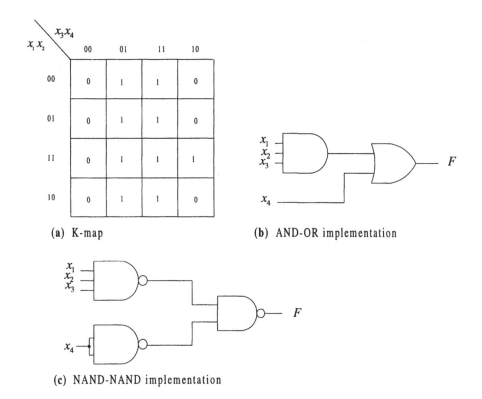

(a) K-map

(b) AND-OR implementation

(c) NAND-NAND implementation

■ **Figure 5.2** NAND implementation of F of Example 5.1

In short, a product-of-sums form for F can be converted to a 2-level NOR implementation by changing all gates to NOR gates. As before, applying this transformation to a minimal product-of-sums form for F results in a minimal 2-level NOR implementation of F.

■ **Example 5.2** Let $F(x_1, x_2, x_3, x_4) = \sum(0, 1, 3, 5, 15)$. By clustering the 0-cells on the K-map of Figure 5.4(a), we obtain

$$F = (x_3' + x_4)(x_1 + x_2' + x_3')(x_1' + x_2)(x_1' + x_3)(x_2' + x_4)$$

which is the minimal product-of-sums for F. The OR-AND implementation of F is shown in Figure 5.4(b). The 2-level NOR-NOR implementation of F, shown in Figure 5.4(c), is obtained by replacing the first-level OR gates and the second-level AND gate by NOR gates.   ■

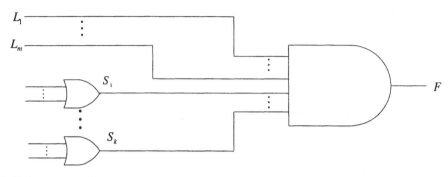

(a) Before conversion

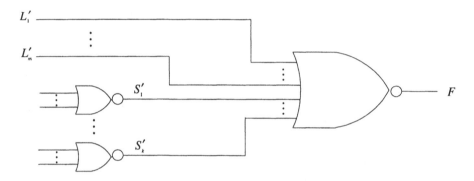

(b) After conversion, with inverted literals

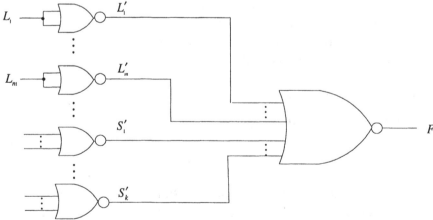

(c) After conversion, with inverting NOR gates

■ **Figure 5.3** OR-AND to NOR-NOR circuit conversion

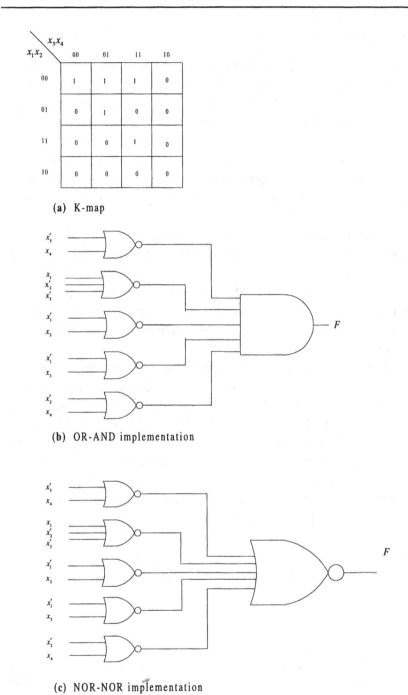

(a) K-map

(b) OR-AND implementation

(c) NOR-NOR implementation

■ **Figure 5.4** NOR implementation of F of Example 5.2

Nondegenerate Forms

The forms considered thus far are not the only ways in which a Boolean expression can be implemented in two levels with the four types of gates: AND, OR, NAND, and NOR. In fact, if we assign one type of gate to the first level and one type of gate to the second level, there are 16 possible 2-level combinations because the same type of gate can be used at both levels (as in the NAND-NAND or in the NOR-NOR implementation). But eight of these 16 forms are **degenerate** since they reduce to a single operation. For example, the AND-AND and OR-OR forms degenerate to just AND and OR, respectively.

The remaining eight **nondegenerate** forms are listed in Figure 5.5. The gate listed first in each of these forms constitutes one of the gates in the first level of the implementation, whereas the gate listed second constitutes a single gate in the second level. Any two forms listed on the same line in Figure 5.5 are duals of each other. The arrows indicate which forms can be obtained from others by successive applications of DeMorgan's laws.

The eight nondegenerate forms are separated into two groups of four related forms each. We call one group the AND-OR family since it contains that form for expressing a function. By the same token, the second group is called the OR-AND family. Conversion between forms in the same family is relatively simple, but interfamily conversion is not as easy. One way that a family-to-family conversion can be accomplished is to go back to the truth table (or K-map) of the original expression. The following example illustrates how we can convert to different nondegenerate forms using DeMorgan's laws.

■ **Example 5.3**   Consider the function $F(a, b, c, d) = \Sigma(3, 4, 5, 8, 9, 10, 11, 12, 13, 14, 15)$. The minimal sum-of-products is

$$F = a + bc' + b'cd \tag{5.7}$$

Hence

$$
\begin{aligned}
F &= [a'(bc')'(b'cd)']' & \text{by Equation (5.1)} && (5.8)\\
&= [a'(b' + c)(b + c' + d')]' & \text{by Equation (5.2)} && (5.9)\\
&= a + (b' + c)' + (b + c' + d')' & \text{by Equation (5.1)} && (5.10)
\end{aligned}
$$

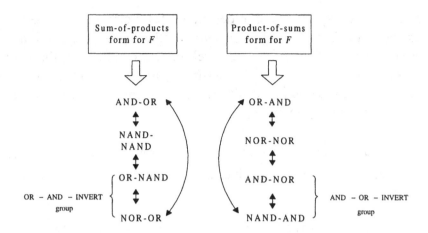

■ **Figure 5.5** The eight nondegenerate forms for Boolean expressions

Equations (5.7) through (5.10) represent, respectively, the AND-OR, NAND-NAND, OR-NAND, and NOR-OR forms for F.

The minimal product-of-sums expression for F is

$$F = (a+b+c)(a+b'+c')(a+c'+d) \tag{5.11}$$

From this expression, we can obtain

$$F = [(a+b+c)' + (a+b'+c')' + (a+c'+d)']'$$
$$\text{by Equation (5.2)} \tag{5.12}$$
$$= [a'b'c' + a'bc + a'cd']' \qquad \text{by Equation (5.1)} \tag{5.13}$$

$$= (a'b'c')'(a'bc)'(a'cd')' \qquad \text{by Equation (5.1)} \tag{5.14}$$

Equations (5.11) through (5.14) represent, respectively, the OR-AND, NOR-NOR, AND-NOR, and NAND-AND forms of F. ■

5.3 MULTILEVEL CIRCUIT IMPLEMENTATION

Since each gate in a circuit has a *propagation time delay* (see Section 1.2), the total time delay between a change in an input and the precipitated change in the output will be proportional to the number of levels in the circuit. Hence, 2-level circuits have the least total time delay.

But it is not always practical to design a 2-level circuit. For one thing, the maximum number of inputs to each gate (the *fan-in*; see Section 1.2) may be limited. Therefore, if a 2-level implementation of the circuit requires more gate inputs than the fan-in capability, we must factor the Boolean expression to obtain a **multilevel** implementation. This procedure is outlined in the following example.

■ **Example 5.4** From the K-map of the function $F(x_1, x_2, x_3, x_4) = \Sigma(7,$ 8, 9, 12, 13, 14, 15), we find that the minimal sum-of-products expression is

$$F = x_1 x_2 + x_1 x_3' + x_2 x_3 x_4$$

This expression requires gates with three inputs. But, if the fan-in is limited to 2, then we must attempt to factor out any common literals from the different product terms. For example, we can factor either x_1 or x_2 to obtain the new expressions

$$x_1(x_2 + x_3') + x_2 x_3 x_4$$

or

$$x_2(x_1 + x_3 x_4) + x_1 x_3'$$

The first expression still requires a gate with three inputs, but the second expression does not require any gate with more than two inputs. The circuit logic diagram of this 4-level circuit with five gates is shown in Figure 5.6(a).

In fact, we can simplify the minimal sum expression even further to a 3-level circuit with only four gates while maintaining the fan-in constraint. Observe that the only troublesome term is $x_2 x_3 x_4$, and then note that it can be written as $x_2 x_3 x_4 + x_3 x_3' x_4$ because $x_3 x_3' = 0$. Therefore,

$$F = x_1 x_2 + x_1 x_3' + x_2 x_3 x_4 = x_1(x_2 + x_3') + x_2 x_3 x_4 + x_3 x_3' x_4$$
$$= x_1(x_2 + x_3') + x_3 x_4(x_2 + x_3') = (x_1 + x_3 x_4)(x_2 + x_3')$$

The circuit diagram for the last expression is shown in Figure 5.6(b) ∎

Note that in both Figures 5.6(a) and (b) each level of the multilevel circuit contains only AND gates or only OR gates. This is always the case for circuits obtained from a factored form of a sum-of-products (or a product-of-sums) expression.

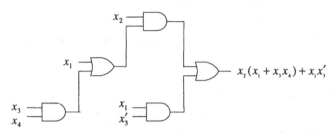

(a) A 4-level implementation with fan-in =2

$$x_2(x_1 + x_3x_4) + x_1x_3'$$

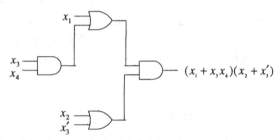

(b) A 3-level implementation with fan-in =2

$$(x_1 + x_3x_4)(x_2 + x_3')$$

■ **Figure 5.6** Different multilevel implementations of a circuit

Design of Multilevel NAND and NOR Circuits

We may use the following procedure to design a *multilevel NAND gate circuit*:

1. Obtain a minimal sum-of-products expression for the function to be implemented.
2. Factor the Boolean expression to satisfy the maximal fan-in requirement.
3. Implement the expression as a multilevel circuit with AND and OR gates so that each level contains only AND gates or only OR gates.

The output gate *must* be an OR gate, but this requirement is satisfied automatically since the expression resulting from step 2 is a sum-of-terms (where each term can be in *mixed form*). Moreover, AND (OR) gate outputs cannot be used as AND (OR) gate inputs.

4. Number the levels starting *backward* from the output OR gate as level 1.
5. Replace all gates with NAND gates, leaving all interconnections between the gates unchanged.
6. Leave the inputs to the *even*-numbered levels unchanged, but complement all the literals that appear as inputs to *odd*-numbered levels.

We can verify that this procedure is valid by partitioning the multilevel circuit into 2-level subcircuits and applying to each the conversion procedure to a NAND-NAND circuit.

The procedure for designing a *multilevel NOR gate circuit* is exactly the same, except that we start with a minimal product-of-sums expression, the output gate of the circuit *must* be an AND gate, and then all gates are replaced with NOR gates. The following examples illustrate the two design procedures.

■ **Example 5.5** Consider the function $F = [(x_1 + x_2')x_3 + x_4' + x_5x_6x_7']x_8 + x_9x_{10}x_{11}' + x_{12}$. We want to implement it using NAND gates having at most three inputs (fan-in = 3).

Note that F is already given in mixed form and that the fan-in requirement is satisfied. Hence, we can skip steps 1 and 2 of the design procedure. Figure 5.7(a) shows the application of steps 3 and 4 while Figure 5.7(b) depicts the NAND gate implementation resulting from steps 5 and 6. ■

■ **Example 5.6** Implement $F(x_1, x_2, x_3, x_4) = \sum (0, 3, 4, 5, 8, 9, 10, 14, 15)$ using NOR gates having at most three inputs (fan-in = 3). The minimal product-of-sums expression for F is

$$F = (x_1' + x_2 + x_3' + x_4')(x_1 + x_2 + x_3 + x_4')(x_1' + x_2' + x_3)$$
$$\cdot (x_1 + x_3' + x_4)(x_1 + x_2' + x_3')$$

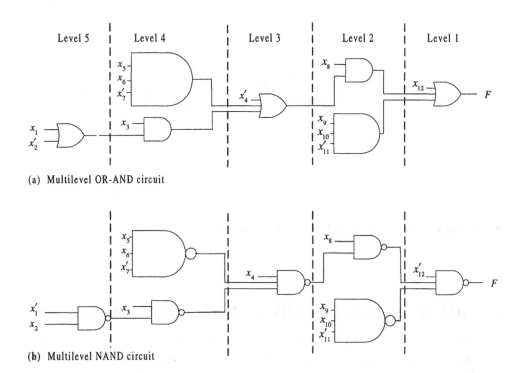

(a) Multilevel OR-AND circuit

(b) Multilevel NAND circuit

■ **Figure 5.7** Multilevel conversion to a NAND gate implementation

Therefore, a 2-level circuit for F would require three 3-input and two 4-input OR gates and one 5-input AND gate.

To reduce the maximum fan-in to 3, we can factor $x_2 + x_4'$ from the first two product terms and $x_1 + x_3'$ from the last two product terms to obtain the following expression:

$$F = [x_2 + x_4' + (x_1 + x_3)(x_1' + x_3')][x_1' + x_2' + x_3][x_1 + x_3' + x_2'x_4]$$

We ask the reader to implement F as a multilevel circuit with AND and OR gates and then to apply the design procedure to convert it to the NOR gate circuit shown in Figure 5.8.   ■

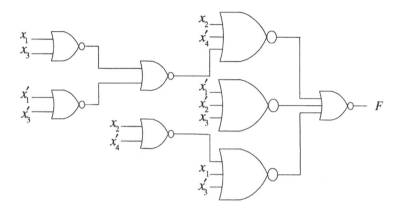

■ **Figure 5.8** Multilevel NOR implementation

5.4 INTEGRATED-CIRCUIT IMPLEMENTATION

The design flow diagram of Figure 4.6 constitutes a general procedure that always produces a combinational circuit to implement any given function. But the truth tables and minimization steps required by this procedure may be too cumbersome if the number of variables is large. Sometimes, an integrated circuit (IC) is already available to implement the desired function; but even if not, the designer may still be able to formulate a method of implementation that incorporates off-the-shelf IC chips.

The basic idea of this approach, referred to as **modular design**, is to use available IC chips as building blocks for the desired circuit. The modular design approach is based on our ability to identify well-defined regularities associated with the circuit functions. If this can be done, then the design problem can be reduced to finding an appropriate inter-connection configuration of IC chips.

In subsequent sections, we will discuss the internal construction of some MSI and LSI chips and illustrate how to use them as modules to build other circuits. In doing so, we limit our discussion to the following five basic categories:

1. Arithmetic circuits
2. Comparators

3. Multiplexers and demultiplexers
4. Decoders and encoders
5. Programmable arrays: read-only memory (ROM), programmable logic array (PLA), and programmable array logic (PAL)

5.5 ARITHMETIC CIRCUITS

Adders

Digital systems are designed to perform a variety of functions on binary numerical data and, among these, arithmetic operations are the most basic. The addition of two 1-bit numbers is accomplished by the **binary full-adder (FA)** circuit first introduced in Example 4.7. Figure 5.9 shows a block diagram of the full-adder circuit where x, y, and z represent the addend bit, augend bit, and a previous carry (referred to as *input carry*), respectively. The outputs, S and C, represent the sum and carry resulting from $x + y + z$. The output C is also called the *output carry*. When the input carry is held constantly at 0, the full-adder circuit specializes to what is called **half-adder (HA)**.

Table 5.1 lists the truth table for C and S, and the K-maps of Figure 5.10 provide the Boolean expressions

$$S = x'y'z + x'yz' + xy'z' + xyz$$
$$C = xy + xz + yz$$

We have already given an AND-OR implementation of the binary full-adder in Figure 4.9. In the following two examples, we consider other implementations.

- **Example 5.7** In Figure 5.11, we show a NAND-NAND implementation of the binary full-adder obtained by applying the technique of Section 5.2 to Figure 4.9. We can also obtain the same results directly from the equations:

$$S = (S')' = [(x'y'z)'(x'yz')'(xy'z')'(xyz)']'$$

and

$$C = (C')' = [(xy)'(xz)'(yz)']'$$

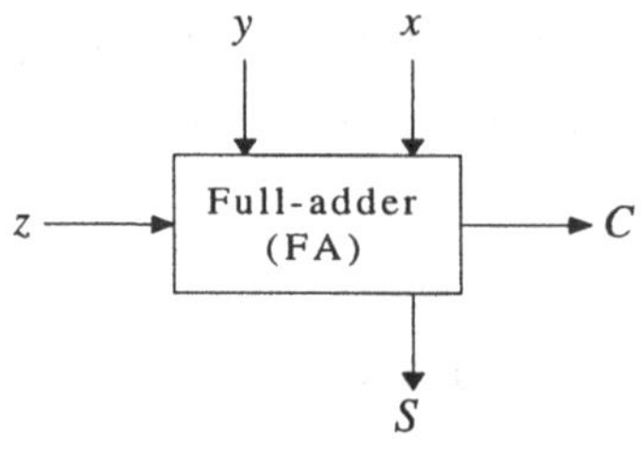

■ **Figure 5.9** Block diagram of a full-adder

x \ yz	00	01	11	10
0	0	1	0	1
1	1	0	1	0

(a) The sum S

x \ yz	00	01	11	10
0	0	0	1	0
1	0	1	1	1

(b) The output carry C

■ **Figure 5.10** K-maps for the full-adder circuit

■ **Table 5.1** Truth table for the full-adder

	Inputs			Outputs	
Decimal	x	y	z	C	S
0	0	0	0	0	0
1	0	0	1	0	1
2	0	1	0	0	1
3	0	1	1	1	0
4	1	0	0	0	1
5	1	0	1	1	0
6	1	1	0	1	0
7	1	1	1	1	1

■ **Example 5.8** Other full-adder configurations can be developed. For instance, S can be rewritten as

$$S = (x'y' + xy)z + (x'y + xy')z'$$

$$= (x \oplus y)'z + (x \oplus y)z' = x \oplus y \oplus z$$

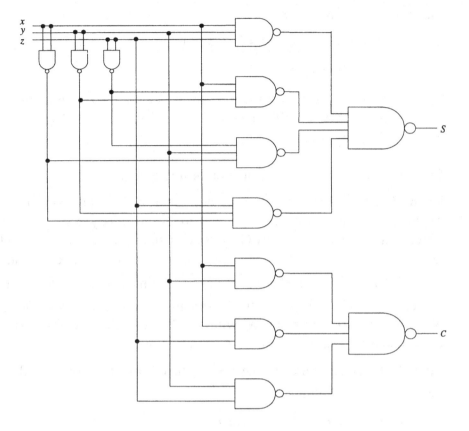

■ **Figure 5.11** NAND implementation of a full-adder

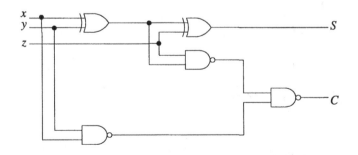

■ **Figure 5.12** Full-adder implemented with NAND and XOR gates

Likewise, C can be written as

$$C = (x'y + xy')z + xy = (x \oplus y)z + xy$$
$$= \{[(x \oplus y)z]'(xy)'\}'$$

Therefore, we can implement the binary full-adder circuit with two 2-input exclusive-OR (XOR) gates and three 2-input NAND gates, as shown in Figure 5.12. ∎

Example 5.9 illustrates an application of full adders.

■ **Example 5.9** Full-adders can be used in the design of code converters. To illustrate this, consider the conversion from binary code to Gray code. (An example of a 4-bit Gray code is shown in Table 2.8.) Let $g_{n-1} \cdots g_2 g_1 g_0$ denote a code word in the n-bit Gray code and $b_{n-1} \cdots b_2 b_1 b_0$ designate the corresponding binary code, where the subscripts 0 and $n-1$ denote the least significant and most significant bits, respectively. For example, with $n = 4$, the Gray code word 1011 corresponds to decimal 13 and, hence, to binary 1101.

The ith Gray code bit g_i can be obtained from the corresponding binary number as follows:

$$g_i = b_i \oplus b_{i+1}, \qquad 0 \le i \le n-2$$

and the most significant bit is given by

$$g_{n-1} = b_{n-1}$$

In other words,

$$g_{n-1} = b_{n-1}$$
$$g_{n-2} = b_{n-1} \oplus b_{n-2}$$
$$g_{n-3} = b_{n-2} \oplus b_{n-3}$$
$$\vdots$$
$$g_1 = b_2 \oplus b_1$$
$$g_0 = b_1 \oplus b_0$$

Hence, all Gray-code bits (except the most significant bit) come from the exclusive-OR sum of successive binary bits. For example, the 6-bit binary number 101101 converts to the Gray-code word 111011 in the manner shown in Figure 5.13(a).

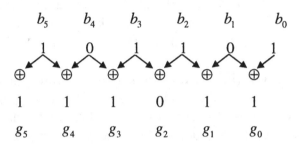

■ **Figure 5.13(a)** Binary-to-Gray conversion process

Since the sum function of a full-adder acts just as an XOR gate if the input carry bit is 0, we can use the full-adder of Figure 5.9 to implement an n-bit binary-to-Gray code converter as illustrated in Figure 5.13(b). ■

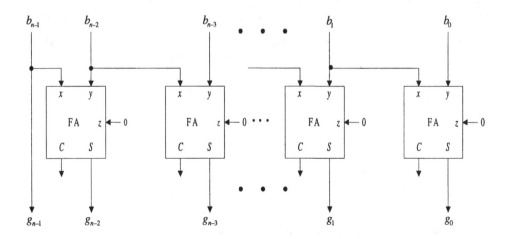

■ **Figure 5.13(b)** n-Bit binary-to-Gray code converter

Subtractors

In digital systems, the difference of two binary numbers x and y is usually computed by adding the 2's complement or the 1's complement of the subtrahend to the minuend, as outlined in Section 2.4. Since the process of subtraction is actually one of addition, subtractors are not usually incorporated into digital systems. Nevertheless, should the need arise, a subtractor can be designed using the design flow diagram of Figure 4.6.

A **binary full-subtractor** (**FS**) is a combinational circuit similar to that of a full-adder except that *borrows*, rather than carries, are propagated to higher-order columns. A full-subtractor consists of three inputs and two outputs. The three inputs, x, y, and z, denote the minuend, subtrahend, and previous (or input) borrow, respectively, while the two outputs, B and D, represent the output borrow and the difference. If z is held constantly at 0, the full-subtractor specializes to **half-subtractor** (**HS**).

Table 5.2 shows the truth table for the subtractor circuit. The rows of the truth table are determined by the subtraction $x - y - z = x - (y + z)$. For instance, the fourth row shows that, whenever the minuend is 0 and both the subtrahend and the input borrow are 1, a net value of -2 results in the column. A borrow is then made, bringing down a value of 2 to offset the -2 value so that the difference $D = 0$ and the borrow $B = 1$. In general, the output borrow B is 1 when $x < y + z$ and the difference D is 1 when the number of 1's in an input combination is *odd*. The K-maps of Figure 5.14 give the corresponding simplified Boolean expressions:

$$D = x'y'z + x'yz' + xy'z' + xyz$$
$$= x \oplus y \oplus z$$

and
$$B = x'y + yz + x'z$$

Note that the output D for the full-subtractor is exactly the same as the output S for the full-adder, except that the variable names for carry and borrow are interchanged. Likewise, the output borrow function B resembles the output carry function; the only difference is that the variable x is complemented in the borrow function. Because of these similarities, a full-adder can be converted to a full-subtractor by merely complementing the input x prior to its application to the gates that ordinarily produce the output carry.

■ **Table 5.2** Truth table for a full-subtractor

Inputs			Outputs	
x	y	z	B	D
0	0	0	0	0
0	0	1	1	1
0	1	0	1	1
0	1	1	1	0
1	0	0	0	1
1	0	1	0	0
1	1	0	0	0
1	1	1	1	1

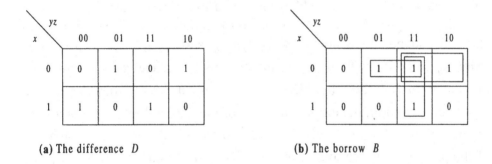

(a) The difference D (b) The borrow B

■ **Figure 5.14** K-maps for a full-subtractor

Binary Parallel Adders

Consider the addition of two 8-bit binary numbers, schematically shown in Figure 5.15. A truth table for such an adder circuit would require $2^{17} = 131,072$ entries. This, no doubt, is one of those times when the designer should choose an alternative design approach. Therefore, let us discuss how we can design a **binary parallel adder** by using full-adder modules.

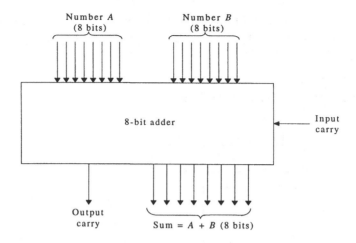

■ **Figure 5.15** Block diagram of an 8-bit binary adder

The n-bit (binary) parallel adder is a device that produces the sum of two n-bit binary numbers by connecting n full adders in **cascade**, with the output carry from one full-adder connected to the input carry of the next full-adder. Figure 5.16 illustrates a 4-bit parallel adder where the augend and the addend are denoted by $A_3A_2A_1A_0$ and $B_3B_2B_1B_0$, respectively. (Bits A_0 and B_0 are the least significant bits.) The carries are connected in a chain through the full-adders (FA). The input carry C_0 may come from a previous addition, the output carry C_1 of the first full-adder becomes the input carry to the second full-adder, and so on. Finally, C_4 is the carry out of the parallel adder circuit. The sum bits are denoted by S_3, S_2, S_1, and S_0. The least significant bit S_0 is produced by the first full-adder, S_1 is produced by the second full-adder, and so on. Thus, the sum of the two 4-bit numbers is the 5-bit number represented by $C_4S_3S_2S_1S_0$.

In general, an n-bit parallel adder will require n full-adders cascaded similarly to the 4-bit parallel adder of Figure 5.16. Alternatively, we can cascade as many 4-bit (or 8-bit) parallel-adder IC chips as required to implement an n-bit binary parallel adder.

Note that the addition of two n-bit numbers can also be accomplished serially rather than in parallel. This calls for a *serial adder*, a sequential circuit described in Chapter 7. The following example illustrates an application of the binary parallel adder.

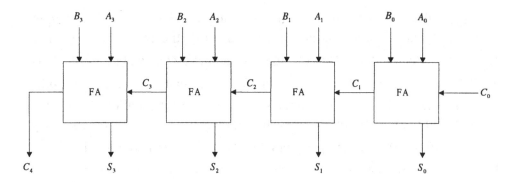

- **Figure 5.16** Four-bit parallel adder

- **Example 5.10** A 4-bit parallel adder can be used to convert from BCD code to excess 3 code (see Section 2.5). The conversion is obtained by adding binary 0011 (decimal 3) to the BCD code word. We can use the 4-bit parallel adder of Figure 5.16, input the BCD number to $A_3 A_2 A_1 A_0$, set $B_3 B_2 B_1 B_0$ to 0011, and assign a 0 to the input carry C_0. Since the largest BCD input is decimal 9, decimal 12 will be the largest possible excess 3 output and, therefore, the outputs can always be represented with 4 bits. Thus, the final carry C_4 of the parallel adder will always be 0. A block diagram of a 4-bit parallel adder used as a BCD-to-excess 3 code converter is shown in Figure 5.17.

We can also design this code converter by following the design flow diagram of Figure 4.6. In this approach, we would have to obtain a truth table for four input variables and four output functions, minimize the Boolean expressions, and then implement the circuit. Clearly, the modular design approach that makes use of an available parallel-adder saves much effort for the designer. ∎

Look-Ahead Carry Adders

The n-bit parallel adder requires fewer logic gates than a comparable 2-level n-bit adder designed by following Figure 4.6, but has one major drawback: The carry signal must *propagate* through the n full-adder modules. Although there are always values at the output terminals, these will not be the correct values unless the signals are given enough time to propagate through the circuit. For example, in Figure 5.16, the output S_3 will not reach a *steady-state* value until after the input carry C_3 has attained its steady state. But C_3 does not achieve steady state until C_2 is available in its steady-state value, and C_2 has to wait on C_1. Since a full-adder is a 2-level circuit, the carry signal of an n-bit parallel adder must ripple through $2n$ gate levels from the input carry C_0 to the output carry C_n. In short, the **carry propagation time** in the parallel adder is a limiting factor on the speed with which two numbers can be added.

Two approaches can be used to speed up the addition process. The first is to use faster gates so that the basic gate propagation time delay is decreased. But this would still leave the carry propagation time dependent on n. A second approach is to bypass the cumulative propagation delays at the expense of increased circuit complexity. Of the several techniques available, the most widely used is that of implementing a **look-ahead carry adder**, resulting in carry propagation time *independent* of n.

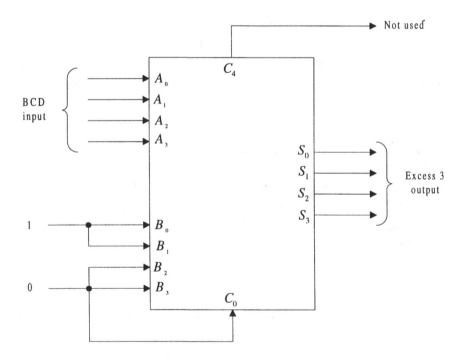

- **Figure 5.17** BCD to excess 3 code converter using a parallel adder

The basic idea of the look-ahead carry adder is that all the carries can be generated *simultaneously* rather than successively. To understand this idea, consider the ith full-adder module of the parallel adder shown in Figure 5.18. The output of the first-level XOR gate is labeled P_i, and the output of the first-level AND gate is labeled G_i. We call G_i the **carry generate** because it generates an output carry when both A_i and B_i are 1, regardless of the value of the input carry (see rows 6 and 7 of Table 5.1). Likewise, P_i is called the **carry propagate** because it is associated with

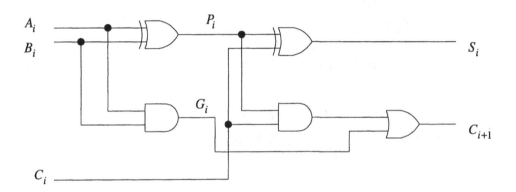

■ **Figure 5.18** The ith full-adder circuit of a parallel adder

the propagation of the carry from C_i to C_{i+1} (see rows 3 and 5 of Table 5.1). In other words, when $A_iB_i = 0$ and the carry of the ith stage is pushed on to the $(i+1)$th stage, then $P_i = 1$.

From Figure 5.18, we obtain

$$P_i = A_i \oplus B_i$$

and

$$G_i = A_iB_i$$

The output sum and carry can now be expressed in terms of P_i and G_i as

$$S_i = P_i \oplus C_i$$

and

$$C_{i+1} = G_i + P_i C_i$$

We can use the last expression to generate the output carry of each stage of the parallel adder in the following *iterative* fashion:

$$
\begin{aligned}
C_1 &= G_0 + P_0 C_0 \\
C_2 &= G_1 + P_1 C_1 \\
    &= G_1 + P_1(G_0 + P_0 C_0) \\
    &= G_1 + P_1 G_0 + P_1 P_0 C_0 \\
C_3 &= G_2 + P_2 C_2 \\
    &= G_2 + P_2(G_1 + P_1 G_0 + P_1 P_0 C_0) \\
    &= G_2 + P_2 G_1 + P_2 P_1 G_0 + P_2 P_1 P_0 C_0
\end{aligned}
$$

and so on.

These expressions reveal that C_3 does not have to wait for C_2 and C_1 to propagate but, instead, C_3 is generated *at the same time* as C_2 and C_1. Furthermore, since each of these expressions is in a sum-of-products form, we can generate *all* the output carries using a 2-level circuit. Hence, all output carries are generated after a delay of only two gate levels, *independently of the number of added bits*.

The circuit that generates the output carries is called a **look-ahead carry generator**. Figure 5.19 shows the circuit logic diagram of a look-ahead carry generator that implements the preceding expressions for C_1, C_2, and C_3. We can incorporate this circuit to design a 3-bit look-ahead carry adder as shown in Figure 5.20. The first-level XOR gates produce the P_i variables, and the first-level AND gates generate the G_i variables. The carries are propagated through the 2-level look-ahead carry generator of Figure 5.19 and are then applied as inputs to the fourth-level XOR gates to produce the sum bits.

Other Examples of Arithmetic Circuits

Decimal Adder

Arithmetic operations in some digital systems (hand-held calculators for example) are performed *directly* in the decimal number system, with the decimal numbers represented in some binary-coded form (see Section 2.5). A **decimal adder** is an arithmetic circuit that accepts binary-coded decimal numbers and produces their sum in the same code.

The actual decimal adder circuit depends on the code used to represent the decimal digits. Therefore, to illustrate the design approach, let us consider as an example the design of a 2-digit decimal adder, where the

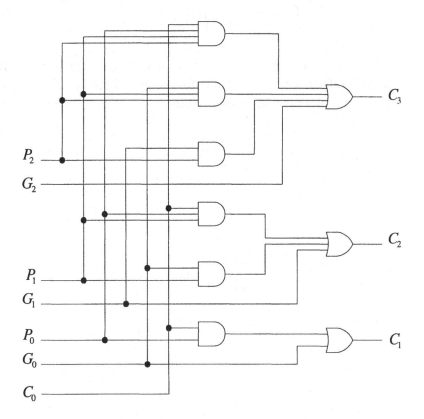

■ **Figure 5.19** Look-ahead carry generator

decimal digits are represented in BCD. Such an adder is called a **BCD adder**.

Assume that the circuit can have an input carry from a previous stage. Since each BCD digit is at most 9, then, with an input carry of 1, the sum of two BCD digits can be at most 19. We can add two BCD digits by using a 4-bit binary parallel adder. The *binary* sum will be as shown in Table 5.3. Note that, if the sum of the two digits and the incoming carry is decimal 9 or less, then the binary sum is the same as the BCD sum. But problems arise if the sum exceeds decimal 9. For example, consider the sum of $8 + 6$:

Decimal	BCD
8	1000
+6	+0110
14	1110

The resulting binary number is *incorrect* because 1110 is not a valid BCD code word. Therefore, a correction must be made whenever a binary sum of more than decimal 9 appears at the outputs of the binary adder.

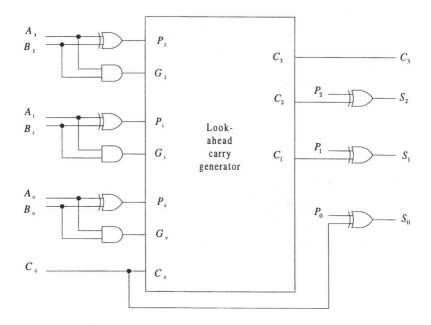

■ **Figure 5.20** Three-bit look-ahead carry adder

The most common correction technique used with the BCD adder is to add decimal 6 (0110) to the binary sum whenever it exceeds 9. This makes the outputs of the parallel adder have the correct BCD value. Thus, the binary sum of 8 + 6 can be corrected to produce the valid BCD sum as follows:

Decimal		BCD	
8		1000	
+ 6		+ 0110	
14		1110	incorrect
		+ 0110	add 6
	0001	0100	correct value
	BCD	BCD	
	for 1	for 4	

■ **Table 5.3** Truth table for a BCD decimal adder

Decimal	Binary Sum					Corrected BCD Sum				
	C_4	S_3	S_2	S_1	S_0	C	T_3	T_2	T_1	T_0
0	0	0	0	0	0	0	0	0	0	0
1	0	0	0	0	1	0	0	0	0	1
2	0	0	0	1	0	0	0	0	1	0
3	0	0	0	1	1	0	0	0	1	1
4	0	0	1	0	0	0	0	1	0	0
5	0	0	1	0	1	0	0	1	0	1
6	0	0	1	1	0	0	0	1	1	0
7	0	0	1	1	1	0	0	1	1	1
8	0	1	0	0	0	0	1	0	0	0
9	0	1	0	0	1	0	1	0	0	1
10	0	1	0	1	0	1	0	0	0	0
11	0	1	0	1	1	1	0	0	0	1
12	0	1	1	0	0	1	0	0	1	0
13	0	1	1	0	1	1	0	0	1	1
14	0	1	1	1	0	1	0	1	0	0
15	0	1	1	1	1	1	0	1	0	1
16	1	0	0	0	0	1	0	1	1	0
17	1	0	0	0	1	1	0	1	1	1
18	1	0	0	1	0	1	1	0	0	0
19	1	0	0	1	1	1	1	0	0	1

The corrected BCD sums are shown in Table 5.3, and Figure 5.21 depicts the BCD adder in block diagram form. The two BCD digits are represented by $A_3 A_2 A_1 A_0$ and $B_3 B_2 B_1 B_0$. The variables T_3, T_2, T_1, T_0, and C are the corrected sum and carry bits. The corrected carry bit C represents the required most significant BCD digit and it is 1 if and only if the binary sum exceeds 9. Note that the most significant BCD digit is, in fact, binary $000X$, where $X = 0$ if the binary sum does not exceed 9 and $X = 1$ if it does. However, since we are only concerned with a 2-digit BCD addition, we can dispose of the three *leading* zeros in representing the most significant digit.

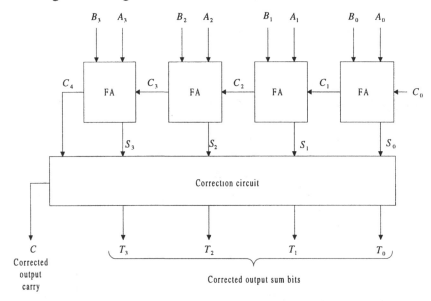

■ **Figure 5.21**   Block diagram of a BCD adder

As can be seen from Table 5.3, the function for the corrected carry bit completely determines the correction circuit. To find C as a function of C_4, S_3, S_2, S_1, and S_0, we can use a 5-variable K-map. If all the don't-cares in the map are taken as 1, the resulting minimal expression for C becomes

$$C = C_4 + S_1 S_3 + S_2 S_3$$

The same expression can also be obtained directly from Table 5.3, and we leave it to the reader to supply the details.

We can, therefore, construct a BCD decimal adder by using two 4-bit binary parallel adders and some gates (required to implement C), as illustrated in Figure 5.22. The two BCD sum digits are indicated by C and $T_3 T_2 T_1 T_0$, respectively. As long as $C = 0$, the binary sum is the same as the BCD sum and is added to 0000 in the lower parallel adder. If $C = 1$, then the lower 4-bit adder adds 0110 (decimal 6) to the outputs of the first adder.

Multiplier

Multiplication can be carried out in two steps (see Section 2.4): (1) Compute the partial products, and (2) sum the partial products (along with the appropriate carries). To illustrate the design of multiplier circuits, let us consider a 2-bit binary multiplier.

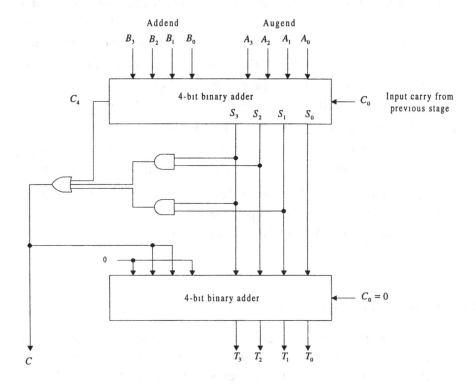

■ **Figure 5.22** BCD decimal adder circuit

The process of multiplying two 2-bit binary numbers, $x_2 x_1$ and $y_2 y_1$, resulting in a 4-bit product $P_4 P_3 P_2 P_1$, is as follows:

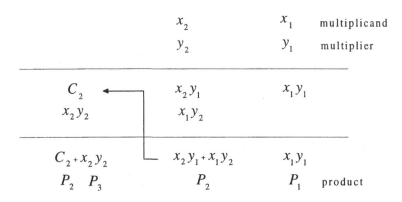

Therefore, we can design the multiplier circuit to consist of two modular parts, as shown in Figure 5.23. One part consists of four AND gates that generate the partial products $P_{11} = x_1 y_1$, $P_{12} = x_2 y_1$, $P_{21} = x_1 y_2$, and $P_{22} = x_2 y_2$; the other part uses two full-adders (FA) to generate the sums of the appropriate partial products. The first full-adder determines the sum $x_2 y_1 + x_1 y_2$, the second bit (P_2) of the product, and carries C_2 to the second full-adder. The second full-adder determines the sum $C_2 + x_2 y_2$ to produce P_3 and P_4 (where the carry results from the addition).

Four-bit multipliers can be designed in the same general modular way, as shown in Figure 5.24. There are 16 partial products $P_{ij} = x_j y_i$ and eight bits in the product output. The full-adders are cascaded to generate the product bits. We will return to the process of binary multiplication after we discuss the read-only memory (ROM) in Section 5.9.

5.6 COMPARATORS

An **n-bit comparator** is a circuit that compares the magnitudes of two n-bit binary numbers A and B. The circuit has three output functions that we designate by L, E, and G, corresponding to $A < B$, $A = B$, and $A > B$. In other words, $L = 1$ if and only if $A < B$, $E = 1$ if and only if $A = B$, and $G = 1$ if and only if $A > B$.

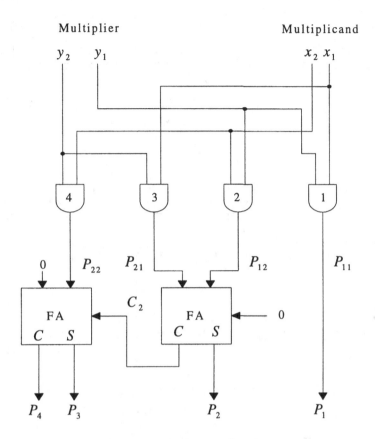

■ **Figure 5.23** Two-bit binary multiplier circuit

Commercially available MSI comparator chips usually include three more input lines, in addition to those representing the two numbers, for connecting comparators in cascade. These cascading inputs correspond to $A < B$, $A = B$, and $A > B$ from a previous comparator stage. Figure 5.25 shows a block diagram of a 4-bit comparator where the two numbers A and B are represented by the bit strings $A_3A_2A_1A_0$ and $B_3B_2B_1B_0$, respectively. The cascade connection of two 4-bit comparators, shown in Figure 5.26, gives an 8-bit comparator. To use this circuit properly to compare two 8-bit numbers, we must assign the following initial conditions to the least significant (right-hand) stage: The $A = B$ input is assigned the value 1, while the $A < B$ and $A > B$ inputs are each assigned the value 0. The reason for this will be understood once we review how two n-bit binary numbers are compared.

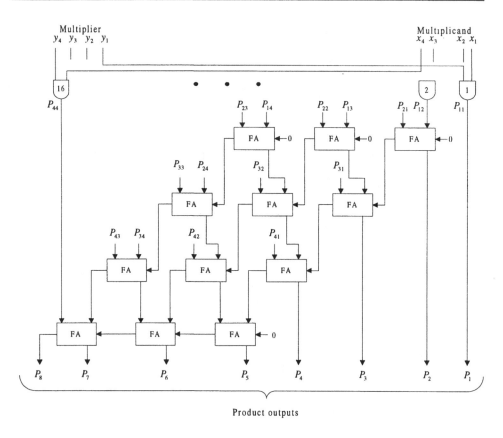

■ **Figure 5.24**   Four-bit binary multiplier circuit

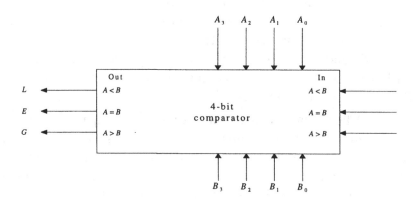

■ **Figure 5.25**   Block diagram of a 4-bit comparator

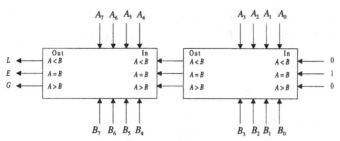

■ **Figure 5.26** Cascading 4-bit comparators to form an 8-bit comparator

Basically, we compare the numbers *lexicographically*, that is, as if they were two *n*-letter words in the dictionary. In other words, we scan the two numbers from left to right (i.e., starting with the most significant bits) until we reach a bit position where they are *not* equal. If the entry of A in this position, say A_i, is less than the corresponding entry B_i of B, then $A < B$. But if, instead, $A_i > B_i$, then we conclude that $A > B$. For example, $A = 100111010$ is greater than $B = 100110011$ since A and B coincide in the first five bit positions from the left, but the sixth bit position of A exceeds that of B.

To describe the design of a comparator circuit, assume that the two numbers are four bits each and consider first the portion of the circuit that generates the output function E corresponding to $A = B$. With $A = A_3 A_2 A_1 A_0$ and $B = B_3 B_2 B_1 B_0$, then $A = B$ if and only if $A_3 = B_3$,

$A_2 = B_2$, $A_1 = B_1$, and $A_0 = B_0$. In other words, $A = B$ if and only if $A_i = B_i$ for each i. The equality of each pair of like bits can be described by the *equivalence* function

$$X_i = A_i B_i + A_i' B_i' = A_i \odot B_i, \quad i = 0,1,2,3$$

Thus, $X_i = 1$ if and only if $A_i = B_i$, and then $A = B$ if and only if $X_3 X_2 X_1 X_0 = 1$. Therefore ,

$$E = X_3 X_2 X_1 X_0$$

describes the output function corresponding to $A = B$.

To obtain the portion of the circuit that generates G, note that $A > B$ if and only if any one of the following conditions is true:

1. $A_3 > B_3$
2. $A_3 = B_3$ and $A_2 > B_2$
3. $A_3 = B_3, A_2 = B_2$, and $A_1 > B_1$
4. $A_3 = B_3, A_2 = B_2, A_1 = B_1$, and $A_0 > B_0$

These separate cases can be described algebraically by the corresponding expressions:

$$A_3 B_3'$$
$$X_3 A_2 B_2'$$
$$X_3 X_2 A_1 B_1'$$
$$X_3 X_2 X_1 A_0 B_0'$$

We can now combine (OR) all four terms and express the function G as

$$G = A_3 B_3' + X_3 A_2 B_2' + X_3 X_2 A_1 B_1' + X_3 X_2 X_1 A_0 B_0'$$

The function L corresponding to $A < B$ can be similarly derived and expressed as

$$L = A_3' B_3 + X_3 A_2' B_2 + X_3 X_2 A_1' B_1 + X_3 X_2 X_1 A_0' B_0$$

The expressions for E, G, and L are implemented in Figure 5.27. Note that we could have designed the circuit from basic principles by following the design flow diagram of Figure 4.6. The truth table corresponding to the comparison of two n-bit numbers has 2^{2n} entries. For example, the truth table of a 4-bit comparator ($n = 4$) has 256 rows. Therefore, the truth table becomes quite cumbersome even for small n. On the other hand, the design process outlined is *modular* and, therefore, can be extended easily to any n. The modular approach, once again, is based on recognizing that a comparator circuit possesses a well-defined regularity that can be exploited to our advantage.

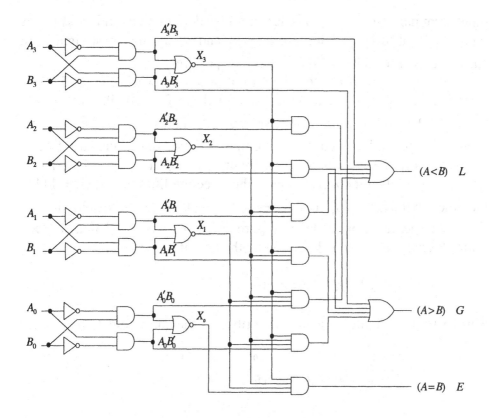

■ **Figure 5.27** Circuit diagram of a 4-bit comparator

5.7 MULTIPLEXERS

A $2^n \times 1$ **multiplexer (MUX)** is a device that selects binary information from one of 2^n input terminals and routes these data to a single output line. For this reason, a multiplexer is also called a **data selector**. The block diagram of a $2^n \times 1$ multiplexer is shown in Figure 5.28. The multiplexer requires n *selection lines*, labeled as $s_{n-1}, s_{n-2}, ..., s_1, s_0$, to select any one of the 2^n inputs, designated by $I_0, I_1, I_2, ..., I_{2^n-1}$. The bit combination of the selection lines determines an n-bit binary number whose decimal equivalent corresponds to the subscript of the selected input terminal. Hence, the n selection lines determine the **address** of the input terminal to be selected, where s_{n-1} and s_0 are the most significant and least significant bits of the address, respectively.

For example, consider the 4×1 multiplexer shown in Figure 5.29(a). As the selection lines s_1 and s_0 range through the four possible values s_1s_0 = 00, 01, 10, and 11, the respective input terminals I_0, I_1, I_2, and I_3 are each selected and routed to the output. These selections are further emphasized in the function table of Figure 5.29(b). Likewise, in a 16×1 multiplexer, if the bit combination of the selection lines is $s_3s_2s_1s_0 = 1110$, then the input terminal I_{14} is selected and its data routed to the output.

From the function table in Figure 5.29(b), it is clear that a 4×1 multiplexer is completely described by the output function

$$F = s_1's_0'I_0 + s_1's_0I_1 + s_1s_0'I_2 + s_1s_0I_3 \tag{5.15}$$

Similarly, an 8×1 multiplexer is described by the output function

$$\begin{aligned} F = {}&s_2's_1's_0'I_0 + s_2's_1's_0I_1 + s_2's_1s_0'I_2 + s_2's_1s_0I_3 \\ &+ s_2s_1's_0'I_4 + s_2s_1's_0I_5 + s_2s_1s_0'I_6 + s_2s_1s_0I_7 \end{aligned} \tag{5.16}$$

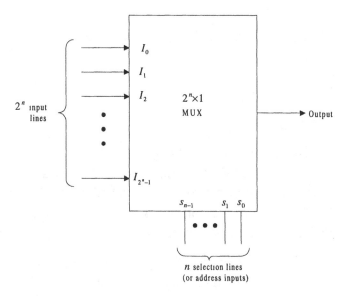

■ **Figure 5.28** Block diagram of a $2^n \times 1$ multiplexer

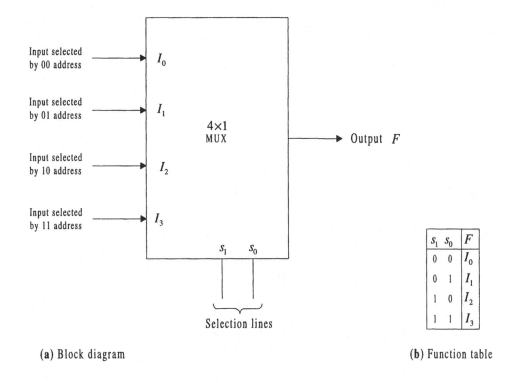

(a) Block diagram   **(b)** Function table

■ **Figure 5.29** A 4×1 multiplexer

In general, the output of a $2^n \times 1$ multiplexer can be expressed by

$$F = \sum_{k=0}^{2^n-1} m_k I_k$$

(5.17)

where m_k is the kth minterm of the variables $s_{n-1}, s_{n-2}, \ldots, s_1, s_0$.

The circuit logic diagram of an 8×1 multiplexer is shown in Figure 5.30. Commercially available multiplexers usually have an additional input line, called **enable** (E) (or **strobe**), to control the operation of the unit. When E is in one binary state, the outputs are disabled, and if E is in the opposite state (the *enable state*), the circuit functions normally as a multiplexer. Thus, Figure 5.30 also illustrates the internal connections necessary to control the operation of the multiplexer with an enable line. We adopt the convention that a multiplexer is enabled when $E = 1$.

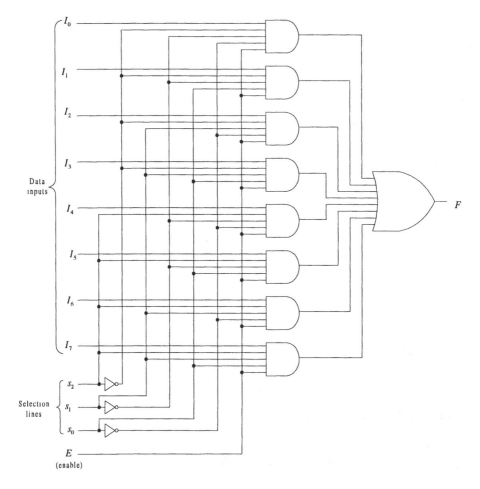

■   **Figure 5.30**   Circuit logic diagram of 8×1 multiplexer with enable line

Multiplexer Trees

Two or more multiplexers can be combined to form a multiplexer with a larger number of inputs. The resulting configuration is called a **multiplexer tree**. There are two basic ways to combine multiplexers into a MUX tree: One uses the enable line and an OR gate, and the other uses an additional multiplexer to select the multiplexer from which data are received. The two methods are illustrated in Figures 5.31(a) and (b).

In Figure 5.31(a), two 4×1 multiplexers are connected to form an 8×1 multiplexer. In this configuration, the ENABLE control line acts like an additional selection line. When ENABLE = 0, the top MUX is enabled (recall that $E = 1$ enables the MUX) and the bottom one is disabled. As the bit combinations of Es_1s_0 range from 000 to 011, one of the inputs I_0, I_1, I_2, or I_3 from the top multiplexer is selected. On the other hand, if ENABLE = 1, the bottom MUX is enabled and the top one is disabled. As the bit combinations of Es_1s_0 range from 100 to 111, one of the four inputs of the bottom multiplexer is selected. Accordingly, we label these inputs as I_4, I_5, I_6, and I_7.

In Figure 5.31(b), five 4×1 multiplexers combine to form a 16×1 multiplexer. To facilitate our explanation, we designate the four multiplexers at the first level of the MUX tree as MUX-0, MUX-1, MUX-2, and MUX-3; the multiplexer at the second level is labeled MUX-4. The selection lines of MUX-4 are labeled s_3 and s_2 to indicate that when

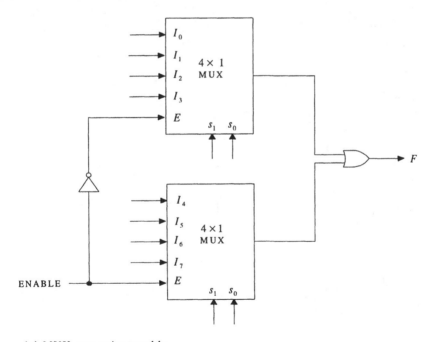

(a) MUX tree using enable

■ **Figure 5.31** Multiplexer trees

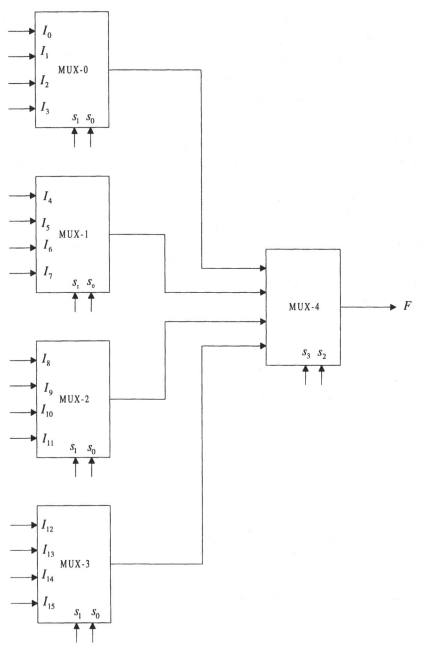

(b) MUX tree using one multiplexer to select other multiplexers

■ **Figure 5.31** Continued

$s_3 s_2 = 00$ the data from MUX-0 are selected; likewise, if $s_3 s_2 = 01$, MUX-1 is selected; and so on. Note that when $s_3 s_2 = 00$ the bit combinations of $s_3 s_2 s_1 s_0$ range from 0000 to 0011 and, for each of these values, one of I_0, I_1, I_2, and I_3 of MUX-0 is selected. If $s_3 s_2 = 01$, then the bit combinations of $s_3 s_2 s_1 s_0$ range from 0100 to 0111 and, for each of these values, one of the four inputs to MUX-1 is selected. Accordingly, we label the inputs to MUX-1 as I_4, I_5, I_6, and I_7. Similarly, we label the inputs to MUX-2 and to MUX-3 as I_8, I_9, I_{10}, and I_{11} and I_{12}, I_{13}, I_{14}, and I_{15}, respectively, to correspond to the bit combinations for $s_3 s_2 s_1 s_0$ when $s_3 s_2 = 10$ and when $s_3 s_2 = 11$.

Some IC chips enclose two or more multiplexers and, therefore, have multiple outputs. For example, Figure 5.32 shows the circuit logic diagram of four 2×1 multiplexers in one IC package (a *quad* 2×1 MUX) used to select one of two 4-bit data words. If the enable line E is 0, then all multiplexers are disabled, whereas if $E = 1$, the inputs are selected according to whether s_0 is 0 or 1. When $E = 1$ and $s_0 = 0$, then the four A inputs are selected; on the other hand, if $E = 1$ and $s_0 = 1$, then the four B inputs are selected.

Multiplexers as Universal Logic Modules

Multiplexers are remarkably versatile implementation devices. A multiplexer can be used as a **universal logic module** in the sense that a $2^n \times 1$ MUX can implement any Boolean function of $n + 1$ or fewer variables. Hence, multiplexing is a *functionally complete* operation (see Section 3.5).

If the Boolean function has $n + 1$ variables, we take n variables and connect them to the selection lines of the multiplexer; the remaining variable, say x, is used as an input variable to the multiplexer. The inputs applied to the multiplexer input terminals are then chosen from the four possibilities x, x', 1, or 0. The following examples illustrate three methods to obtain the implementation information.

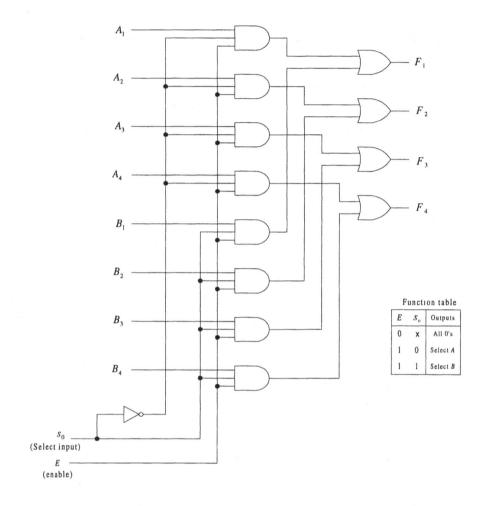

Function table		
E	S_0	Outputs
0	x	All 0's
1	0	Select A
1	1	Select B

■ **Figure 5.32** Quad multiplexer used for data selection

■ **Example 5.11** To implement the Boolean function $F(x_1, x_2, x_3, x_4) = \sum(0, 1, 3, 6, 7, 8, 11, 12, 14)$, we require an 8×1 multiplexer because $n+1 = 4$ and, therefore, $2^n = 2^3 = 8$. We arbitrarily choose x_1, x_2, and x_3 as inputs to the selection lines s_2, s_1, and s_0, respectively, and use x_4 as the input variable to the multiplexer.

We can take an *algebraic approach* and express F as in Equation

(5.17) with minterms in the three selection variables x_1, x_2, and x_3:

$$F = (x_1'x_2'x_3')(x_4' + x_4) + (x_1'x_2'x_3)(x_4) + (x_1'x_2x_3)(x_4' + x_4)$$
$$+ (x_1x_2'x_3')(x_4') + (x_1x_2'x_3)(x_4) + (x_1x_2x_3')(x_4') + (x_1x_2x_3)(x_4')$$

Then we rewrite F so that *every* minterm in x_1, x_2, and x_3 has some input symbol appended to it (actually, this is Shannon's expansion theorem; see Table 3.1):

$$F = (x_1'x_2'x_3')(1) + (x_1'x_2'x_3)(x_4) + (x_1'x_2x_3')(0) + (x_1'x_2x_3)(1)$$
$$+ (x_1x_2'x_3')(x_4') + (x_1x_2'x_3)(x_4) + (x_1x_2x_3')(x_4') + (x_1x_2x_3)(x_4')$$
$$= m_0(1) + m_1(x_4) + m_2(0) + m_3(1) + m_4(x_4') + m_5(x_4)$$
$$+ m_6(x_4') + m_7(x_4')$$

This last equation tells us to assign 1 to I_0, x_4 to I_1, 0 to I_2, 1 to I_3, x_4' to I_4, x_4 to I_5, x_4' to I_6, and x_4' to I_7, as illustrated in Figure 5.33(a). ■

■ **Example 5.12** The same implementation information for F of Example 5.11 can be obtained from the *K-map* shown in Figure 5.33(b). Since we have chosen in Example 5.11 to use x_1, x_2, and x_3 as the selection variables, the bit combination $x_1x_2x_3 = 000$ selects I_0. But, when $x_1x_2x_3 = 000$, the bit combination of the four variables $x_1x_2x_3x_4$ determines two cells on the K-map, those cells with coordinates 0000 and 0001. Since the functional values in these two cells are both 1, the input for I_0 is 1. Likewise, the bit combination $x_1x_2x_3 = 001$ selects I_1 and determines the two cells with coordinates 0011 and 0010 on the K-map. Here the functional values are 1 and 0, respectively, but we notice that the functional value is 1 or 0 exactly where the x_4 coordinate is, respectively, 1 or 0. Thus, the input for I_1 is x_4.

By contrast, we observe that each of the two cells on the K-map determined by $x_1x_2x_3 = 010$ has the functional value 0. Thus, the I_2

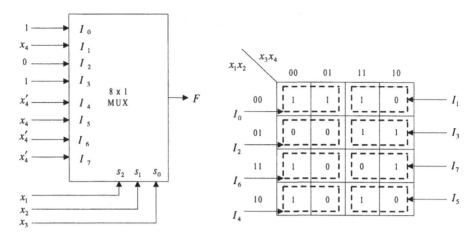

(a) Block diagram   (b) K-map indicating inputs to MUX

(c) Implementation table

■ **Figure 5.33** An 8×1 MUX implementation of F of Example 5.12

input must be 0. On the other hand, the two cells determined by $x_1 x_2 x_3 = 100$ have the value 1 where the x_4 coordinate is 0 and the value 0 where the x_4 coordinate is 1. This means that the input to I_4 must be x_4'. The other inputs are determined similarly, resulting in the 8×1 MUX of Figure 5.33(a).   ■

■ **Example 5.13** The implementation information for F of Example 5.11 can also be obtained by using an **implementation table**. Observe that the variable x_4 is uncomplemented in exactly one-half of the 16 minterms of x_1, x_2, x_3, and x_4, while the other half have x_4 complemented. Therefore, in the implementation table, we list the complemented and uncomplemented input variable (x_4 in our case) as the headings of two rows of minterm numbers, as shown in Figure 5.33(c). The first row lists all minterms in which x_4 is complemented, and the second row lists all minterms with x_4 uncomplemented. The column headings are the multiplexer input terminals $I_0, I_1, \cdots, I_7$.

Clearly, the entries in each column are determined by the bit combinations of $x_1 x_2 x_3 x_4$. For example, under the column for I_3 will be the decimal equivalents of 0110 and 0111. The first three bits 011 are determined, in our case, by the selections variables x_1, x_2, and x_3 and constitute the address of I_3. The fourth bit is determined by the input variable x_4 so that, when x_4 is either 0 or 1, the corresponding decimal number is placed either in the row for x_4' or in the row for x_4, respectively.

Next we circle all the minterms for which $F = 1$ and inspect each column separately. We adhere to the following rules in order to determine what is to be applied to the input terminals of the multiplexer:

(1) If the two minterms in a column are not circled, apply 0 to the corresponding multiplexer input.

(2) If the two minterms in a column are circled, apply 1 to the corresponding multiplexer input.

(3) If the top minterm in a column is circled and the bottom one is not, apply the complement of the input variable to the corresponding multiplexer input.

(4) If the bottom minterm in a column is circled and the top one is not, apply the uncomplemented variable to the corresponding multiplexer input.

Applying these rules to our implementation table, we obtain the last row shown in Figure 5.33(c). As can be seen, the implementation table provides the same information depicted in Figure 5.33(a). ∎

Note that we are not constrained to any one choice for the selection variables and input variable. We can similarly derive an implementation

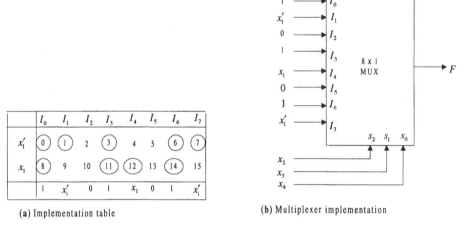

(a) Implementation table   (b) Multiplexer implementation

■ **Figure 5.34** Alternate MUX implementation of F of Example 5.12

table for F of Example 5.12, as shown in Figure 5.34(a), where now we use x_2, x_3, and x_4 as the selection variables assigned to s_2, s_1, and s_0. By circling the minterms of the function on the implementation table and applying the rules stated previously, we obtain the implementation shown in Figure 5.34(b).

The implementation table has the advantage over the algebraic and K-map methods because it provides a more systematic and straightforward design procedure. In fact, as the number of variables increases, the implementation table becomes the only feasible method of the three.

Although it is not an efficient design, it is possible to implement a function of *n or fewer* variables with a $2^n \times 1$ multiplexer (rather than with a $2^{n-1} \times 1$ multiplexer). The following example illustrates this approach.

■ **Example 5.14**   The function $F(x_1, x_2, x_3, x_4) = \Sigma(0, 1, 3, 6, 7, 8, 11, 12, 14)$ can be implemented by a 16×1 multiplexer. In this case, we let x_1, x_2, x_3, and x_4 be assigned to the selection terminals s_3, s_2, s_1, and s_0, respectively, and then all we need do is set I_0, I_1, I_3, I_6, I_7, I_8, I_{11}, I_{12}, and I_{14} to 1 and all the other inputs to 0, as shown in Figure 5.35.   ■

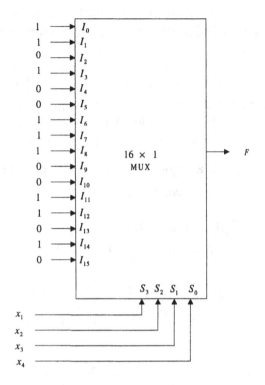

■ **Figure 5.35** A 16×1 MUX implementation of *F* of Example 5.14

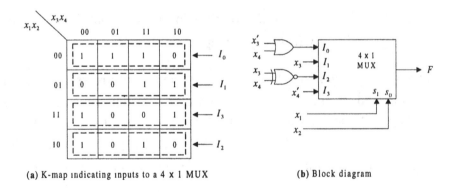

(a) K-map indicating inputs to a 4 x 1 MUX   (b) Block diagram

■ **Figure 5.36** Implementation of *F* of Example 5.15 with 4×1 MUX and added gates

A function of *more* than $n+1$ variables can be implemented by a $2^n \times 1$ multiplexer, but additional gates must be employed as shown in the following example.

■ **Example 5.15** The function $F(x_1, x_2, x_3, x_4) = \sum(0,1,3,6,7,8,11,12,14)$ can be implemented with an 8×1 multiplier following any of the procedures outlined in Examples 5.11 through 5.13. But the function can also be implemented with a 4×1 multiplexer and additional gates.

Let x_1 and x_2 be assigned to the selection terminals s_1 and s_0, respectively. We could use the expression

$$F = \sum(0, 1, 3, 6, 7, 8, 11, 12, 14)$$
$$= x_1'x_2'I_0 + x_1'x_2I_1 + x_1x_2'I_2 + x_1x_2I_3$$

and solve for I_0, I_1, I_2, and I_3 by collecting terms algebraically. For example, only the three minterms m_0, m_1, and m_3 have the factor $x_1'x_2'$, so

$$m_0 + m_1 + m_3 = x_1'x_2'I_0 = x_1'x_2'(x_3'x_4' + x_3'x_4 + x_3x_4)$$
$$= x_1'x_2'(x_3' + x_3x_4) = x_1'x_2'(x_3' + x_3)(x_3' + x_4)$$
$$= x_1'x_2'(x_3' + x_4)$$

Thus, $I_0 = x_3' + x_4$. We could continue in this vein, but the information can be gained easily from the K-map.

The 4-bit combinations for x_1x_2 determine the four rows of the K-map as indicated in Figure 5.36(a). Each row then represents a 2-variable K-map expressing each I_i in terms of x_3 and x_4. Therefore, $I_0 = x_3' + x_4$, $I_1 = x_3$, $I_2 = x_3'x_4' + x_3x_4 = (x_3 \oplus x_4)'$ and $I_3 = x_4'$. These input values determine the implementation shown in Figure 5.36(b). ■

A multiple-output system can be implemented by interconnecting several multiplexers, as illustrated in Example 5.16.

■ **Example 5.16**   Consider the three output functions

$$F_1(x_1, x_2, x_3, x_4) = \sum(0, 1, 3, 6, 7, 8, 11, 12, 14)$$
$$F_2(x_1, x_2, x_3, x_4) = \sum(4, 5, 10, 11, 13)$$
$$F_3(x_1, x_2, x_3, x_4) = \sum(1, 2, 3, 8, 11, 12)$$

F_1 was already considered in Examples 5.11 through 5.13. Therefore, if we use x_1, x_2, and x_3 as inputs to the selection terminals s_2, s_1, and s_0, respectively, then the implementation for F_1 is the same as that of Figure 5.33(a). The implementation tables for F_2 and F_3 are given in Figures 5.37(a) and (b). The block diagram of the multiple-output circuit implemented with three 8×1 multiplexers is shown in Figure 5.37(c).   ■

Multiplexers are used mostly to provide selected routes between multiple sources and a single destination. Nevertheless, they can be used in the design of small, special combinational circuits not otherwise available as MSI functions. For circuits with a large number of inputs and outputs, a more suitable IC device, the read-only memory (ROM), is available; it is discussed in Section 5.9.

5.8   DECODERS

The decoder is another versatile MSI device. This device can, in fact, operate in a dual mode: It can function either as a **decoder** or as a **demultiplexer**. Let us first consider the decoder function and then show how a decoder can be used to perform the opposite function to that of a multiplexer.

An $n \times m$ **decoder** is a combinational circuit that converts binary information from n input lines to m output lines, where $m \leq 2^n$. The decoder will have less than 2^n active outputs only if the n-bit decoded information has unused (or don't-care) combinations; otherwise, the decoder has exactly 2^n active output lines and is called a *complete*

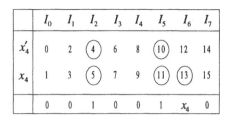

(a) Implementation table for F_2 (b) Implementation table for F_3

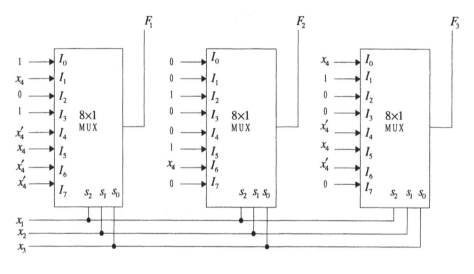

(c) Block diagram

■ **Figure 5.37** MUX implementation of a multiple-output system

decoder. A fundamental property of a decoder is that the output values are *mutually exclusive*; that is, only one output is 1 for any given input combination. Hence, the decoder generates a unique output code for each input combination. The bit combinations of the input variables form the **address** of the output that is 1 for that input.

The block diagram of an $n \times m$ decoder is shown in Figure 5.38(a). To decode the input binary information uniquely, each of the output functions is defined as

$$D_i = m_i$$

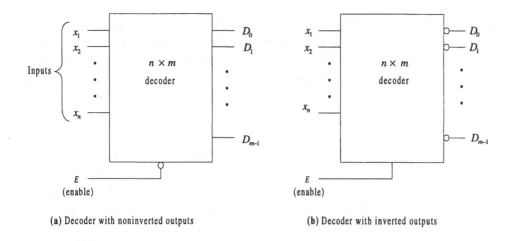

(a) Decoder with noninverted outputs   (b) Decoder with inverted outputs

■ **Figure 5.38** Block diagrams of $n \times m$ decoders

where m_i is the ith minterm in the n input variables. Hence, $D_i = 1$ if and only if the binary combination corresponding to m_i is present on the input lines and all the other outputs are 0. Therefore, this decoder generates the first m minterms of n variables. If $m = 2^n$, then the decoder generates *all* minterms of n variables.

We can also decode the input binary information by defining each output function as

$$D_i = M_i$$

where M_i is the ith maxterm in the n input variables. In this case, $D_i = 0$ if and only if the binary combination corresponding to M_i is present on the input lines and all the other outputs are 1. The block diagram of this decoder is shown in Figure 5.38(b). The small circles on the output lines indicate that the decoder generates maxterms (active-LOW outputs).

Commercial decoders have an **enable** (E) input to control the operation of the circuit. The small circle on the enable line in Figure 5.38(a) indicates that the decoder is enabled when $E = 0$. On the other hand, the absence of the small circle from the block diagram of Figure 5.38(b) indicates that the decoder is enabled with $E = 1$.

■ **Example 5.17**   The 3×8 decoder of Figure 5.39(a) generates all eight minterms of the three input variables. Exactly one of the eight output lines will be 1 for each bit combination of the input variables. The circuit is disabled when $E = 1$ and, in that case, the outputs are all 0. If $E = 0$, then the decoder is enabled and the truth table of Figure 5.39(b) shows the input/output relationships.

If each AND gate in Figure 5.39(a) is replaced by a NAND gate, then each output will be complemented (and hence will be a maxterm). If, in addition, the inverter is removed from the enable line, then the circuit will be disabled when $E = 0$ and, in that case, the output of each NAND gate will be 1 for each input combination.   ■

In general, an $n \times m$ decoder requires n inverters and m n-input decoding gates. If $m < 2^n$, the circuit can be designed with some decoding gates having less than n inputs. This is illustrated in Example 5.18.

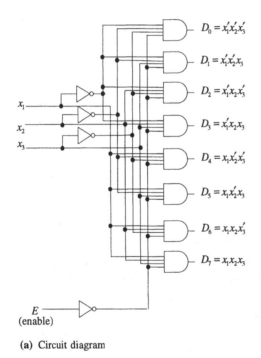

(a) Circuit diagram

Inputs			Outputs							
x_1	x_2	x_3	D_0	D_1	D_2	D_3	D_4	D_5	D_6	D_7
0	0	0	1	0	0	0	0	0	0	0
0	0	1	0	1	0	0	0	0	0	0
0	1	0	0	0	1	0	0	0	0	0
0	1	1	0	0	0	1	0	0	0	0
1	0	0	0	0	0	0	1	0	0	0
1	0	1	0	0	0	0	0	1	0	0
1	1	0	0	0	0	0	0	0	1	0
1	1	1	0	0	0	0	0	0	0	1

(b) Truth table

■ **Figure 5.39**   A 3×8 decoder

■ **Example 5.18** A 4 × 10 decoder is frequently used to convert from BCD code to decimal. The inputs are the ten decimal digits represented in BCD code. Since four input variables can represent up to 16 combinations, there are six don't-care conditions (corresponding to decimals 10 through 15).

The circuit has ten outputs, so ten K-maps would be needed to simplify the output functions. In our case, however, all can be drawn on one map as shown in Figure 5.40(a). We label the appropriate cells with the output variables $D_0, \ldots, D_9$ and mark the other six cells with don't care $\times$'s.

The designer must decide how to treat the don't-care conditions. Should the designer decide to minimize each output function, then the following output functions will be obtained:

$$D_0 = x_1' x_2' x_3' x_4'$$
$$D_1 = x_1' x_2' x_3' x_4$$
$$D_2 = x_2' x_3 x_4' \qquad \text{(combine cells for } m_2, m_{10})$$
$$D_3 = x_2' x_3 x_4 \qquad \text{(combine cells for } m_3, m_{11})$$
$$D_4 = x_2 x_3' x_4' \qquad \text{(combine cells for } m_4, m_{12})$$
$$D_5 = x_2 x_3' x_4 \qquad \text{(combine cells for } m_5, m_{13})$$
$$D_6 = x_2 x_3 x_4' \qquad \text{(combine cells for } m_6, m_{14})$$
$$D_7 = x_2 x_3 x_4 \qquad \text{(combine cells for } m_7, m_{15})$$
$$D_8 = x_1 x_4' \qquad \text{(combine cells for } m_8, m_{10}, m_{12}, m_{14})$$
$$D_9 = x_1 x_4 \qquad \text{(combine cells for } m_9, m_{11}, m_{13}, m_{15})$$

These equations determine the circuit logic diagram of Figure 5.40(b). In this circuit, only two of the ten AND gates require four input lines. Note, however, that while we have reduced the complexity of the circuit in terms of the number of input lines to the decoding gates an *erroneous* output will appear, as indicated in Table 5.4, should one of the six invalid (don't-care) input combinations occur.

To prevent the occurrence of false outputs, we can redesign the circuit by making all outputs 0 when an invalid input combination occurs. This is the same as assigning each don't-care the value 0 on the K-map of Figure 5.40(a). Hence, this second design has an advantage over the first in that a zero value at all outputs detects an invalid input. The circuit now has a *false data rejection* capability but requires 4-input AND gates to implement. Other assignments of don't-cares may be considered, but the designer should investigate their effect once the circuit is in operation. ∎

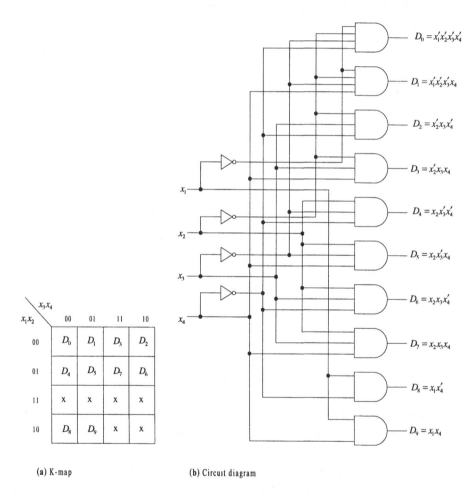

(a) K-map   (b) Circuit diagram

■ **Figure 5.40** BCD-to-decimal decoder

■ **Table 5.4** Partial truth table for the circuit of Figure 5.40(b)

Inputs				Outputs									
x_1	x_2	x_3	x_4	D_0	D_1	D_2	D_3	D_4	D_5	D_6	D_7	D_8	D_9
1	0	1	0	0	0	1	0	0	0	0	0	1	0
1	0	1	1	0	0	0	1	0	0	0	0	0	1
1	1	0	0	0	0	0	0	1	0	0	0	1	0
1	1	0	1	0	0	0	0	0	1	0	0	0	1
1	1	1	0	0	0	0	0	0	0	1	0	1	0
1	1	1	1	0	0	0	0	0	0	0	1	0	1

Decoders can be combined to form a larger decoder by using the enable input. An example of a 4×16 decoder constructed from two 3×8 decoders is shown in Figure 5.41. The input variables are designated as x_1, x_2, x_3, x_4, with x_1 being the most significant bit. Hence, when $x_1 = 0$, the top decoder is enabled while the bottom decoder is disabled; the bit combinations of $x_1 x_2 x_3 x_4$ range from 0000 to 0111 and determine the addresses of the outputs $D_0, D_1, \ldots, D_7$. If $x_1 = 1$, the bottom decoder is enabled and the top decoder disabled; the bit combinations now range from 1000 to 1111 and determine the addresses of the outputs of the bottom decoder, designated accordingly as $D_8, D_9, \ldots, D_{15}$.

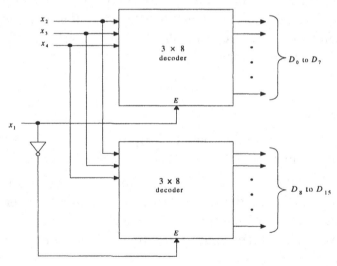

■ **Figure 5.41** A 4×16 decoder constructed with two 3×8 decoders

Decoder Implementation of Arbitrary Functions

A decoder may be used to convert from one code to another as illustrated in Example 5.18. But a decoder has other uses as well, not the least of which is the implementation of arbitrary Boolean functions. Since an n-input decoder can generate all minterms of n variables, any Boolean function of n variables can be implemented by connecting an OR gate to the appropriate minterm outputs of a decoder. (If the outputs of the decoder are maxterms, then we connect a NAND gate to the appropriate outputs.) Similarly, any combinational circuit with n inputs and m outputs can be implemented with an $n \times 2^n$ decoder and m OR gates.

■ **Example 5.19** The circuit of Example 5.16 with output functions

$$F_1(x_1, x_2, x_3, x_4) = \sum(0, 1, 3, 6, 7, 8, 11, 12, 14)$$
$$F_2(x_1, x_2, x_3, x_4) = \sum(4, 5, 10, 11, 13)$$
$$F_3(x_1, x_2, x_3, x_4) = \sum(1, 2, 3, 8, 11, 12)$$

is implemented in Figure 5.42 with a 4×16 decoder and three OR gates. ■

To implement a single-output function, a multiplexer is more economical than a decoder with an OR gate. But, in general, for multiple-output circuits, a single decoder with several OR gates is likely to be a more economical design than one with several multiplexers.

Demultiplexers

A decoder can be utilized to route data from the enable line to an address specified by the input variables $x_1, \dots, x_n$. Since such a circuit routes data from a single source to one of several outputs, it performs the opposite function of that of a multiplexer and is called a **demultiplexer**. The dual role of a decoder/demultiplexer is shown in Figure 5.43. In a sense, all that we have to do is change our perspective by "reflecting" the decoder as indicated.

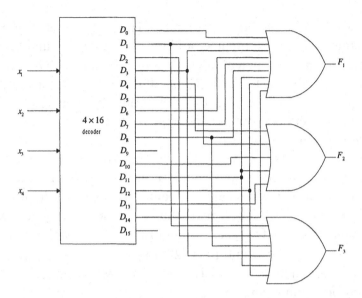

■ **Figure 5.42** Decoder and added OR gates implementing a multiple-output system

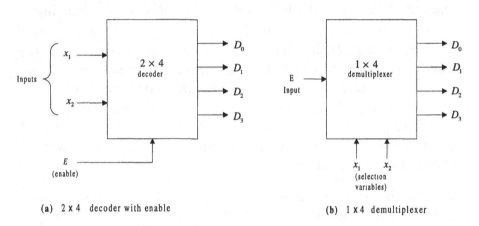

(a) 2 x 4 decoder with enable

(b) 1 x 4 demultiplexer

■ **Figure 5.43** Block diagrams indicating the dual role of a decoder/demultiplexer

Encoders

An encoder is a circuit that performs the function of a decoder in reverse. It has m inputs and n outputs, where $m \leq 2^n$. The outputs generate the binary codes for the m input variables. The inputs must be *mutually exclusive*; that is, only one input should be 1 at a time. To understand this, think of the m inputs as potential customers, where an active input (equal to 1) means that the customer is demanding service. Then the mutually exclusive input condition is essentially requiring that only one customer at a time demand service. But what happens if more than one customer *simultaneously* demands service?

There are two basic ways to handle this situation. In the first, we *assume* that external conditions monitor the customer demands so that only one customer request comes to the encoder at a time. In this case, the design of the encoder circuit is particularly simple; it is just a one-level circuit consisting of n OR gates. Usually, the truth table of a Boolean function of m variables would have 2^m rows, but since we are assuming that the inputs are mutually exclusive, only m rows are required; the other input conditions cannot happen.

- **Example 5.20** An octal-to-binary (8×3) encoder has eight inputs, one for each of the eight octal digits, and three outputs that generate the corresponding binary numbers. We assume that external conditions prevent any input other than those that represent the eight octal digits. Therefore, the truth table has only eight input conditions that have any meaning, while all the remaining $2^8 - 8 = 248$ input combinations cannot happen. The truth table for the 8×3 encoder is given in Table 5.5. From the truth table, we can construct the encoder circuit with OR gates as shown in Figure 5.44. ∎

The other basic way to design an encoder circuit is to impose a *priority assignment* on the customers so that, if two or more customers request service simultaneously, the priority assignment will determine which customer receives service. With priority assignment, the encoder circuit is called a **priority encoder**. One commonly used priority assignment scheme assigns higher priority to the input with the higher subscript number. In other words, if D_i and D_j are both 1 simultaneously and $i > j$, then D_i has priority over D_j. For example, if D_1, D_3, and D_5 are 1 simultaneously, then the output will be the binary equivalent of decimal 5 since D_5 has the highest priority.

■ **Table 5.5** Truth table for the octal-to-binary encoder

Inputs								Outputs		
D_0	D_1	D_2	D_3	D_4	D_5	D_6	D_7	x_1	x_2	x_3
1	0	0	0	0	0	0	0	0	0	0
0	1	0	0	0	0	0	0	0	0	1
0	0	1	0	0	0	0	0	0	1	0
0	0	0	1	0	0	0	0	0	1	1
0	0	0	0	1	0	0	0	1	0	0
0	0	0	0	0	1	0	0	1	0	1
0	0	0	0	0	0	1	0	1	1	0
0	0	0	0	0	0	0	1	1	1	1

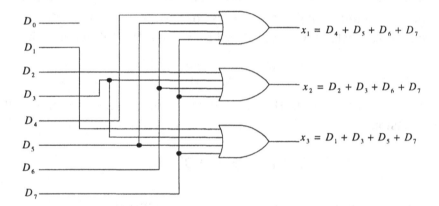

■ **Figure 5.44** Octal-to-binary encoder

■ **Example 5.21** The truth table for an 8×3 priority encoder is shown in Table 5.6. The outputs x, y, and z generate a binary number corresponding to the highest-priority input. Hence, if only D_0 is 1, the output will be $xyz = 000$. On the other hand, the third row of the truth table describes a situation where $D_2 = 1$ while D_0 and D_1 may or may not be 1 simultaneously. This situation is indicated by the entry 1 in column D_2 and the don't-cares in columns D_0 and D_1; no request of a higher priority than D_2 is present at this time. Therefore, the output code must be the binary equivalent for decimal 2, that is, $xyz = 010$. The rest of the table is filled out in a similar fashion.

■ **Table 5.6** Truth table for an 8×3 priority encoder

Inputs								Outputs		
D_0	D_1	D_2	D_3	D_4	D_5	D_6	D_7	x	y	z
1	0	0	0	0	0	0	0	0	0	0
$\times$	1	0	0	0	0	0	0	0	0	1
$\times$	$\times$	1	0	0	0	0	0	0	1	0
$\times$	$\times$	$\times$	1	0	0	0	0	0	1	1
$\times$	$\times$	$\times$	$\times$	1	0	0	0	1	0	0
$\times$	$\times$	$\times$	$\times$	$\times$	1	0	0	1	0	1
$\times$	$\times$	$\times$	$\times$	$\times$	$\times$	1	0	1	1	0
$\times$	$\times$	$\times$	$\times$	$\times$	$\times$	$\times$	1	1	1	1

We can derive the Boolean expressions for x, y, and z from Table 5.6. We could simplify the expressions using the Quine-McCluskey method but, instead, we will appeal to the simple algebraic identity

$$z + z'w = (z + z')(z + w) = z + w$$

and use it repeatedly. Hence, from the truth table, we obtain the expression

$$x = D_4 D_5' D_6' D_7' + D_5 D_6' D_7' + D_6 D_7' + D_7$$

Using the identity, this expression can be simplified to

$$x = D_4 + D_5 + D_6 + D_7$$

Likewise, we obtain the expressions

$$y = D_2 D_3' D_4' D_5' D_6' D_7' + D_3 D_4' D_5' D_6' D_7' + D_6 D_7' + D_7$$
$$= D_2 D_4' D_5' + D_3 D_4' D_5' + D_6 + D_7$$

and

$$z = D_1 D_2' D_3' D_4' D_5' D_6' D_7' + D_3 D_4' D_5' D_6' D_7' + D_5 D_6' D_7' + D_7$$
$$= D_1 D_2' D_4' D_6' + D_3 D_4' D_6' + D_5 D_6' + D_7$$

The circuit implementation of these expressions is straight forward and is, therefore, omitted.   ■

5.9 PROGRAMMABLE ARRAYS

Read-Only Memory

A $2^n \times m$ **read-only memory (ROM)** is an LSI circuit chip containing an $n \times 2^n$ decoder connected to an *array* of m OR gates, as shown in the block diagram of Figure 5.45. Each combination of the input variables is called an **address**. With n input variables, the number of distinct addresses is 2^n. Each combination on the output lines is called a **word**, and the number of output lines determines the **word length**. Since there are 2^n distinct addresses, there are 2^n distinct output words. The number of bits $(2^n \times m)$ that a ROM contains is referred to as the **capacity** of the ROM and completely specifies it. For example, a 4096-bit ROM may be

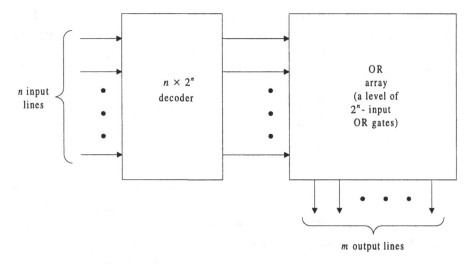

■ **Figure 5.45** Basic ROM structure

organized as 2048 words of 2 bits each, meaning that it has 2 output lines and requires 11 input lines to specify $2^{11} = 2048$ words. The same ROM can be organized also as 1024 words of 4 bits each, requiring 4 output lines and 10 input lines, or as 512 words of 8 bits each with 8 outputs and 9 inputs.

All the 2^n outputs of the decoder in Figure 5.45 are connected through **fusible links** to *each* OR gate; therefore, the number of links is $2^n \times m$. When we specify a truth table for the ROM, the binary information is embedded in the ROM by "blowing" those links that are not part of the required interconnection pattern. The process of opening links is referred to as **programming** the ROM. Hence, we refer to a ROM as a **programmable array** in which the OR array is programmable. Once a ROM has been programmed, the interconnection pattern *cannot* be changed and the information content of the ROM can only be read; hence the term *read-only*.

Recall that a decoder generates 2^n minterms of n input variables and that adding OR gates to it (see Example 5.19) enables us to implement any combinational circuit. Obviously, therefore, we can also program a ROM to implement any combinational circuit, as illustrated in the following example.

- **Example 5.22**   Consider the truth table in Figure 5.46(a), which specifies a combinational circuit with two inputs and three outputs. The corresponding Boolean functions expressed in canonical sum-of-products forms are

$$F_1(x_1, x_0) = \sum(1, 2, 3)$$
$$F_2(x_1, x_0) = \sum(0, 3)$$
$$F_3(x_1, x_0) = \sum(0, 1)$$

  The ROM that implements this circuit must have two inputs and three outputs. Therefore, its capacity is 4×3 bits, calling for a 2×4 decoder and three OR gates, each having four fusible links. Figure 5.46(b) shows the internal construction of this ROM. For each OR gate, those minterms that correspond to an output of 0 should not have a path to the output through the gate. Consequently, in programming the ROM, their corresponding links are blown, leaving open inputs to the OR gates, which are interpreted as 0 inputs. Only those links associated with the minterms that specify the functions are left intact, as shown in Figure 5.46(b).   ∎

In the actual construction of commercial ROMs, switching devices such as diodes or transistors may be used instead of OR gates. A switching device is placed at the intersection of a minterm line and an output line if the corresponding minterm is to be included in the output function; otherwise, the switching element is omitted.

In Example 5.22, each output of the ROM is considered *separately* and represents a Boolean function expressed in canonical sum-of-products form. But we can reinterpret the binary information provided by the truth

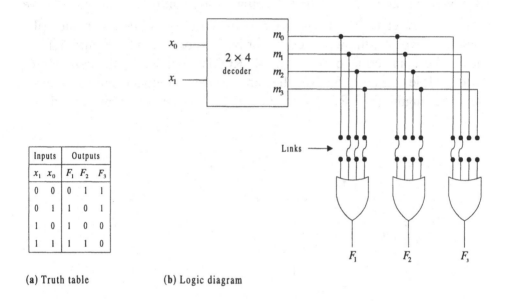

Inputs		Outputs		
x_1	x_0	F_1	F_2	F_3
0	0	0	1	1
0	1	1	0	1
1	0	1	0	0
1	1	1	1	0

(a) Truth table   (b) Logic diagram

∎ **Figure 5.46** ROM implementation of a combinational circuit

table in Figure 5.46(a) by considering the three outputs *collectively*, rather than treating them separately. In this way, for each input combination, the three outputs form a 3-bit *word* whose *address* is specified by that input combination. For example, each time the address 10 is placed on the input lines, the ROM will produce on its output lines the 3-bit word 100. The four 3-bit words produced by the ROM of Figure 5.46 are said to be **stored** in the ROM; hence the term *memory*.

Consequently, a ROM is a memory device with *fixed* word patterns that can only be read upon application of a given address. The bit patterns stored in the ROM are *permanent* and cannot be changed during normal circuit operation. As a result, the information stored will not be lost even when power is turned off or interrupted. A memory device having this feature is called **nonvolatile**. A different type of memory is required if the stored information needs to be changed during normal circuit operation; it is called a **read/write memory (RWM)** and is introduced in Chapter 7. Unlike ROMs, RWMs implemented with semiconductor ICs are **volatile** devices; they lose the information when power is interrupted or removed.

Like most MSI and LSI chips, the operation of a ROM can be controlled through an *enable* (*E*) input. This control input can also be used to cascade ROMs in order to increase the storage capacity (the number of words), the word length, or both. For example, Figure 5.47(a) shows the connection of two 16×4 ROMs to increase the number of words, while in Figure 5.47(b) the same ROMs are connected to increase the word length. (The ROM outputs in Figure 5.47(a) are wired-ORed.)

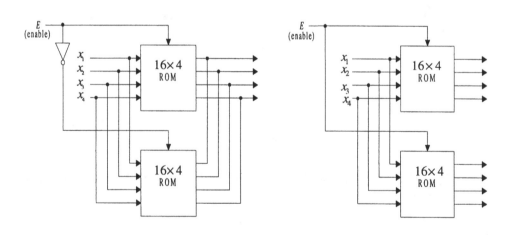

(a) Two 16 x 4 ROMs connected to form a 32 x 4 ROM   (b) Two 16 x 4 ROMs connected to form a 16 x 8 ROM

■ **Figure 5.47** Cascading ROM modules

Types of ROM

Three basic types of ROMs are available commercially: mask-programmable ROM, programmable ROM (PROM), and erasable PROM (EPROM). **Mask-programmable ROMs** are custom made by the device manufacturer to specifications (a truth table) submitted by the customer. The binary information is permanently stored in the ROM at the time of manufacture by a special process called *masking*. But, since masking is expensive, mask-programmable ROMs are economically feasible only if large quantities (typically a thousand or more) of the same ROM are produced.

For small quantities, it is more economical to use a **programmable ROM (PROM)**. PROMs are typically manufactured with all the fusible links present and can be programmed by the user using a special device called a *PROM programmer*.

But programming a ROM or a PROM is an irreversible process. Each time the information stored has to be changed, a new device must be produced. In contrast, the third type of ROM, the **erasable PROM (EPROM)**, is reprogrammable and enables the user to modify the stored data. EPROMs are programmed with a PROM programmer and are available in two types distinguished by the method used for erasure. In one EPROM type, the data are erased by placing the device under a special ultraviolet light source for some time. In the other EPROM type, called *electrically alterable ROM (EAROM)*, the data are erased on the application of special electrical signals.

ROM Look-Up Table

Besides being able to implement any combinational circuit and to store information, a ROM may be used as a look-up table. To illustrate this application, consider, for example, the process of binary multiplication shown schematically in Figure 5.48. The product value of two digits, X and Y, is stored in the ROM and can be accessed by addressing the correct location. In other words, the multiplication table stored in the ROM contains the results of all possible combinations of the two input operands.

If the multiplicand and multiplier contain s and t bits, respectively then the total number of inputs to the ROM is $s+t$, and the product will have at most $s+t$ bits. Therefore, the capacity of the ROM is $2^{s+t}(s+t)$

bits. For example, the product of two 8-bit binary numbers would require a memory capacity of $2^{16} \times 16 = 2^{20}$ bits. This means that over 1 million bits is required to store the multiplication table for two 8-bit numbers. By contrast, the multiplication table for two 4-bit numbers only requires a 256×8 (2048-bit) ROM.

Programmable Logic Array

Implementing a combinational circuit having a large number of don't-care conditions with a ROM may be quite wasteful because the don't-cares represent input addresses that will never occur and, consequently, not all of the interconnection patterns are used. In such cases, it is more economical to use another LSI device, called a **programmable logic array (PLA)** and shown in block diagram form in Figure 5.49.

A PLA performs the same functions as a ROM: With n inputs and m outputs, a PLA can implement m functions of n variables. Structurally, however, the PLA differs from the ROM in that it does not generate all the

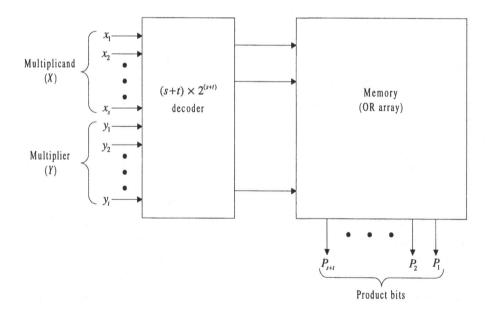

■ **Figure 5.48** ROM used as a look-up multiplication table

minterms of the input variables. Rather, the decoder of the ROM is replaced in the PLA by an AND (gate) array that can be programmed to produce only selected product terms. This difference eliminates the inefficient storage of unused minterms required in a ROM. The OR array in both devices plays the same basic role: Certain product terms are ORed together to form the output functions. Moreover, both devices can implement any combinational circuit. Therefore, a PLA implements functions in their *simplified* sum-of-products form, rather than the *canonical* sum-of-products form required by a ROM. Consequently, a PLA is best employed to implement Boolean functions described by a small number of product terms, whereas a ROM is best suited for implementing functions requiring a large number of minterms.

As shown in Figure 5.49, both the AND and OR arrays of the PLA are programmable. To program the AND array, fusible links are inserted between all the n inputs and their complements and each of the AND gates. To program the OR array, links are also provided between the outputs of the AND gates and each of the OR gates. Hence, if the PLA consists of n inputs, p product terms, and m outputs, it requires p

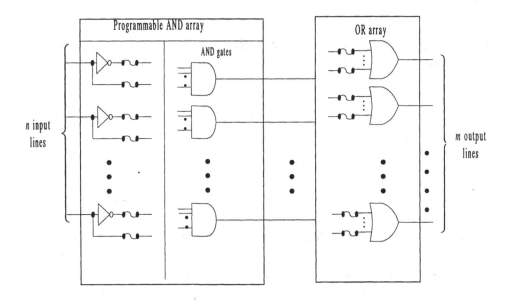

■ **Figure 5.49** Basic PLA structure

AND gates, m OR gates, and a total of $2np + pm$ fusible links. Some PLAs also come with programmable output inverters to generate either complemented or uncomplemented output functions. Each output line is connected to an inverter having a link placed across it. If we leave the link intact, the inverter is bypassed and the PLA output is uncomplemented; with the link broken, the inverter is enabled and generates the complemented output. This arrangement calls for an additional set of m programmable links.

The size of a PLA is specified by the number of inputs, the number of product terms generated by the AND array, and the number of outputs. Commercially available PLAs typically have 16 inputs, 48 product terms, and 8 outputs, and the outputs may be complemented or uncomplemented. In contrast, a ROM with 16 inputs generates 2^{16} minterms, but if at most 48 products terms are needed, then $2^{16} - 48$ words of the ROM would be unused. This inefficiency points out one of the major drawbacks to ROM implementation. Typically, a PLA can accommodate many more input variables than a ROM without stretching the storage capacity of the chip. In contrast, the capacity of a ROM *doubles* with the addition of each new input variable.

■ **Example 5.23**   To compare a ROM and a PLA implementation, consider the following 3-input, 4-output circuit:

$$F_1(x_1, x_2, x_3) = \sum(0, 1, 6, 7) = x_1'x_2' + x_1x_2$$
$$F_2(x_1, x_2, x_3) = \sum(1, 3, 5, 6, 7) = x_3 + x_1x_2$$
$$F_3(x_1, x_2, x_3) = \sum(1, 2, 3) = x_1'x_3 + x_1'x_2$$
$$F_4(x_1, x_2, x_3) = \sum(0, 1, 3, 5, 7) = x_3 + x_1'x_2'$$

The canonical forms are implemented with a ROM and the simplified expressions with a PLA, as shown in Figures 5.50(a) and (b). The ROM circuit requires eight lines for all possible minterms in the three variables, while the PLA circuit requires only five lines for the *distinct* product terms. Since the technologies have similar costs, we could say that the cost of implementation is proportional, among other things, to the number of lines. Therefore, the cost of the PLA circuit of Figure 5.50(b) would be about five-eighths of the ROM cost.   ■

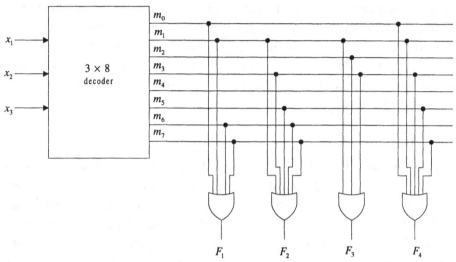

(a) ROM implementation

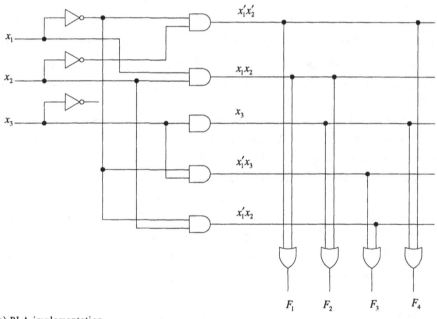

(b) PLA implementation

■ **Figure 5.50** Comparison of ROM and PLA implementations

There are two basic types of PLAs: **mask-programmable PLA** and **field-programmable PLA (FPLA)**. With a mask-programmable PLA, the customer submits a PLA program table (introduced in the following section) describing the output functions to be implemented, and then the manufacturer uses the table to program the PLA with the desired interconnection patterns. As the name indicates, a FPLA, like the PROM, can be programmed in the field by the user.

PLA Program Table

A PLA is specified by a table called the **PLA program table**. For example, the PLA program table for the circuit of Figure 5.50(b) is shown in Table 5.7. All that the program table specifies are the internal patterns of the PLA circuit. The first column lists the product terms, the second group of columns specifies the required connections between the inputs and the AND gates, and the third group of columns specifies the connections between the AND gates and the OR gates.

The connections between the inputs and the AND gates are specified under the heading *inputs*. For each product term, an input x_i is marked with 1, 0, or dash$(-)$ according to whether x_i appears, x_i' appears, or neither appears in the product term. Therefore, a 1 in any input column indicates that the input is connected to the AND gate that forms the corresponding product term; on the other hand, a 0 indicates that the complemented input is connected to the AND gate. A dash specifies no connection at all. For example, the first row in Table 5.7 under the heading *inputs* is labeled $00(-)$ to indicate the product term $x_1'x_2'$.

Likewise, the connections between the AND and OR gates are specified under the heading *outputs*. A 1 (or a dash) is placed at the intersection of a row and a column headed by a function if the corresponding product term appears (or does not appear) in the expression for the function. For example, the column F_1 in Table 5.7 is labeled $11(-)(-)(-)$ to indicate that the first two product terms appear in the expression for F_1, but the other three product terms do not appear.

When we design a combinational circuit with a PLA, we need not show the internal circuit of the PLA. All that is required is the PLA program table from which the manufacturer can program the PLA to supply the appropriate connections.

■ **Table 5.7** PLA program table for Figure 5.50(b)

Product Term	Inputs			Outputs			
	x_1	x_2	x_3	F_1	F_2	F_3	F_4
$x_1'x_2'$	0	0	—	1	—	—	1
x_1x_2	1	1	—	1	1	—	—
x_3	—	—	1	—	1	—	1
$x_1'x_3$	0	—	1	—	—	1	—
$x_1'x_3$	0	1	—	—	—	1	—

Frequently, the design or analysis of a PLA circuit is also carried out in terms of a symbolic (matrix) representation, like that of Figure 5.51, rather than with a circuit logic diagram. In the symbolic representation, a dot at the intersection of a product term line and an input line or an output line indicates the presence of a switching element in the array. For example, Figure 5.51 shows the symbolic representation of the PLA implementation of the circuit of Example 5.23. To be sure, the correspondence between this symbolic representation and the circuit logic diagram of Figure 5.50(b) is clear.

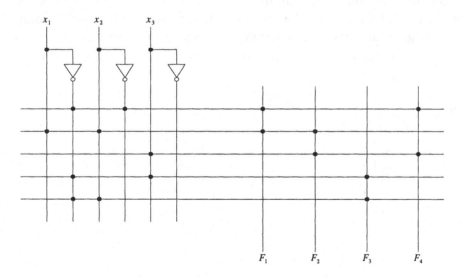

■ **Figure 5.51** Symbolic representation of a PLA implementation

PLA Minimization

Since a given PLA has a limited number of product terms, we must design the circuit so that the number of *distinct* product terms required will not exceed that of the PLA chip at hand. If the number of distinct product terms exceeds that of the chip, we can use the familiar simplification techniques (the K-map or Quine-McCluskey method) to reduce it. Note, however, that we need only simplify to the point where the number of distinct product terms is less than or equal to that of the PLA chip.

To illustrate this point, suppose that the PLA chip available has four inputs, six product terms, and three outputs, and consider using it to implement the multiple-output system

$$F_1(x_1, x_2, x_3, x_4) = \sum(2, 3, 5, 7, 8, 9, 10, 11, 13, 15)$$
$$F_2(x_1, x_2, x_3, x_4) = \sum(2, 3, 5, 6, 7, 10, 11, 14, 15)$$
$$F_3(x_1, x_2, x_3, x_4) = \sum(6, 7, 8, 9, 13, 14, 15)$$

(This system was used in Section 4.10 to illustrate the multiple-output minimization procedure.) We cannot implement the canonical forms of the functions with the given PLA because they require 12 distinct minterms; nor can we implement the minimal sum-of-products expressions, obtained by minimizing each function separately, because they require 7 distinct product terms. We can apply the multiple-output minimization procedure and obtain an implementation suitable for our PLA that requires only 5 distinct product terms (see Section 4.10). But

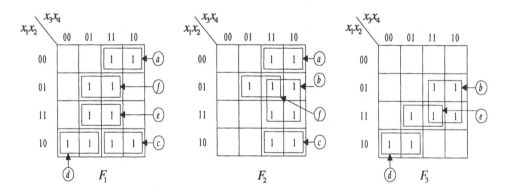

- **Figure 5.52** K-maps for a multiple-output system

this method is rather lengthy and time consuming and can be avoided in some cases because we are only interested in a simplification that can be accommodated by a given PLA.

We follow the same basic idea of multiple-output minimization: We search for *shared* product terms by selectively combining minterms of each function. To illustrate this, consider the K-maps for the three output functions shown in Figure 5.52. By judiciously clustering minterms, we can express the three functions in terms of only six distinct products: $a = x_1'x_2'x_3$, $b = x_2x_3$, $c = x_1x_2'x_3$, $d = x_1x_2'x_3'$, $e = x_1x_2x_4$, and $f = x_1'x_2x_4$. The resulting PLA implementation is shown in symbolic form in Figure 5.53.

In actual practice, it may not be possible to fit a given multiple-output circuit on a single chip. In such cases, we should attempt to reduce the number of product terms to fit the circuit on as few chips as possible. Note, however, that minimizing PLA designs can be a very difficult problem for circuits with large numbers of inputs and outputs.

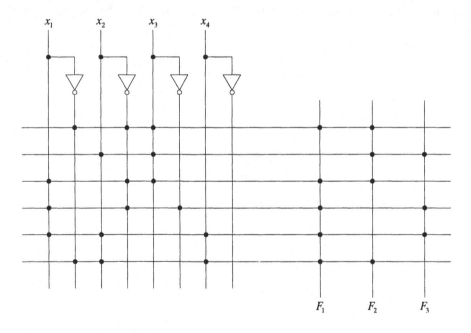

■ **Figure 5.53**   PLA implementation of the multiple-output system of Figure 5.52

Programmable Array Logic

The **programmable array logic (PAL)** is a special case of the PLA of Figure 5.49 in which the AND array is programmable and the OR array is fixed. (In some PAL types, the OR array is replaced by a NOR array.) Since only the AND array is programmable, the PAL is less expensive and easier to program than the more general PLA.

Figure 5.54(a) shows a portion of an unprogrammed PAL. The accepted notation is to use an × to represent an intact link. Although one common line is shown in Figure 5.54(a) for each AND gate, it represents four fusible-link inputs to the gate. When the PAL is programmed, the appropriate links are selectively "blown" to leave the desired connections to the AND gates inputs. As an example, we use this PAL in Figure 5.54(b) to implement the function $F = x_1 x_2 + x_1' x_2'$.

Since the OR array is fixed, the size of the PAL is determined not only by the number of inputs, product terms, and outputs, but also by the number of inputs per OR gate. Typical PAL integrated circuits are available with 10 to 35 inputs, 1 to 30 outputs, and 2 to 20 inputs per OR gate.

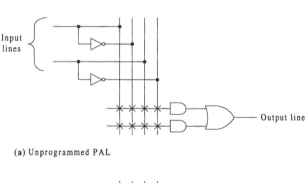

(a) Unprogrammed PAL

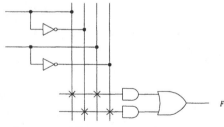

(b) PAL implementation of F

■ **Figure 5.54** Basic PAL structure

The PAL allows the designer to specify the nature of the product terms, but the ways in which the product terms may be formed into sums is fixed in the chip. Unlike the PLA, product terms (AND gates) in a PAL implementation cannot be shared among two or more OR gates. Therefore, to implement a multiple-output circuit with a PAL, each output function must be simplified separately without regard to sharing common gates. For a given type of PAL, the number of product terms feeding each OR gate is fixed and limited. Consequently, if the number of product terms in a simplified function is too large, we must choose a different PAL with more OR gate inputs or fit the function on more than one PAL.

PROBLEMS

5.1   Find the minimal sum for $F = \sum(0, 2, 8, 9, 10, 11, 14, 15)$ by means of a K-map. Draw the simplified circuit logic diagram with
        (a) AND-OR gates   **(b)** NAND gates

5.2   Find the minimal product-of-sums form for $F = \sum(2, 3, 4, 5, 6, 7, 11, 14, 15)$. Draw the OR-AND circuit logic diagram and then implement the circuit with NOR gates.

5.3   Implement the minimal sum of $F = \sum(0, 2, 4, 5)$ with NAND gates. Use inverters to complement inputs.

5.4   **(a)**   Show that the circuit in Figure P5.4 produces the exclusive-OR (XOR) function.

        (b)   Replace the NAND gates in Figure P5.4 by NOR gates and show that the exclusive-NOR (equivalence) function is obtained.

5.5   Obtain the circuit logic diagram for a two-input equivalence function using NAND gates.

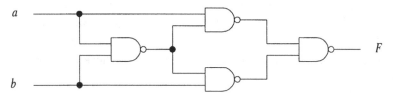

■ **Figure P5.4**

5.6   Convert the circuit logic diagram of Figure P5.6 to a:
        (a) NAND implementation   **(b)** NOR implementation

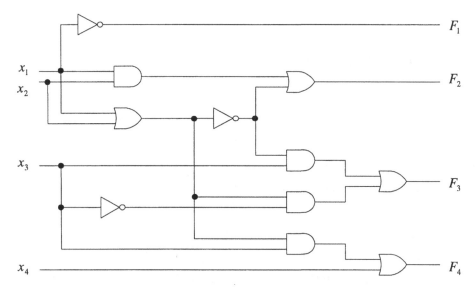

■ **Figure P5.6**

5.7   Obtain a NAND implementation of the circuit in Figure P5.7.

5.8   Derive the Boolean output function for the circuit logic diagram in Figure P5.7.

5.9   **(a)**   Find the minimal sum-of-products expression for an even-parity generator G for 3-bit messages. (See Problem 4.15).

        (b)   Draw a NAND implementation of the minimal sum for G.

        (c)   Obtain a minimal product-of-sums expression for G and implement it with NOR gates.

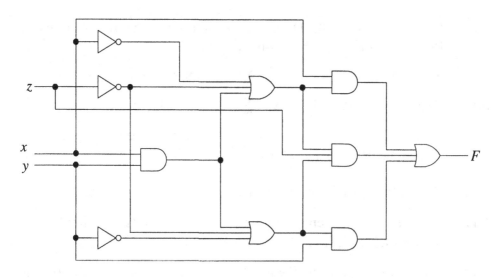

- **Figure P5.7**

5.10 Convert the circuit diagram of Figure P5.10 to one implemented with AND, OR, and NOT gates.

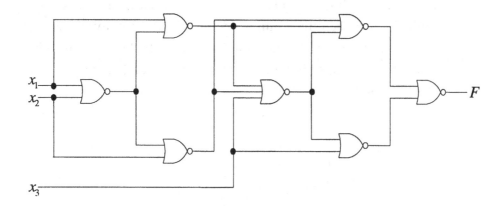

- **Figure P5.10**

5.11 Implement $F = \sum(1,5,6,10,13,14)$ as a 3-level OR-AND-OR circuit where the fan-in capability of each gate is at most 3.

5.12   Obtain a 3-level AND-OR-AND circuit for the function F in Problem 5.11 where the fan-in is at most 3.

5.13   Repeat Problem 5.11 for $G = \sum(7,11,13,14,15)$.

5.14   Repeat Problem 5.11 for $H = \sum(0,1,2,13,14,15)$.

5.15   Convert $F = \sum(1,2)$ into its eight nondegenerate forms involving AND, OR, NAND, or NOR gates.

5.16   Repeat Problem 5.15 for $G = \left[(x_1 + x_2')' + x_3\right]'$.

5.17   Find eight different minimal 2-level circuits to implement $F = \sum(0,2,6,8,12)$.

5.18   The block diagram in Figure P5.18 shows a base 3 (ternary) full-adder that receives two base 3 digits $X = (x_2 x_1)$ and $Y = (y_2 y_1)$ plus an input carry C_i and produces the sum $S = (s_2 s_1)$ in base 3, as well as an output carry C. The base 3 digits 0, 1, and 2 are coded in binary: 0 by 00, 1 by 01, and 2 by 10. Addition is carried out in base 3. For example, if x and y are equal to 2 in base 3, and if C_i is 1, then their sum is decimal 5, which converts in base 3 to $(12)_3$. Accordingly, the sum S must be 2 while C must be 1 (and 2 must be coded in binary). Design the circuit using gates and binary full-adders.

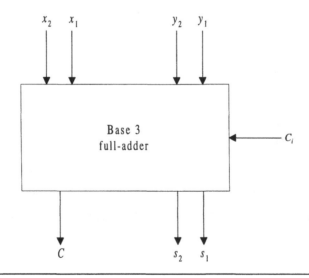

5.19 Design a base 5 full-adder circuit using binary full-adders and gates. Assume that the circuit receives two base 5 digits and an input carry C_i, all coded in binary. The circuit produces the binary-coded base 5 sum S and an output carry C.

5.20 Design a 2-bit binary multiplier combinational circuit that multiplies two 2-bit binary numbers $x_2 x_1$ and $y_2 y_1$ and produces a product $P_4 P_3 P_2 P_1$. First derive the truth table and minimize the output functions P_4, P_3, P_2, and P_1 with K-maps. (Compare this circuit with the circuit of Figure 5.23, which uses full-adders.)

5.21 Design an excess 3-to-BCD code converter using 4-bit binary full-adders.

5.22 Using four MSI 4-bit parallel adders, construct a binary parallel adder that can add two 16-bit binary numbers. Label all carries between the MSI circuits.

5.23 Design a circuit that adds four 1-bit binary numbers simultaneously by using full-adder circuits.

5.24 Derive the 2-level equation for the output carry C_5 in a look-ahead carry generator.

5.25 Design a binary multiplier circuit that multiplies a 4-bit number $B = b_4 b_3 b_2 b_1$ by a 3-bit number $A = a_3 a_2 a_1$ to form a product $P = P_7 P_6 P_5 P_4 P_3 P_2 P_1$. This can be done with 12 AND gates and two 4-bit parallel adders. The AND gates are used to form the partial products of pairs of bits, and then these partial products are summed by the parallel adders.

5.26 Implement a 12-bit comparator using 4-bit comparator chips.

5.27 Design a circuit that compares two 4-bit binary numbers, $A = A_3 A_2 A_1 A_0$ and $B = B_3 B_2 B_1 B_0$, to check if they are equal. The circuit has one output E, so $E = 1$ if $A = B$ and $E = 0$ otherwise.

5.28 Suppose that we are given a logic device X that compares two 3-bit numbers $A_{(3)} = a_1 a_2 a_3$ and $B_{(3)} = b_1 b_2 b_3$, where a_3 and b_3 are the least significant bits. Device X has two outputs $G_{(3)}$ and $L_{(3)}$ that behave as follows:

$$G_{(3)} = 1 \quad \text{if and only if} \quad A_{(3)} > B_{(3)}$$

$$L_{(3)} = 1 \quad \text{if and only if} \quad A_{(3)} < B_{(3)}$$

and

$$G_{(3)} = L_{(3)} = 0 \quad \text{if and only if} \quad A_{(3)} = B_{(3)}$$

(a) Design a device Y so that, together with device X, both units will combine as in Figure P5.28 to serve as a comparator for two 4-bit binary numbers $A_{(4)} = a_1 a_2 a_3 a_4$ and $B_{(4)} = b_1 b_2 b_3 b_4$.

(b) Find expressions for $G_{(4)}$ and $L_{(4)}$ in terms of the inputs to unit Y and show an implementation of these expressions using only NAND gates.

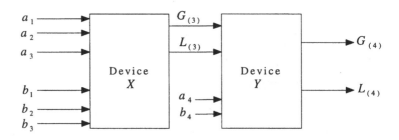

■ **Figure P5.28**

5.29 Implement the function $F(x_1, x_2, x_3, x_4) = \sum(0, 1, 3, 4, 8, 9, 15)$ with an 8×1 multiplexer where the following variables are connected in the specified order to selection lines s_2, s_1, and s_0, respectively:

(a) x_1, x_2, x_3 (b) x_2, x_3, x_4

(c) x_1, x_2, x_4 (d) x_1, x_3, x_4

5.30 Show how two 2×1 MUXs (with no added gates) can be connected to form a 3×1 multiplexer. Use the inputs to the selection lines as follows:

if $x_1 x_2 = 00$, select I_0

if $x_1 x_2 = 01$, select I_1

if $x_1 x_2 = 1 \times (x_2$ is a don't-care), select I_2

5.31 Implement a 16×1 multiplexer with five 4×1 multiplexers.

5.32 Implement a 32×1 multiplexer with:

(a)   Two 16×1 MUXs, each with an enable line

(b)   Two 16×1 MUXs and one 2×1 MUX

(c)   Two 16×1 MUXs and one 4×1 MUX

(d)   Five 8×1 MUXs

5.33 (a)   Draw the circuit logic diagram of a 4×1 MUX with an enable line.

(b)   Show how two 4×1 multiplexers, each with an enable input, can be connected to form an 8×1 MUX.

5.34 Implement a full-adder circuit with multiplexers.

5.35 Implement $F(x_1, x_2, x_3) = \sum(0, 2, 3, 5, 7)$ with an 8×1 MUX by connecting x_1, x_2, and x_3 to the selection lines s_2, s_1, and s_0, respectively.

5.36 Use 4×1 multiplexers to implement the combinational circuit with output functions
$$F_1(x_1, x_2, x_3) = \sum(0, 1, 6)$$
$$F_2(x_1, x_2, x_3) = x_1' + x_2$$
$$F_3(x_1, x_2, x_3) = \sum(0, 1, 6, 7)$$

5.37 Use 8×1 multiplexers to implement the combinational circuit with output functions
$$F_1(x_1, x_2, x_3, x_4) = \sum(2, 4, 10, 11, 12, 13)$$
$$F_2(x_1, x_2, x_3, x_4) = \sum(4, 5, 10, 11, 13)$$
$$F_3(x_1, x_2, x_3, x_4) = \sum(1, 2, 3, 10, 11, 12)$$

5.38   Draw the logic diagram of a 2×4 decoder with an enable input using:

    (a) NAND gates   **(b)** NOR gates

5.39   Implement a full-adder circuit with a decoder and two OR gates.

5.40   Implement $F_1(x_1, x_2, x_3, x_4) = \sum(1, 3, 4)$ and $F_2(x_1, x_2, x_3, x_4) = \sum(4, 6, 7, 9)$ with a 4×10 decoder and two NAND gates.

5.41   Implement a BCD-to-excess 3 code converter with a 4×10 decoder and four NAND gates.

5.42   Show how to construct a 5×32 decoder with four 3×8 decoders (with enable inputs) and one 2×4 decoder.

5.43   With a decoder and external gates, implement the circuit defined by

$$F_1(x_1, x_2, x_3) = x_1'x_2' + x_1x_2x_3'$$
$$F_2(x_1, x_2, x_3) = x_1' + x_2$$
$$F_3(x_1, x_2, x_3) \;\; = \sum(0, 1, 6, 7)$$

5.44   With a 4×16 decoder and external gates, implement the circuit of Problem 5.37.

5.45   Design a 4×2 priority encoder. Include an output E to indicate that at least one input is a 1.

5.46   **(a)**   Draw the block diagram of a 32×8 ROM with enable input. How many address lines and output lines are needed?

       (b)   Show the external connections of two 32×8 ROMs in order to produce a 64×8 ROM.

5.47   Implement a circuit that converts from the 4-bit BCD code to the 4-bit Gray code by using a ROM.

5.48   Design a 32×8 ROM by cascading a four 16×4 ROMs.

5.49   List the truth table of a 16×4 ROM that multiplies two 2-bit binary numbers and produces a 4-bit product.

5.50   Compare a ROM implementation with a PLA implementation of the circuit with output functions

$$F_1(x_1, x_2, x_3) = \sum(0, 2, 3, 5, 6)$$
$$F_2(x_1, x_2, x_3) = \sum(1, 2, 3, 4, 7)$$
$$F_3(x_1, x_2, x_3) = \sum(2, 3, 4, 5, 6)$$

5.51 List the PLA program table for the BCD-to-excess 3 code converter circuit.

5.52 The internal connection of a PLA is given in Figure P5.52.

   (a) Write the equations implemented by the PLA.

   (b) Specify the truth table of a ROM that would implement the same circuit as the PLA.

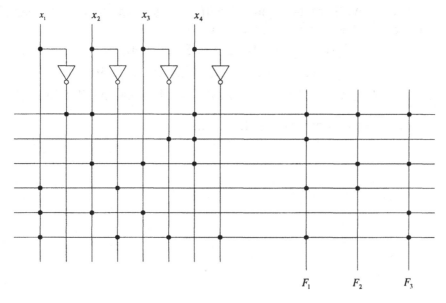

■ **Figure P5.52**

5.53 Implement with a PLA the circuit with output functions

$$F_1(x_1, x_2, x_3, x_4) = \sum(4, 5, 10, 11, 12)$$
$$F_2(x_1, x_2, x_3, x_4) = \sum(0, 1, 3, 4, 8, 11)$$
$$F_3(x_1, x_2, x_3, x_4) = \sum(0, 4, 10, 12, 14)$$

Specify the PLA program table and draw a symbolic representation of the internal circuit of the PLA.

5.54   Implement the seven-segment display circuit whose truth table is shown in Figure 3.3 using a PAL with 5 inputs, 8 outputs, and 4 inputs per OR gate. (Assume that all don't-cares are zero).

5.55   Given a PAL with 12 inputs and 6 outputs, where two OR gates have 4 inputs each and the other four OR gates have 2 inputs each, implement a circuit that converts a 4-bit binary number to a hexadecimnal digit and outputs the 7-bit ASCII code for the hexadecimal digit.

5.56   Given a PAL with 14 inputs, 4 outputs, and 4 inputs per OR gate, implement the circuit with the output functions

$$F_1 = \sum(0,2,4,6,7,8,10,12,14,15,22,23,30,31)$$
$$F_2 = \sum(0,1,4,5,13,15,20,21,22,23,24,26,28,30,31)$$
$$F_3 = \sum(0,1,4,5,16,17,20,21,37,42,43,46,47,53,58,59,62,63)$$

(*Hint*:   See Examples 4.22, 4.23, and 4.24.)

6 Asynchronous Sequential Circuits

6.1 INTRODUCTION

Recall that, in a combinational circuit, the Boolean functions relating the outputs $(z_1, z_2, \ldots, z_m)$ to the inputs $(x_1, x_2, \ldots, x_n)$ are given by

$$z_i = F_i(x_1, x_2, \ldots, x_n), \qquad i = 1, 2, \ldots, m \tag{6.1}$$

We have said that the outputs at any instant of time are functions only of the inputs at that time. But, in practice, this statement is not absolutely true because every physical device introduces a *propagation time delay*, albeit small, in the signal path between its inputs and outputs. Therefore, in a combinational circuit, the output at time t is actually a function of the inputs at time $t - t_{pd}$, where t_{pd} denotes the signal propagation time delay.

Nevertheless, we say that a combinational circuit is **memoryless**: Given an output value at time t_n, the output value at $t_{n+1} > t_n$ depends *only* on the input values at t_{n+1}.

By contrast, in a circuit with **memory**, an output value at t_{n+1} must be a function not only of the inputs at t_{n+1} but also of the outputs at t_n. To achieve this, the circuit must have some **feedback** connections from its outputs to its inputs. These feedback connections are what differentiates a memoryless circuit from one with memory. But note that *a circuit with memory is a combinational circuit incorporating some feedback connections.*

Programmable arrays (see Section 5.9) may seem exceptions to this rule because they are said to have memory, yet they are combinational circuits. But they do not have memory in the present sense because (1) they have only *feed-forward* connections between their inputs and outputs, and (2) they are *read-only* memories, so their contents cannot be changed.

To implement feedback, signals are fed back from the outputs to the inputs of the combinational circuit through **memory devices**. A memory device *stores* an output value at time t_n so that it can be input to the combinational circuit and processed, together with the external inputs, to produce the output value at t_{n+1}. But, then, the output at t_n depends on input values at $t_{n-1} < t_n$, the output at t_{n-1} depends on input values at $t_{n-2} < t_{n-1}$, and so on. Therefore, the circuit outputs actually depend not only on the *present* inputs but also on *past* inputs. In other words, the circuit maps *input sequences* to *output sequences*, rather than input values to output values as in a combinational circuit.

Circuits with memory are called **sequential circuits**[*] and are represented by the model shown in Figure 6.1. The circuit consists of n independent input variables (called *circuit inputs*), m output variables (called *circuit outputs*), and k memory devices. The outputs of the memory devices $(y_1, y_2, ..., y_k)$ collectively define the **present state** of the sequential circuit, and the inputs $(Y_1, Y_2, ..., Y_k)$ to the memory devices specify its **next state**. The binary variables are referred to as **state variables**, and the prefix *present* or *next* is added to distinguish between them. The circuit inputs and the present state determine the values of the circuit outputs, as well as the inputs to the memory devices, which, in turn, determine the next present state of the circuit (hence the term *next state*). The passage from a present state to the next state is called **state transition**.

With k state variables (memory devices), the circuit can be in any one of at most 2^k *distinct* states. Alternatively, if the circuit has r distinct states, then the number of state variables required must be such that $2^k \geq r$ (or $k \geq \lceil \log_2 r \rceil$, where $\lceil a \rceil$ denotes the smallest integer that is greater than or equal to a).

The behavior of any sequential circuit can be represented by

---

[*]Terms such as *sequential machine, finite automaton,* or *finite-state machine* are also used interchangeably to describe the model of sequential circuits.

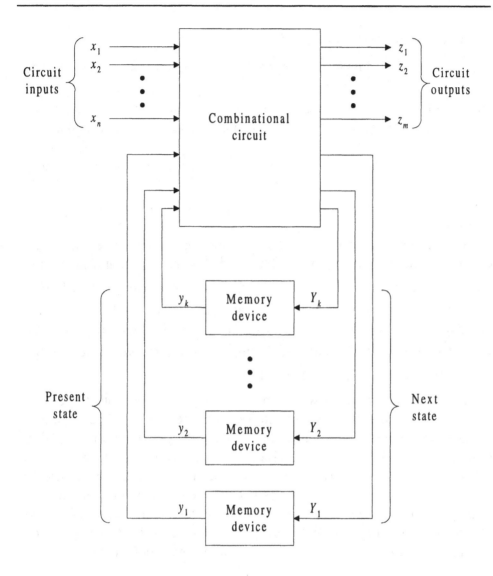

■ **Figure 6.1** Sequential circuit model

mathematical relations between the three sets of variables: those for the inputs, the outputs, and the present state. These relations, described by Boolean functions, can be expressed as

$$Y_i = G_i(x_1, x_2, \ldots, x_n, y_1, y_2, \ldots, y_k), \quad i = 1, 2, \ldots, k \qquad (6.2)$$

referred to as the *next-state equation*, and

$$z_i = F_i(x_1, x_2, \ldots, x_n, y_1, y_2, \ldots, y_k), \quad i = 1, 2, \ldots, m \qquad (6.3)$$

that is called the *output equation*. In some sequential circuits, the outputs are functions of only the present state and are, therefore, represented as

$$z_i = F_i(y_1, y_2, \ldots, y_k), \quad i = 1, 2, \ldots, m \qquad (6.4)$$

Note that, in either case, *the outputs are always associated with the present state*.

The timing element enters these equations in the following way: Given the present state and the values of the circuit inputs at time t_n, the output z_i and the next state variable Y_i (for every i) are produced at t_n (because they are outputs of a combinational circuit). But the ith state variable Y_i will be converted by the ith memory device into the ith present state some time later (the Y_i information must propagate through the memory device). Hence, $Y_i(t_n)$ determines $y_i(t_n + t_{pd})$.

A sequential circuit in which the outputs depend on both the circuit inputs and the present state [as in Equation (6.3)] is called a **Mealy circuit**. When the outputs depend only on the present state [as in Equation (6.4)], the circuit is called a **Moore circuit**. However, there are no substantial differences between the two circuit types. In fact, every Mealy circuit with m outputs and k states is equivalent to a Moore circuit with m outputs and no more than $mk+1$ states. As we will see later, either type can be used in the design of sequential circuits, and the choice between them is usually made based on the particular design problem at hand.

Classification of Sequential Circuits

Synchronous versus Asynchronous

The timing of the signals in the circuit determine two types of sequential circuits: synchronous and asynchronous. In a **synchronous** sequential circuit, the state can change *only* at discrete instants of time. To achieve that, the circuit uses a timing device, called a **clock generator**, that produces trains of periodic or aperiodic **clock pulses** (see Section 1.2). The clock pulses are input to the memory devices so that they can change state *only* in response to the arrival of a pulse and *only once* for each pulse occurrence. We say that the operation of the circuit is *synchronized* with the clock pulse input; hence the name synchronous circuit. We discuss the analysis and design of synchronous sequential circuits in Chapter 7.

The behavior of an **asynchronous** sequential circuit, on the other hand, depends *only* on the order in which the inputs change and can be affected at any instant of time. We use unclocked memory devices so that state transitions can occur only when the circuit inputs change. As a result, the correct operation of an asynchronous circuit depends critically on the timing of the inputs. Since there is no timing device to synchronize all the changes, the circuit must have a chance to settle down (become *stable*) from the last input change; otherwise, the circuit may operate improperly. Consequently, it is necessary to restrict the way in which the inputs can change; only one input is allowed to change at any given time, and any change must occur after the circuit has become stable. We consider the details of these timing issues in this chapter.

The design of asynchronous circuits is somewhat more difficult than that of synchronous circuits because they present a number of problems (such as instability, critical races, and essential hazards) that are eliminated from synchronous circuits due to their clocked operation. Because of this, many texts discuss synchronous circuits first. By contrast, we chose to treat asynchronous circuits first because of two main reasons: (1) The operation of sequential circuits depends on proper timing and is best understood in the context of asynchronous circuits, and (2) since asynchronous circuits include synchronous circuits as a special case, we can present a unified design methodology for both circuit types and avoid duplication of similar topics.

Completely Specified versus Incompletely Specified

If all the next states and circuit outputs are specified for every combination of the circuit inputs, the sequential circuit is said to be **completely specified**. We know, however, that in many situations some input combinations may not occur or, for one reason or another, are not allowed to occur. In such cases, some state transitions (next states) may be undefined. But even if all the next states are defined, we might not care about some output values and leave them unspecified. Such sequential circuits are called **incompletely specified**.

We will see that asynchronous sequential circuits are *inherently* incompletely specified. On the other hand, synchronous sequential circuits can be either completely or incompletely specified.

Representation of Sequential Circuits

The logic equations, (6.2) and (6.3) or (6.4), completely characterize the behavior of the sequential circuit shown in Figure 6.1. Other tools, however, represent the operation of the circuit in pictorial and tabular forms and offer better insights into its behavior. We discuss in this section three commonly used tools: state diagram, state table, and timing sequences and diagrams.

State Diagram

The state diagram depicts graphically the operation of a sequential circuit. Figure 6.2 shows two examples of state diagrams of single-input, single-output circuits. The states are represented by circles (vertices) labeled by (state) names, and the vertices are connected by *directed* branches (edges) that indicate the transitions from one state to another. To reflect the difference between Mealy and Moore circuits, we use two different labeling schemes. For a Mealy circuit, we label each branch with the circuit input values that cause the transition *and* the corresponding circuit output values (using a slash to separate them). The state diagram shown in Figure 6.2(a) represents a Mealy circuit. By contrast, for a Moore circuit, we insert the output values into the vertices and use a slash to separate them from the state names. The state diagram of Figure 6.2(b) corresponds to a Moore circuit.

To illustrate how the state diagram depicts the operation of a sequential circuit, let us apply the input sequence 0101 to state a of the Mealy circuit in Figure 6.2(a). The first input value (0) in the sequence is applied when the circuit is in state a and, in response, a 0 output is generated and the circuit makes a transition to state b. (Hence, state b is the resulting next state.) In this state, the second input (1) in the sequence causes a 0 output and the circuit goes to (the next) state c. The third input (0) applied to state c generates a 0 output and the circuit goes to state d. The fourth input (1) applied to state d generates a 1 output and the circuit

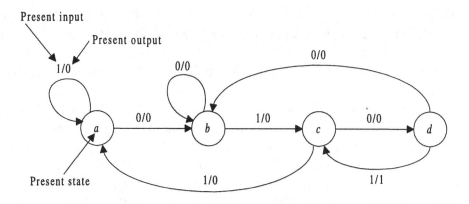

(a) Mealy state diagram

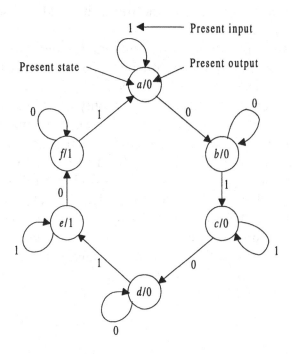

(b) Moore state diagram

■ **Figure 6.2** Examples of state diagrams

makes a transition to state c. Thus, in response to the input sequence 0101 applied to state a, the output sequence 0001 is generated, and the *final* state of the circuit is state c. Note that if the same input sequence is applied to a different *initial* state, then a different output sequence may be produced.

If we now apply the same input sequence (0101) to state a of the Moore circuit in Figure 6.2(b), then the output sequence 0000 will be generated, and the final state of the circuit will be state e. To understand how this is determined, recall that the output in a Moore circuit depends only on the state. Thus, when the first input (0) is applied to state a, the output generated is 0, and the circuit goes to (the next) state b. In state b, the second output (0) is generated, and the second input (1) causes the circuit to go to state c. In this state, the third output (0) is generated, and the third input (0) causes the circuit to make a transition to state d. In state d, the fourth output (0) is generated, and the fourth input (1) causes the circuit to go to the final state e.

Figures 6.2(a) and (b) differ in one other important aspect: The state diagrams represent, respectively, a synchronous and an asynchronous sequential circuit. To understand this distinction, recall that state transitions in an asynchronous circuit can occur only when the circuit inputs are changed. As we can see in Figure 6.2(b), once the circuit reaches any state, it will remain in that state as long as the input value is not changed. (Note that the transition leading from any state to itself is labeled with the same input value of the transition leading to that state.) By contrast, Figure 6.2(a) shows that the circuit can leave a state under the same input value that led to that state. For example, a 0 input applied to state c causes the circuit to go to state d, but, in state d, a subsequent 0 input causes the circuit to go to state b. Since the timing in a synchronous circuit is controlled by a clock, the 0 input leading to d and the 0 input leading out of d are two different events.

State Table

The state table presents in a tabular form the same information contained in the state diagram. Table 6.1 shows the state table for the Mealy circuit of Figure 6.2(a). In general, if the Mealy circuit has k memory devices, then the state table will contain 2^k rows, one for each of the present states

(PS). With n circuit inputs, the next state (NS) portion of the table contains 2^n columns, one for each input combination. The entry at the intersection of each row and column indicates the corresponding next state. The output portion of the table also contains 2^n columns, one for each circuit input combination. The entry at the intersection of each row and column indicates the corresponding output values.

■ **Table 6.1** Mealy state table for Figure 6.2(a)

PS	NS		Output (z)	
	$x = 0$	$x = 1$	$x = 0$	$x = 1$
a	b	a	0	0
b	b	c	0	0
c	d	a	0	0
d	b	c	0	1

■ **Table 6.2** Alternative representation of Table 6.1

PS	NS/Output (z)	
	$x = 0$	$x = 1$
a	$b/0$	$a/0$
b	$b/0$	$c/0$
c	$d/0$	$a/0$
d	$b/0$	$c/1$

Because the outputs in a Mealy circuit are associated with state transitions, the output part of the table can be merged with the next-state portion, resulting in 2^n (rather than a total of 2×2^n) columns. Table 6.2 provides this alternative representation of Table 6.1. In this case, the entry at the intersection of each row and column indicates the next state *and* the output values (using a slash to distinguish between them).

Table 6.3 is the state table for the Moore circuit of Figure 6.2(b). In contrast with a Mealy circuit, the output portion of a Moore state table always contains a *single* column. (The next state part is the same as that for a Mealy circuit.) The entry at the intersection of any row with the output column indicates the output values corresponding to the present state associated with that row.

- **Table 6.3**   Moore state table for Figure 6.2(b)

PS	NS $x = 0$	$x = 1$	Output z
a	b	a	0
b	b	c	0
c	d	c	0
d	d	e	0
e	f	e	1
f	f	a	1

Table 6.4 shows another example of a state table. This table represents an incompletely specified (Mealy) sequential circuit with two inputs (x_1 and x_2) and a single output (z). For example, if the present state (PS) is either a or e, then the corresponding next states and outputs are both undefined when the input combination is $x_1 x_2 = 00$. On the other hand, if PS $= d$, then it is only the output associated with the transition to state a that is unspecified when $x_1 x_2 = 00$.

Timing Sequences and Diagrams

The operation of a sequential circuit can also be described using a **timing sequence** or a **timing diagram**. Both depict the input/output behavior of the circuit along the time axis. They can be intricately detailed by providing information pertaining to the evolution in time of various internal signals in the circuit, or they can be kept concise and show only the behavior of the outputs in response to the circuit inputs.

These two representations may not be useful if the number of possible input sequences is relatively large or if we want to *fully* describe the

■ **Table 6.4** State table of an incompletely specified sequential circuit

PS	NS/Output (z) Input Combination ($x_1 x_2$)			
	00	01	11	10
a	$-/-$	$c/1$	$b/-$	$e/1$
b	$e/0$	$-/-$	$-/-$	$-/-$
c	$f/0$	$f/1$	$-/-$	$-/-$
d	$a/-$	$-/-$	$e/-$	$b/1$
e	$-/-$	$f/0$	$d/1$	$a/0$
f	$c/0$	$-/-$	$c/1$	$b/0$

behavior of the circuit. Often, however, even a partial description is still quite helpful. We use the timing sequence to get an idea of how a circuit is supposed to operate, and we can use the timing diagram when we want to accentuate the timing issues involved in the operation of the circuit.

6.2 MEMORY DEVICES

The simplest device that can provide short-term memory is the delay element shown in Figure 6.3(a). It operates according to the timing diagram of Figure 6.3(b) and is characterized by the equation

$$y_i(t + \Delta T) = Y_i(t)$$

The memory capability of a delay element is due to the fact that it takes a finite time for the signal to propagate through the element.

The idea of a delay element is quite useful in explaining the operation of asynchronous sequential circuits. If we replace the memory devices in Figure 6.1 with delay elements, then when a circuit input changes value, the y variables do not change instantaneously; rather, the signals propagate through the combinational circuit and generate new values for the next-state variables (the Y's). These values then propagate through the delay elements and become the new present state. In the transition period, some

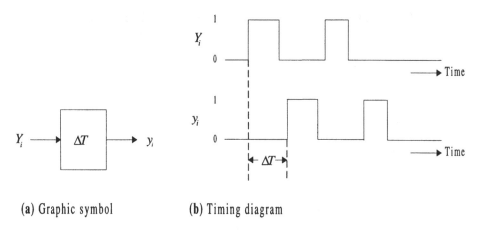

(a) Graphic symbol   (b) Timing diagram

■ **Figure 6.3** Delay element

or all y's differ in value from the corresponding Y's, but after the expiration of the delay, all y's become equal to the corresponding Y's.

Thus, it is only a change in an input variable that causes a transition from one state to another.

In practice, we do not have to actually insert delay elements because the propagation time delays between the inputs and the outputs of the combinational part of the circuit provide sufficient delay across the feedback loops. Besides that, the memory part in sequential circuits is mostly implemented with *bistable* devices. As the name implies, a bistable device has two stable states. It can remain in either one of them indefinitely until directed by an input signal to change state. Usually, the device has two complementary outputs designated by Q and Q' so that the two stable states are $Q = 1\,(Q' = 0)$ and $Q = 0\,(Q' = 1)$. When $Q = 1$, we say that the device is **set** (or is in the **set state**); if $Q = 0$, the device is **reset** (or is in the **reset state**).

We distinguish between two types of bistable devices: **latches** and **flip-flops**. They are similar in many ways, but differ in the method used to change their state. A latch changes state when the input values change. The new output value is delayed *only* by the propagation time delays of the gates between the inputs and outputs of the latch. This property is called the *transparency* property and is common to all latches. In contrast, flip-flops do not have the transparency property. Rather, a flip-flop has a control (triggering) input, called a **clock**, and can change state *only* in response to a *transition* of a clock pulse at this input.

We will use latches to implement the memory part of asynchronous circuits and use flip-flops in synchronous circuits. To introduce latches, we consider only two types in this section: the *SR* latch and the *D* latch. We will identify two other latch types, the *JK* latch and the *T* latch, when we discuss flip-flops in Section 7.2.

SR Latch

The graphic symbol for the *SR* latch, shown in Figure 6.4(a), consists of two inputs, *S* (set) and *R* (reset), and two complementary outputs *Q* and *Q'*. It can be constructed with two cross-coupled NOR gates as shown in Figure 6.4(b). We will see shortly that this latch operates with both inputs *normally* at 0 and requires a logic 1 at the inputs to change its state. For this reason, it is called an **active-HIGH** latch. On the other hand, an **active-LOW** latch operates with both inputs normally at 1 and requires a logic 0 at the inputs to change its state. An active-LOW *SR* latch, designated by the graphic symbol shown in Figure 6.5(a), differs from that of Figure 6.4(a) only in that the *S* and *R* inputs are preceded by small circles. An active-LOW latch can be implemented with two cross-coupled NAND gates as shown in Figure 6.5(b).

The **characteristic table** shown in Figure 6.6(a) details the operation of the active-HIGH latch. The *Q* and *Q⁺* columns designate the present state and the next state, respectively. With $S = R = 0$, the latch maintains its state so that it remains *set* or *reset* according to whether it previously

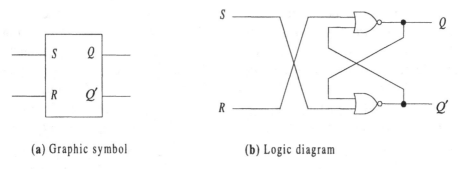

(a) Graphic symbol   (b) Logic diagram

■ **Figure 6.4** Active-HIGH *SR* latch

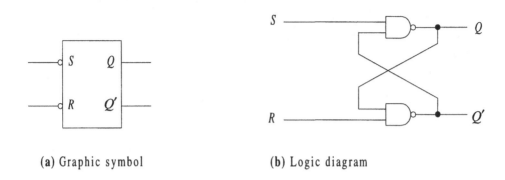

(a) Graphic symbol   (b) Logic diagram

■   **Figure 6.5**   Active-LOW SR latch

was in the set or reset state. When $SR = 10$, the latch is set if it was previously reset or remains set if it was previously set; changing S back to 0 leaves the latch in the set state. If $SR = 01$, the latch is reset if it was previously set or else its state is left unchanged; returning R back to 0 leaves the latch in the reset state. When both S and R are equal to 1, the outputs are no longer complementary (both Q and Q' become 0). Furthermore, by changing S and R back to 00, Q will either stay 0 or change to 1, depending on whether S or R is the first to go to 0. Therefore, with $SR = 11$, the next state of the latch is indeterminate, as indicated in Figure 6.6(a). Hence, to ensure normal operation, S and R should not be 1 simultaneously. This condition implies that the logic expression $S \cdot R$ must always be 0.

The characteristic table can be easily reduced to the equivalent table shown in Figure 6.6(b). Since the latch operates asynchronously, the characteristic table should be interpreted as a sequence table in the sense that the inputs are always returned to 0 after changing them. In deriving either characteristic table, we must allow the signals time to propagate from the output of one NOR gate back to the input of the other NOR gate through the feedback connection. In particular, the S and R inputs should be held constant during that time. To show the feedback connection, we redraw the circuit as shown in Figure 6.6(c). The output Q is identical to the next state variable Y (which we have designated by Q^+) and to the present state variable y at the "output" of the feedback loop.

S	R	Q	Q⁺
0	0	0	0
0	0	1	1
0	1	0	0
0	1	1	0
1	0	0	1
1	0	1	1
1	1	0	Indeterminate
1	1	1	

S	R	Q⁺
0	0	Q
0	1	0
1	0	1
1	1	Indeterminate

(a) Characteristic table   (b) Equivalent characteristic table

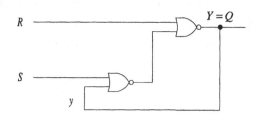

Q	Q⁺	S	R
0	0	0	×
0	1	1	0
1	0	0	1
1	1	×	0

(c) Circuit showing feedback   (d) Excitation table

■ **Figure 6.6** Operation of *SR* latch with NOR gates

We do not have to insert a delay element in the feedback loop because the propagation time delays of the two NOR gates between the input y and the output Y provide the necessary delay.

From Figure 6.6(c), the Boolean expression for the output of the active-HIGH *SR* latch is given by

$$Q^+ = R'Q + R'S$$

We can OR the term SR to this expression because the condition $SR = 0$ must be satisfied and obtain the **characteristic equation**

$$Q^+ = R'Q + S \quad (SR = 0) \tag{6.5a}$$

Notice that the condition $SR = 0$ is part of the characteristic equation of the active-HIGH *SR* latch.

It is a simple matter to show that the characteristic equation of the active-LOW SR latch of Figure 6.5(b) is given by

$$Q^+ = RQ + S' \quad (S'R' = 0) \tag{6.5b}$$

Here the condition to be avoided is that S and R should not be 0 simultaneously. Comparing Equations (6.5a) and (6.5b), we see that the input variables for the NAND latch require the complemented values of those used in the NOR latch.

The characteristic table is useful in the *analysis* of sequential circuits. In this case, the inputs and the present state are known (given), and we can use the table to find the resulting next state. On the other hand, in the *design* of sequential circuits, the present state and the desired next state are specified, and we must establish the values at the latch inputs that will achieve the required state transition. This information, shown in the **excitation table** of Figure 6.6(d), is obtained simply by manipulating the characteristic table.

D Latch

The graphic symbol of the D latch is shown in Figure 6.7(a). The D latch differs from the SR latch in that it has only one input, but can be obtained from the SR latch as shown in Figure 6.7(b). Hence, we can derive the characteristic table and the excitation table provided in Figures 6.7(c) and (d) simply by substituting $S = R'$ into the corresponding tables of the SR latch.

The characteristic equation of the D latch is given by

$$Q^+ = D$$

From this equation we conclude that, after a delay, the output Q follows (latches onto) the input D; when $D = 0$, the latch will reset, and if $D = 1$, the latch will set.

Gated Latches

In many applications, data must be entered into the latch only when some control signal becomes active. For example, Figure 6.8 shows how this can be accomplished with an SR latch. The **enable (E)** input controls the

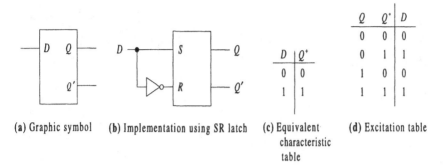

(a) Graphic symbol   (b) Implementation using SR latch   (c) Equivalent   (d) Excitation table
                                                              characteristic
                                                              table

■ **Figure 6.7** D latch

operation of the circuit so that the latch will not change state as long as
$E = 0$. But when $E = 1$, the latch operates normally, and the S and R
inputs can affect its state. Hence, anytime S is 1 and R is 0, a 1 on the E
input sets the latch; anytime S is 0 and R is 1, a 1 on the E input resets
the latch. Since the operation of the latch is, in fact, synchronized with the
enable input, such input is often called a *synchronous input*. A latch that
has a synchronous input is called a **gated latch**.

To ensure the correct operation of a gated latch, it is necessary to
control the relative timing of the inputs. The data must be present at the
latch inputs some time *prior* to activating E. This time interval is called
the **setup time** and is necessary to ensure that the correct data are
recognized by the latch by the time E is activated. In other words, the
previous data inputs must be allowed to propagate through the latch before
E is activated to accept the new input values. Furthermore, the data must
also be held at the latch inputs for a period of time, called the **hold time**,
after E has been activated. This time interval is required to ensure the
continued recognition of the data while the information propagates
through the latch. In other words, the new latch state is allowed to
propagate through the circuit before removing the input signal that
produced that state.

Waveforms illustrating the gated SR latch timing are shown in Figure
6.8(c). It is assumed that $Q = 0$ initially. We have also included in the
timing diagram the propagation time delays associated with the time taken
for a change in the enable input to affect the Q output of the latch. These
delays are designated as t_{on} and t_{off} and represent the times taken by Q to
change from 0 to 1 and from 1 to 0, respectively (see Section 1.2).

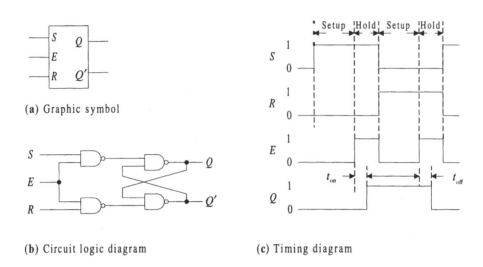

(a) Graphic symbol

(b) Circuit logic diagram

(c) Timing diagram

■ **Figure 6.8** Gated *SR* latch

6.3 ANALYSIS

The analysis of asynchronous sequential circuits begins with a circuit logic diagram or a set of next-state and output equations and terminates when we obtain a state table (or a state diagram) describing the sequence of states and outputs as a function of changes in the input variables. However, unlike combinational circuits, the analysis of sequential circuits is more complex due to the presence of feedback loops. We illustrate this by means of two examples. The first example introduces some of the basic concepts associated with the operation of asynchronous circuits. The second example, presented in the following section, carries the analysis procedure to its conclusion, but also illustrates some of the pitfalls encountered with asynchronous circuits.

Let us consider the 2-input circuit of Figure 6.9(a). The diagram shows a single feedback loop so that the circuit has one present state variable (y) and one next state variable (Y). The delay associated with the feedback loop is obtained from the propagation time delays of the gates between the y input and the Y output. From the circuit diagram, we obtain readily the Boolean expression for the next-state variable as a function of the present state and the input variables:

$$Y = x_1 x_2 + x_2' y$$

We can also represent this expression in a tabular form, called the **next-state table**, as shown in Figure 6.9(b).

To determine how the circuit operates, assume that $x_1 x_2 = 00$ and $y = 1$. Using these values in the circuit diagram, we can verify that $Y = 1$. If we now change the inputs to $x_1 x_2 = 01$, y is still 1, but Y changes to 0 after some delay (the propagation time delay through the NAND gates). In other words, we have come upon an *inconsistency* in the operation of the circuit since $Y \neq y$. But this inconsistency is short-lived because $Y = 0$ *forces* y to 0, and the circuit is "locked in" with $Y = y$.

Let us summarize the lessons learned from this example. For a given binary combination at the circuit inputs, the circuit is in a **stable state** if and only if $y_i = Y_i$ for every i. When we change the value of one or more circuit inputs, the combinational part of the circuit may or may not produce a new set of values for the next-state variables. If it does, then the next state will differ from the present state ($Y_i \neq y_i$ for some i), and the circuit enters an **unstable state**. If the circuit inputs are kept at their present values, the y's will become equal to the Y's, and the circuit will be in the *next* stable state. Therefore, for proper operation, we must ensure that the circuit inputs are allowed to change *only* when the circuit is in a stable state.

We can detect the presence of stable and unstable states directly from the next-state table. (As we will see later, they can also be detected in the state table.) For a state to be stable, the value of Y must be the same as that of y. We indicate a stable condition by circling those entries in the table where $Y = y$. An uncircled entry represents an unstable state. Figure 6.9(c) shows the next-state table with stable states indicated.

Let us consider another problem that arises with the operation of asynchronous circuits. Assume that $x_1 x_2 = 00$ and $y = 0$ in the circuit of Figure 6.9. Now suppose that we change the input values to $x_1 x_2 = 11$. From Figure 6.9(c), we see that the circuit should reach the stable state $Y = y = 1$ under this input combination. However, since we can never ensure that both variables will change *simultaneously*, they may first change from 00 to 01 or from 00 to 10 before reaching their final values. These intermediate values will be obtained depending on whether x_2

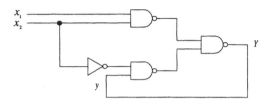

(**a**) Logic diagram

	NS (Y)			
PS	Input Combination $(x_1 x_2)$			
y	00	01	11	10
0	0	0	1	0
1	1	0	1	1

(**b**) Next-state table

	NS (Y)			
PS	Input Combination $(x_1 x_2)$			
y	00	01	11	10
0	⓪	⓪	1	⓪
1	①	0	①	①

(**c**) Next-state table with stable
states indicated

■ **Figure 6.9**   Example of asynchronous sequential circuit

changes slightly *before* or *after* x_1. In either case, we see in Figure 6.9(c) that the circuit will "lock in" the stable state $Y = y = 0$, which is not the state intended. In other words, the intermediate input values may cause an indeterminate circuit operation.

Therefore, for proper operation, we must ensure that *the circuit inputs change only one at a time and only when the circuit is in a stable state.* Asynchronous sequential circuits operating under these constraints are called **fundamental-mode circuits**.

Analysis Example

To carry the analysis procedure to its conclusion, let us introduce another example. Consider a fundamental-mode asynchronous circuit with two inputs (x_1 and x_2) and a single output (z) represented by the following next-state and output equations:

$$Y_1 = y_2 x_1 x_2' + y_2 x_1' x_2 + y_2 y_3 x_1' + y_1 y_2 + y_1 y_3 \tag{6.6a}$$

$$Y_2 = y_1' y_3' x_1' x_2 + y_1 y_2 + y_2 x_1' + y_2 x_2' + y_1 y_3 + y_1' y_3 x_1 + y_1' y_3 x_2' \tag{6.6b}$$

$$Y_3 = y_1 y_2 x_2' + y_1' y_2' x_2 + y_1' x_1 x_2' + y_1 y_3 + y_3 x_1' \tag{6.6c}$$

and

$$z = y_1 x_1 + y_2' x_2' \tag{6.7}$$

Since the circuit is quite complex, we chose to omit its logic diagram.

Next-State and Output Tables

The *internal* behavior of the circuit is represented by the **next-state table**, which specifies the next state $(NS = Y_1 Y_2 Y_3)$ as a function of the present state $(PS = y_1 y_2 y_3)$ and the circuit inputs (x_1 and x_2). The next-state table, shown in Table 6.5, is obtained directly from the next-state equations. As in Figure 6.9(c), those entries in the table where $PS = NS$ are circled to indicate a stable condition, while uncircled entries represent unstable states where $PS \neq NS$.

On the other hand, the *external* behavior of the circuit is represented by an **output table** that specifies the outputs as a function of the present state and the circuit inputs. Shown in Table 6.6, the output table is obtained directly from Equation (6.7).

Race Conditions and Stability Considerations

To consider the internal behavior of the circuit, assume that the present state is $y_1 y_2 y_3 = 000$ and the input combination is changed to $x_1 x_2 = 01$. Table 6.5 indicates that the next state should be $Y_1 Y_2 Y_3 = 011$. However, since the specified state transition requires that two state variables change *simultaneously*, the timing of these changes may become an important factor if we consider the following three possibilities:

1. Y_3 is the first to change, leading to $y_1 y_2 y_3 = 001$, which is stable under $x_1 x_2 = 01$.

2. Y_2 changes first, leading to $y_1 y_2 y_3 = 010$, which is unstable under $x_1 x_2 = 01$. In turn, $PS = 010$ leads to $NS = 110$, which is a temporary unstable condition that leads to the stable state 110.

3. Both Y_2 and Y_3 change at the same time, leading to the unstable condition $y_1 y_2 y_3 = 011$, which, in turn, leads to the stable state $y_1 y_2 y_3 = 111$.

■ **Table 6.5** Next-state table

PS $y_1y_2y_3$	NS $(Y_1Y_2Y_3)$ Input Combination (x_1x_2)			
	00	01	11	10
000	000	011	001	001
001	011	001	011	011
011	111	111	010	111
010	010	110	000	111
110	111	110	110	111
111	111	111	111	111
101	111	111	111	111
100	000	000	000	000

■ **Table 6.6** Output table

PS $y_1y_2y_3$	Output (z) Input Combination (x_1x_2)			
	00	01	11	10
000	1	0	0	1
001	1	0	0	1
011	0	0	0	0
010	0	0	0	0
110	0	0	1	1
111	0	0	1	1
101	1	0	1	1
100	1	0	1	1

As we can see, the resulting next state depends on the *relative* propagation time delays in the combinational part of the circuit that generates the Y variables.

We say that a **race** condition exists in an asynchronous sequential circuit when two or more state variables change value in response to a change in an input variable. If the circuit reaches two or more different stable states depending on the order in which the state variables change, the race is said to be **critical**. Hence, there is a critical race condition in our circuit when $PS = 000$ and the inputs are changed to $x_1x_2 = 01$.

In some cases, there may be race conditions whose outcome does not depend on the order in which the state variables change value (the relative propagation time delays). Such races are called **noncricital** races. For example, if $PS = 010$ and $x_1x_2 = 10$, the next state indicated in Table 6.5 should be 111. We can easily verify that, regardless of whether Y_1 changes first, Y_3 is the first to change, or both change at the same time, the next state will be the specified stable state 111.

There is yet another possibility, known as **buzzer**, that leads to a totally *unstable* circuit operation. In this case, the circuit responds by going from one unstable state to another without ever reaching a stable state. For example, if $PS = 001$ and $x_1x_2 = 11$, the circuit enters a sequence of unstable states, 011, 010, 000, 001, 011, ..., and continuously repeats it. The same buzzer can be entered also if $PS = 100$ and $x_1x_2 = 11$.

Clearly, critical races and buzzers must be avoided in designing asynchronous circuits. Noncritical races, on the other hand, do not affect the operation of the circuit and can be left unattended.

Flow Table

We can combine the next-state and output tables to form the **transition table**. To get the state table from the transition table, we use symbols, such as letters or numerals, to designate the *stable* states without making specific reference to their binary values. In the context of asynchronous sequential circuits, a state table is commonly referred to as a **flow table**.

The flow table shows only the transitions from one stable state to another. Therefore, any row in the transition table without a stable state need not be represented in the flow table. However, if a row contains more than one stable state, we can designate all of them by the same letter, because they are distinguished by their corresponding input combination. For a Mealy circuit, the entries in each row can be associated with different output values, because the outputs are functions of both the present state and the input combination. On the other hand, for a Moore circuit, *all* the entries in each row must be associated with the same output values, because the outputs are functions of the present state only.

The flow table for our example is shown in Table 6.7. To derive this table, we use the following letter assignments to designate stable states:

$$a = 000$$
$$b = 001$$
$$c = 010$$
$$d = 110$$
$$e = 111$$

Note that rows 011, 101, and 100 in the next-state table (also in the transition table) are not represented in the flow table because they contain no stable states.

To obtain a Moore flow table for our circuit, we have to split each row in the transition table containing stable states with different outputs into one or more rows. For our circuit, we can do so using the following letter assignments to designate the stable states:

$$A = 000 \text{ under } x_1 x_2 = 00$$
$$B = 001 \text{ under } x_1 x_2 = 01$$
$$C = 010 \text{ under } x_1 x_2 = 00$$
$$D = 110 \text{ under } x_1 x_2 = 01$$
$$E = 110 \text{ under } x_1 x_2 = 11$$
$$F = 111 \text{ under } x_1 x_2 = 00 \text{ and under } x_1 x_2 = 01$$
$$G = 111 \text{ under } x_1 x_2 = 11 \text{ and under } x_1 x_2 = 10$$

■ **Table 6.7**   Mealy flow table for the analysis example

	NS				Output (z)			
	Input Combination $(x_1 x_2)$				Input Combination $(x_1 x_2)$			
PS	00	01	11	10	00	01	11	10
a	ⓐ	b, d, e	–	e	1	0	–	1
b	e	ⓑ	Buzzer	–	1	0	–	–
c	ⓒ	d	–	e	0	0	–	0
d	e	ⓓ	ⓓ	e	0	0	1	1
e	ⓔ	ⓔ	ⓔ	ⓔ	0	0	1	1

The resulting Moore flow table is shown in Table 6.8, and we encourage the reader to supply the details involved in obtaining it.

Since we are dealing with a fundamental-mode circuit, we must interpret properly the information provided by the flow table. For example, consider in Table 6.8 stable state G under $x_1 x_2 = 11$. We can change the input combination to 10 or 01 but *not* to 00.

■ **Table 6.8** Moore flow table for the analysis example

	NS				
	Input Combination $(x_1 x_2)$				
PS	00	01	11	10	Output (z)
A	Ⓐ	B, D, F	—	G	1
B	F	Ⓑ	Buzzer	—	0
C	Ⓒ	D	—	G	0
D	F	Ⓓ	E	—	0
E	—	D	Ⓔ	G	1
F	Ⓕ	Ⓕ	G	G	0
G	F	F	Ⓖ	Ⓖ	1

State Diagram

The state diagram shows only the transitions between stable states and does not specify the unstable states that are passed along the way. The Mealy state diagram for our analysis example, shown in Figure 6.10, contains five vertices corresponding to the five rows in Table 6.7. A state like *a* or *c* is called an *initial state* because, once it leaves such a state, the circuit can never return to it (as indicated by the outgoing arrows). A state like *e* is referred to as a *terminal state* because, once it enters such a state, the circuit cannot leave it (as indicated by the incoming arrows). The dashed branches represent the races shown in the flow table, but are usually omitted from the state diagram. We leave to the reader the task of obtaining the Moore state diagram from Table 6.8.

Primitive Flow Tables

We will see in Section 6.6 that design problem are often formulated in terms of a verbal description of the desired circuit performance. To design the circuit, we have to transform the verbal statement into a precise

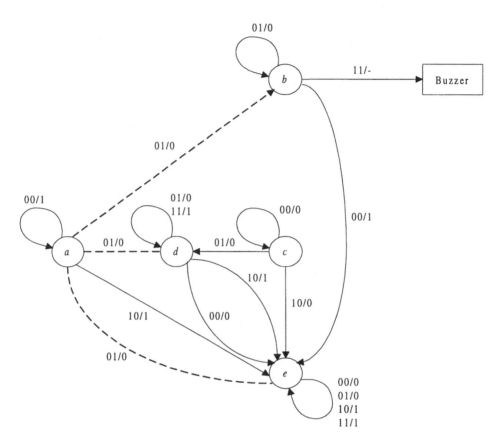

■ **Figure 6.10** Mealy state diagram for the analysis example

description that specifies the circuit operation for every applicable input sequence. This process entails the derivation of a flow table, which, as we have just seen, is also the last step in the analysis procedure.

In most cases, the initial flow table can be represented conveniently in the form of a **primitive flow table**. Such a table is characterized by having only *one* stable state in each row, as shown in Table 6.9. We interpret a primitive flow table and a flow table in the same manner: The circled entries designate stable states, while uncircled entries represent unstable states and indicate the stable states into which they lead. The unspecified next-state entries result either from the fundamental-mode assumption or because some input combinations are not allowed.

The specification of the outputs in a primitive flow table needs to be considered carefully due to the distinction between stable and unstable states. The outputs are of interest only when the circuit is in a stable state, although some output values may not be of interest and can be left unspecified. For example, stable state j in Table 6.9 has an unspecified output. On the other hand, the outputs associated with unstable states are specified or left unspecified according to the following situations:

1. If we start with a Moore primitive flow table and want to implement a Moore circuit, the outputs of the unstable states in each row *must* be made equal to those associated with the row's stable state. For example, row a in Table 6.9 would read in this case,

$$\boxed{a\,/\,00}\quad b\,/\,00\quad -\,/\,-\quad e\,/\,00$$

- **Table 6.9** A primitive flow table

PS	NS/Output $(z_1 z_2)$ Input Combination $(x_1 x_2)$			
	00	01	11	10
a	$a\,/\,00$	$b\,/\,-$	$-\,/\,-$	$e\,/\,-$
b	$c\,/\,-$	$b\,/\,10$	$f\,/\,-$	$-\,/\,-$
c	$c\,/\,00$	$h\,/\,-$	$-\,/\,-$	$d\,/\,-$
d	$a\,/\,-$	$-\,/\,-$	$f\,/\,-$	$d\,/\,11$
e	$g\,/\,-$	$-\,/\,-$	$f\,/\,-$	$e\,/\,11$
f	$-\,/\,-$	$h\,/\,-$	$f\,/\,01$	$e\,/\,-$
g	$g\,/\,00$	$h\,/\,-$	$-\,/\,-$	$e\,/\,-$
h	$c\,/\,-$	$h\,/\,10$	$i\,/\,-$	$-\,/\,-$
i	$-\,/\,-$	$j\,/\,-$	$i\,/\,01$	$e\,/\,-$
j	$a\,/\,-$	$j\,/\,-$	$i\,/\,-$	$-\,/\,-$

2. If we start with a Mealy primitive flow table and want to implement a Mealy circuit, the outputs of the unstable states in each row are left unspecified initially. (We will show in Section 6.6 how to specify them at some stage in the design process.)

3. Sometimes it may be preferable to implement a Mealy circuit, even though the design specifications lend themselves more readily to a Moore model. In such cases, the outputs of the unstable states are left unspecified initially.

6.4 STATE REDUCTION

In transforming the design specifications into a state table, often we define states that carry information already associated with some other states. Such states are referred to as **redundant states**. In general, we should not be too concerned about adding states that may turn out to be redundant. We need only include enough states so that the state table will represent the circuit completely for every allowable input sequence. We can then use the **state reduction** process to establish whether the initially derived state table contains redundant information and, if it does, to eliminate the redundancies.

The idea behind state reduction is quite simple: If the given state table has redundant states, then eliminating redundancies enables us to derive a **reduced** state table **equivalent** to the original one. By reduced we mean that the resulting state table contains fewer rows than the original one, and by equivalent we mean that both tables completely characterize the behavior of the sequential circuit. There are two main reasons for finding and eliminating redundant states:

1. The number of states is related to the number of memory devices used to implement the circuit; therefore, eliminating redundant states *may* result in a circuit with a *smaller* number of memory devices.
2. Even if the number of memory devices is not reduced, the combinational part of the circuit derived from the reduced state table will *always* be less complex than the one derived from the original state table.

We begin our discussion of state reduction with completely specified tables. Then we show that this procedure can be applied to fundamental-mode tables (primitive flow tables) even though they are incompletely specified. Finally, we introduce a reduction procedure for incompletely specified state tables.

Reduction of Completely Specified State Tables

The essence of state reduction of completely specified tables is the ability to identify redundant states. A state s_i is redundant if it is **indistinguishable** from another state s_j. To establish whether two states s_i and s_j are indistinguishable, we can perform the following experiment:

1. Start the circuit in s_i, apply to it every possible input sequence, and observe the resulting output sequences.

2. Repeat step 1 with the circuit started in s_j.

□ **Definition 6.1**   If identical output sequences are generated from steps 1 and 2, the two states are indistinguishable. In this case, they are referred to as **equivalent** and we write $s_i = s_j$. Furthermore, if there is at least one input sequence that generates different output sequences in steps 1 and 2, the two states are **distinguishable** and we write $s_i \neq s_j$. □

If a state is redundant and can be eliminated, the resulting state table will have one less row. If more than two states are indistinguishable, the state table can be reduced even further. To establish if several states are equivalent, we extend the previous experiment as follows:

□ **Definition 6.2**   States $s_1, s_2, \ldots, s_r$ are equivalent if and only if, for every possible input sequence, the same output sequence will be generated regardless of whether $s_1, s_2, \ldots,$ or s_r is the initial state.   □

State equivalence is an *equivalence relation* since the following properties can be verified:

1. **Reflexivity:** $s_i = s_i$ for any state.

2. **Symmetry:** If $s_i = s_j$, then $s_j = s_i$.

3. **Transitivity:** If $s_i = s_j$ and $s_j = s_k$, then we are guaranteed without even checking that $s_i = s_k$ (and, therefore, $s_i = s_j = s_k$).

Moreover, the set of states can be partitioned into nonempty disjoint **equivalence classes**. The importance of this partition is that it allows us to identify redundant states. In fact, if each equivalence class contains only a single state, then the circuit has no redundant states. But if one or more equivalence classes contain more than one state, then all we need do is keep any one state from each equivalence class and eliminate all the others.

But this plan is easier described than implemented. The difficulty is that we have to apply *every possible input sequence* in order to determine equivalent states. This may require a large number of input sequences and is impractical in many cases. In contrast, the following result, stated without a proof, provides a more practical solution to this problem.

□ **Theorem 6.1**   Two states s_i and s_j of a completely specified state table are equivalent if and only if, for every input combination, (1) the outputs produced by s_i and s_j are the same, and (2) their next states are equivalent.   □

Note what is now required from us: We have to check the two conditions for every input *combination* and *not* for every input *sequence*. Therefore, our task has been considerably simplified.

Before showing (in Example 6.1) how this result can be applied, let us consider some types of relations encountered in the course of using Theorem 6.1. Assume that two states s_i and s_j have the same output under some input combination, and let their respective next states for that input be s_m and s_n. Further assume that the two conditions of Theorem 6.1 are satisfied for *all* the other input combinations. Hence, if we can show that the pair of states (s_m, s_n) is equivalent, then the pair of states (s_i, s_j) will also be equivalent. In this case, we write

$$s_i = s_j \text{ if } s_m = s_n$$

and say that the pair (s_i, s_j) **implies** the pair (s_m, s_n). Basically, what this accomplishes is to shift the problem from the pair (s_i, s_j) to the pair (s_m, s_n). In this context, a relation such as

$$s_i = s_j \text{ if } s_i = s_j$$

is a *tautology* implying that $s_i = s_j$. Another relation encountered is

$$s_i = s_j \ \text{ if } \ s_m = s_n \ \text{ and } \ s_m = s_n \ \text{ if } \ s_i = s_j$$

which can be shown to yield

$$s_i = s_j \ \text{ and } \ s_m = s_n$$

In other words, the characteristic of equivalent states is such that, if (s_i, s_j) implies (s_m, s_n) and (s_m, s_n) implies (s_i, s_j), then s_i and s_j are equivalent, as well as s_m and s_n.

The following example illustrates the application of Theorem 6.1.

■ **Example 6.1** Consider the state table shown in Table 6.10, representing a single-input (x), single-output (z), completely specified sequential circuit.

■ **Table 6.10** State table for Example 6.1

PS	NS/Output (z)	
	$x = 0$	$x = 1$
a	$c/0$	$b/1$
b	$d/0$	$b/1$
c	$a/1$	$d/0$
d	$b/1$	$c/0$

We see that the present states a and b have the same output for the same input combination. Under $x = 0$, their next states are c and d, respectively, while under $x = 1$ their next states are b. If we can show that the pair (c, d) is equivalent, then the pair (a, b) will be equivalent because they will have the same or equivalent next states. In other words,

$$a = b \ \text{ if } \ c = d \ \text{ and } \ b = b$$

But, since b is obviously equivalent to itself, the problem reduces to

$$a = b \quad \text{if} \quad c = d \tag{6.8}$$

We can continue in this fashion and apply Theorem 6.1 repeatedly to every pair of present states in an orderly manner to obtain the following statements:

$$
\begin{aligned}
&a \neq c \quad \text{(due to different outputs)}\\
&a \neq d \quad \text{(due to different outputs)}\\
&b \neq c \quad \text{(due to different outputs)}\\
&b \neq d \quad \text{(due to different outputs)}\\
&c = d \quad \text{if} \quad a = b \quad \text{and} \quad c = d \tag{6.9}
\end{aligned}
$$

From the relations in (6.8) and (6.9), we conclude that

$$a = b \quad \text{and} \quad c = d$$

Hence, there are two redundant states, and we can obtain the reduced state table containing only two rows as shown in Table 6.11. States b and c have been eliminated, and every occurrence of b has been replaced by a, while every occurrence of c has been replaced by d. ∎

■ **Table 6.11**   Reduced state table for Example 6.1

PS	NS/Output (z)	
	$x = 0$	$x = 1$
a	$d\,/\,0$	$a\,/\,1$
d	$a\,/\,1$	$d\,/\,0$

Note that, due to property 2 of state equivalence, we need only traverse Table 6.10 in one direction. Nevertheless, with a large number of states, the process illustrated in Example 6.1 can become rather tedious. In such cases, state reduction can be accomplished by using the implication table.

The **implication table** is a lower-triangle matrix whose rows are labeled by all the states except the first and whose columns are labeled by all the states except the last. The entry at the intersection of each row and column is marked by an $\times$ if the corresponding states are not equivalent and by a check mark ($\checkmark$) if they are. If equivalence is implied but not yet verified one way or the other, the implied pairs of states are entered. We make successive passes through the implication table to determine if any additional entries should be marked with a cross. In particular, an entry is crossed out if it contains at least one implied pair of states that is not equivalent. This process is repeated until no additional entry can be crossed out. The remaining entries constitute pairs of equivalent states and can be recorded with check marks. When we have finished, the information provided by the implication table indicates those states that cannot be eliminated and those that are pairwise equivalent. The following example illustrates this procedure.

■   **Example 6.2**   The implication table for the state table of Table 6.12 is shown in Figure 6.11. We start with the top square in the left column and proceed down that column, marking each square with a cross ($\times$), a check ($\checkmark$), or with an implied pair of states. We then proceed to the next column to the right and continue in this vein until we have investigated every column of the implication table.

After constructing the implication table, we make successive passes to determine which entries should be crossed out or checked. For example, (e, f) implies (c, d), and (c, d) implies (e, f) and (c, d). Hence, $e = f$ and $c = d$, and we check the corresponding entries. But, then, (f, g) and (e, g) imply (c, d) and, therefore, since $c = d$, $f = g$ and $e = g$, so these entries are also checked. Since $d \neq f$, then $b \neq e$, and this entry is crossed out. Similarly, because $b \neq f$, then $a \neq f$, and we conclude that $d \neq h$.

Continuing in this manner, we can verify Figure 6.11 in its entirety and conclude:

$$c = d, \ \ e = f, \ \ e = g, \ \text{and} \ f = g \hspace{3cm} ■$$

The results obtained from the implication table can be stated in terms

■ **Table 6.12** State table for Example 6.2

PS	NS/Output (z)	
	$x = 0$	$x = 1$
a	$b / 1$	$h / 1$
b	$f / 1$	$d / 1$
c	$d / 0$	$e / 1$
d	$c / 0$	$f / 1$
e	$d / 1$	$c / 1$
f	$c / 1$	$c / 1$
g	$c / 1$	$d / 1$
h	$c / 0$	$a / 1$

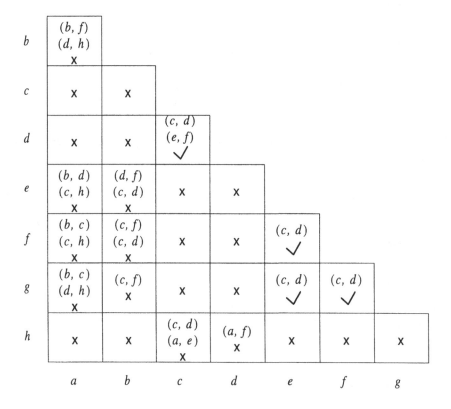

■ **Figure 6.11** Implication table for Example 6.2

of *equivalence classes* of states. Equivalence classes are sets in which *all* states are equivalent. *Distinguishable* states must always be in *different* equivalence classes. Conversely, if states x and y are in distinct equivalence classes, then x and y are distinguishable. Thus, the equivalence classes are *mutually exclusive (disjoint)*. Moreover, the set of equivalence classes **covers** (includes) all the states of the original state table in the sense that every state is in some equivalence class. If an equivalence class contains a single state, then that state is *nonredundant*. Two states are in the same equivalence class if and only if they are equivalent.

The set of equivalent pairs for Example 6.2 is

$$\{a\},\{b\},\{c,d\},\{e,f\},\{e,g\},\{f,g\},\{h\} \tag{6.10}$$

But note that $e = f$, $e = g$, and $f = g$ imply $e = f = g$. Therefore, the pairs $\{e,f\}$, $\{e,g\}$, and $\{f,g\}$ can be combined into a larger set to yield

$$\{a\},\{b\},\{c,d\},\{e,f,g\},\{h\} \tag{6.11}$$

No combination into larger sets is possible because to do so would combine distinguishable states into the same equivalence class. Consequently, the five sets are the equivalence classes for Example 6.2, and Table 6.12 can be reduced to a state table that contains only five rows. We can either assign a new name to each equivalence class in (6.11) or follow the procedure used in Example 6.1 of striking out redundant states. The former approach is demonstrated in the following example.

■ **Example 6.3** To derive the reduced state table for Example 6.2, let us rename the equivalence classes in the set (6.11) as follows:

$$A = \{a\}$$
$$B = \{b\}$$
$$C = \{c,d\}$$
$$D = \{e,f,g\}$$
$$E = \{h\}$$

With these assignments, the reduced state table equivalent of Table

6.12 is shown in Table 6.13.   ■

To determine the set of states representing the separate equivalence classes, we can use a graphical aid called the **implication graph**. Each vertex in the implication graph is labeled with a state name. If two states are equivalent, their vertices are connected by a branch. (Note that this information is obtained from the implication table.) We search the implication graph for **complete (fully connected) subgraphs** not contained within any larger complete subgraph. (A collection of vertices and branches is a fully connected subgraph if *every* pair of vertices is connected by a branch. In this context, an isolated vertex is a fully connected subgraph that contains no branches.) Since the states covered by a complete subgraph are all pairwise equivalent, those complete subgraphs not contained in larger complete subgraphs determine the equivalence classes. Example 6.4 illustrates this procedure with respect to Table 6.12 of Example 6.2.

■ **Table 6.13** Reduced state table equivalent of Table 6.12

	NS/Output (z)	
PS	$x = 0$	$x = 1$
A	$B/1$	$E/1$
B	$D/1$	$C/1$
C	$C/0$	$D/1$
D	$C/1$	$C/1$
E	$C/0$	$A/1$

■ **Example 6.4** The implication graph for Table 6.12, obtained from the implication table in Figure 6.11 [or the set (6.10)], is shown in Figure 6.12. The set of complete subgraphs consists of three isolated vertices, a, b, and h; the single branch (c, d); and the triangle (e, f, g). The correspondence with the set (6.11) is obvious.   ■

Reduction of Flow Tables

Even though a fundamental-mode asynchronous circuit is incompletely specified, we can still reduce a flow table (or a primitive flow table) using the method of the previous section. This calls for a two-step procedure: (1) Eliminate redundant states using the technique established for completely specified state tables, and (2) merge rows. From the first step we obtain a reduced flow table in which all redundant stable states have been removed. In the second step, we "condense" the table by eliminating intermediate unstable states.

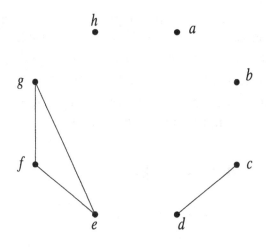

■ **Figure 6.12** Implication graph for Table 6.12

The reduction method for incompletely specified tables, discussed in the following section, can also be applied to fundamental-mode tables. In this case, state reduction is carried out in one step. In general, however, the application of the two-step method outlined here is more straightforward and requires less work to carry out.

Let us discuss the first step of the procedure. Since we are only interested in stable states, any two states that are *candidates* for equivalency must be stable under the same inputs; that is, they must appear in the *same* column of the flow table. To be equivalent, the two states must show the same output values, and their next states must be equivalent under *any* input combination. In other words, the indications in any other entry (next-state and output values) in their respective rows *must not conflict*. Hence, we can apply the state-reduction process of completely specified state tables to flow tables. We illustrate this procedure in the following example.

■ **Example 6.5** Consider the primitive flow table shown in Table 6.9. The only candidates for equivalence are the stable states a, c, and g in column 00; b, h, and j in column 01; f and i in column 11; and d and e in column 10.

The implication table is shown in Figure 6.13. Consider, for example, the intersection of column a and row c. In Table 6.9, we see that stable states a and c have the same outputs; unstable states $\{b,h\}$ and $\{d,e\}$ have unspecified outputs; and both next states and outputs are unspecified under column 11. Hence, $a = c$ if $b = h$ and $d = e$, and we enter the two implied pairs $\{b,h\}$ and $\{d,e\}$ in the implication table.

Note that if row c had looked like

$$\widehat{c / 00} \quad -/- \quad -/- \quad d/-$$

or

$$\widehat{c / 00} \quad h/10 \quad -/- \quad d/-$$

or any other form that deviates from what is shown in Table 6.9, then $a \neq c$. Note also that, when two stable states are equivalent but the output of one of them is unspecified, their combination would result in a stable state whose output is specified.

By making successive passes through the implication table to determine if any additional entries should be crossed out, we can verify that all the implied entries of the table receive a check mark.

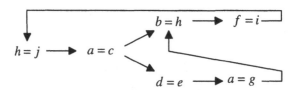

	a	b	c	d	e	f	g	h	i
b	X								
c	(b, h) (d, e) ✓	X							
d	X	X	X						
e	X	X	X	(a, g) ✓					
f	X	X	X	X	X				
g	(b, h) ✓	X	(d, e) ✓	X	X	X			
h	X	(f, i) ✓	X	X	X	X	X		
i	X	X	X	X	X	(h, j) ✓	X	X	
j	X	(a, c) (f, i) ✓	X	X	X	X	X	(a, c) ✓	X

- **Figure 6.13** Implication table for Example 6.5

For example, from the implication table we get for the pair (h, j) (the arrows stand for "implies"):

Hence, $h=j$, $a=c$, $b=h$, $f=i$, $d=e$, and $a=g$.

Thus, the equivalent pairs obtained from the implication table of Figure 6.13 are

$$a = c, \quad a = g, \quad b = h, \quad b = j, \quad c = g, \quad d = e, \quad f = i, \quad h = j$$

and the corresponding implication graph, shown in Figure 6.14, reveals the following set of equivalence classes:

$$\{a,c,g\},\{b,h,j\},\{d,e\},\{f,i\}$$

Therefore, the original 10-row table can be reduced to the flow table shown in Table 6.14, which consists of four rows. ∎

Note that eliminating redundant states in a fundamental-mode state table does not change its format. Table 6.14 shows that, if we start with a primitive flow table, the reduced table will also be primitive.

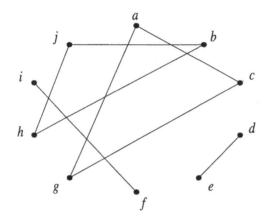

■ **Figure 6.14** Implication graph for Example 6.5

■ **Table 6.14** Reduced primitive flow table for Example 6.5

	NS/Output $(z_1 z_2)$			
	Input Combination $(x_1 x_2)$			
PS	00	01	11	10
A	(A/00)	B/–	–/–	C/–
B	A/–	(B/10)	D/–	–/–
C	A/–	–/–	D/–	(C/11)
D	–/–	B/–	(D/01)	C/–

$$A=\{a,\,c,\,g\};\ B=\{b,\,h,\,j\};\ C=\{d,\,e\};\ D=\{f,\,i\}$$

Merger Process

A reduced flow table is obtained after removing all the redundant stable states. We can, however, condense the table even further by eliminating intermediate unstable states through the **merger process**. This is the second step of our two-step reduction process.

To understand the merger process, consider the first two rows in Table 6.14. Under the input combination $x_1x_2 = 00$, the circuit is in stable state A. If we change the inputs to 01, the circuit will make a transition to stable state B, as indicated by the uncircled entry $B/-$. (Recall that an uncircled entry represents an unstable state labeled similarly to the stable state to which it leads.) Likewise, if we change the inputs to 10 while the circuit is in stable state A, the transition to stable state C will pass through unstable state $C/-$. The input combination 11 is not allowed if the present state (PS) is A. A similar situation occurs when the circuit is in stable state B under $x_1x_2 = 01$. If we change the input combination to either 00 or 11, the circuit will go to either stable state A or D through unstable state $A/-$ or $D/-$, respectively. Of course, the input combination 10 is not allowed if PS $= B$.

Let us now consider the following row:

PS	Inputs 00	01	11	10
(A,B)	A/00	B/10	D/–	C/–

This row conveys the *same* information contained in the first two rows of Table 6.14, except that, now, the transitions through unstable states $A/-$ and $B/-$ have been eliminated. We say that the original rows A and B have been *merged* into the new row and note that, since the new row represents both of them, Table 6.14 can be reduced to three rows. But, following the same process, we can merge rows A and C, A and D, B and C, B and D, and C and D, and then we can merge all four into the single row:

PS	Inputs			
	00	01	11	10
(A, B, C, D)	$A/00$	$B/10$	$D/01$	$C/11$

In other words, Table 6.14 can be actually represented by only a single row. All the information contained in the table is preserved through the merger process, but, instead of having to implement the circuit with two memory devices, we see now that none is required. [Recall that a state table with m rows (states) requires $k \geq \lceil \log_2 m \rceil$ memory devices.] Therefore, the circuit described by Table 6.14 is actually a combinational circuit. Note, however, that we must properly interpret the information provided by the single row representing Table 6.14. We should remember that it pertains to a fundamental-mode circuit and, therefore, transitions are not allowed from stable state A to D under 11 or from B to C under 10.

Let us turn now to the definition of the merger operation.

□ **Definition 6.3**   Any two rows, r_i and r_j, in a flow table can be merged into a single row, (r_i, r_j), if and only if there are no *conflicts* between their entries in any column.

□

The following comments explain our use of the word *conflict* in defining the merger operation:

1. An unspecified entry is always nonconflicting.
2. In merging two rows, an unstable state is merged into the stable state, and an unspecified next-state entry is merged into the specified next-state entry (if one is indeed specified). Obviously, a row can be merged with itself, so the merger operation is *reflexive*.
3. The merger operation is *symmetric*; if r_i can be merged with r_j, then r_j can be merged with r_i.
4. The merger operation need not be *transitive*; that is, (r_i, r_j) and (r_j, r_k) need not imply (r_i, r_k), as the following example shows:

	Inputs			
PS	00	01	11	10
a	ⓐ	b	–	c
b	a	ⓑ	d	–
c	a	–	e	ⓒ

We conclude that b can be merged with a, a can be merged with c, but b cannot be merged with c.

All that we have said in regard to the merger of two rows can be extended to more than two rows.

□ **Definition 6.4** Rows $r_1, r_2, \ldots, r_j$ can be merged into a single row, $(r_1, r_2, \ldots, r_j)$, if and only if each pair of rows r_i and r_k can be merged.

□

Consider a collection of pairwise mergeable rows and call it a **merger group**. (In other words, these rows are related by transitivity.) Then a **maximal merger group** is a merger group for which no other row is mergeable with each row in the group. Hence, to reduce a flow table to a *minimal* number of rows, we must select a set of *mutually exclusive* maximal merger groups that *covers* all the rows of the flow table. Note, however, that there may be more than one set of maximal merger groups for a given flow table, resulting in alternative flow tables, each with the same (minimal) number of rows.

To facilitate the selection of maximal merger groups, we use the **merger graph**. This graph is similar to the implication graph in that each vertex represents a row in the flow table, but any two vertices in the merger graph are connected by a branch if and only if their corresponding rows can be merged. To obtain the maximal merger groups, we search the merger graph for *maximal complete subgraphs*. (A maximal complete subgraph is a complete subgraph not contained in any larger complete subgraph.) The following example illustrates this procedure.

■   **Example 6.6**   Consider the flow table shown in Table 6.15 and let us investigate the following two cases:

(a) If the flow table represents a *Moore circuit*, the output corresponding to any row (stable state) must be associated with *each* entry in that row. The merger graph for this case is shown in Figure 6.15. The set of maximal merger groups obtained from the graph is

$$(a,b,f), (c), (d,e)$$

As can be seen, the merger groups are disjoint and cover all the rows of Table 6.15. We can rename these groups A, B, and C, respectively, and reduce Table 6.15 to the 3-row flow table of Table 6.16.

(b) If the flow table represents a *Mealy circuit*, the outputs are identical to those of the Moore table when the circuit is in any stable state, but the outputs associated with the remaining entries are left unspecified. In contrast to the Moore table, this allows us more degrees of freedom in merging the flow table. The merger graph for this case is shown in Figure 6.16, and the set of maximal merger groups obtained is

$$(a,b,f), (c,d,e)$$

■   **Table 6.15**   Flow table for Example 6.6

	NS				Output (z)
	Input Combination $(x_1 x_2)$				
PS	00	01	11	10	
a	ⓐ	b	—	f	0
b	a	ⓑ	d	—	0
c	ⓒ	b	—	—	0
d	—	b	ⓓ	e	1
e	c	—	d	ⓔ	1
f	a	—	d	ⓕ	0

As can be seen, the merger groups are disjoint and cover all the rows of Table 6.15. Renaming these groups A and B, respectively, we obtain the minimal-row Mealy flow table shown in Table 6.17.

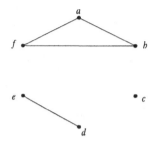

■ **Figure 6.15** Merger graph for Table 6.15 considered as a Moore flow table

■ **Table 6.16** Minimal-row Moore flow table for Example 6.6

PS	NS				Output (z)
	Input Combination $(x_1 x_2)$				
	00	01	11	10	
A	Ⓐ	Ⓐ	C	Ⓐ	0
B	Ⓑ	A	—	—	0
C	B	A	Ⓒ	Ⓒ	1

Note that we could have selected the following merger groups in Figure 6.16:

$$(a, f), (b, d), (c, e)$$

These groups are disjoint and cover all the rows of Table 6.15, but they are *not* maximal. The reduced flow table resulting from this choice is equivalent to Table 6.15 (as well as to table 6.17) but requires two memory devices, as opposed to the single memory device required due to the selection leading to Table 6.17. ■

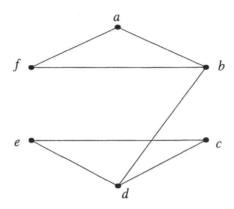

■ **Figure 6.16** Merger graph for Table 6.15 considered as a Mealy flow table

■ **Table 6.17** Minimal-row Mealy flow table for Example 6.6

PS	NS			
	Input Combination $(x_1 x_2)$			
	00	01	11	10
A	$A/0$	$A/0$	$B/-$	$A/0$
B	$B/0$	$A/-$	$B/1$	$B/1$

We did not retain the original letter symbols in Tables 6.16 and 6.17. Instead, we assigned a common letter symbol to all the stable states in each merger group. Clearly, there is no ambiguity in this assignment because each stable state in each row is distinguished by an input combination.

In general, if we being with a flow table representing a Moore circuit, we can apply the merger process to obtain a minimal-row Moore table [as in Example 6.6(a)], or we can first convert the flow table to a Mealy table and then proceed to obtain a minimal-row Mealy table [as in Example 6.6(b)]. As we see from Example 6.6, the latter approach may be preferable since it often leads to a merged flow table with fewer rows.

* Reduction of Incompletely Specified State Tables

In an incompletely specified sequential circuit, some outputs and/or state transitions are not defined. They can be utilized as don't-cares and can be assigned in any way we choose. Once the don't-cares are specified, however, the state table is no longer incompletely specified, and the state-reduction technique of the previous sections can be applied to it. But, then, as the following example demonstrates, don't-care assignment *prior* to state reduction may result in an unnecessarily complex circuit.

■ **Example 6.7** Consider the incompletely specified state table shown in Table 6.18(a). Since some of the outputs are not defined, we can specify them arbitrarily. Two such examples are shown in Tables 6.18(b) and (c). As we see, Table 6.18(b) contains no redundant states, while Table 6.18(c) can be reduced to two rows because there are two equivalence classes $\{a,b\}$, $\{c,d\}$. ■

Instead of specifying the don't-cares before state reduction, we should keep them unspecified for as long as possible and take advantage of the degrees of freedom afforded by them. Consequently, incompletely specified states can be combined to reduce the number of rows in the state table, but they are not equivalent in the sense of Definition 6.1 because of lack of specification in at least one occurrence of the inputs. Rather, two incompletely specified states can be combined if they are compatible.

☐ **Definition 6.5** Two states s_i and s_j of an incompletely specified sequential circuit are **compatible** if and only if, for each possible input sequence applied to both, s_i and s_j produce the same output sequence when the outputs are specified. ☐

The following theorem reduces the task of checking for compatibility.

☐ **Theorem 6.2** Two states s_i and s_j of an incompletely specified sequential circuit are compatible if and only if, for each possible input

---

* This section is optional.

combination, (1) the outputs produced by s_i and s_j are the same when both are specified, and (2) the next states are compatible when both are specified.   □

We can use the implication table to find compatible states in the same manner as it is used to determine equivalent states. The only difference is that now the dashes (which represent the unspecified conditions or don't-cares) have no effect when we compare the rows of the state table. From the implication table, we obtain the set of **compatible pairs**, as illustrated by the following example.

■ **Example 6.8**   Consider the incompletely specified state table originally shown in Table 6.4 and reproduced in Figure 6.17(a) for convenience. The implication table in Figure 6.17(b) is derived using Theorem 6.2. For example, rows a and b are compatible, but rows b and d will be compatible only if a and e are compatible. However, rows a and e are incompatible because they have different outputs under 01 and 10. Continuing in this vein, we obtain the following set of compatible pairs from Figure 6.17(b):

$$\{a,b\}, \{a,c\}, \{a,d\}, \{b,e\}, \{c,f\} \tag{6.12}$$

■

■ **Table 6.18** State tables for Example 6.7

(a)	Original table		(b)	Don't-care assignment 1		(c)	Don't-care assignment 2	
	NS/Output (z)			NS/Output (z)			NS/Output (z)	
PS	$x=0$	$x=1$	PS	$x=0$	$x=1$	PS	$x=0$	$x=1$
a	$c/-$	$b/1$	a	$c/1$	$b/1$	a	$c/0$	$b/1$
b	$d/0$	$b/-$	b	$d/0$	$b/0$	b	$d/0$	$b/1$
c	$a/-$	$d/-$	c	$a/1$	$d/0$	c	$a/1$	$d/0$
d	$b/1$	$c/-$	d	$b/1$	$c/1$	d	$b/1$	$c/0$

As we see, the derivation of the implication table for incompletely specified sequential circuits is similar to that of completely specified tables, except that, here, we have more flexibility due to the don't-care conditions. Nevertheless, by inspecting the set (6.12), we note the fundamental difference between compatibility and equivalence. If (6.12) was a set of equivalence pairs, then, in addition to $\{a,b\}$ and $\{a,c\}$, we would also have obtained $\{b,c\}$ because the equivalence relation is transitive. In contrast, the *compatibility relation* is *not* transitive due to the don't-cares. Indeed, the compatible pairs $\{a,b\}$ and $\{a,c\}$ do not automatically imply the compatibility of b and c.

(a)

PS	NS/Output (z) Input Combination $(x_1 x_2)$			
	00	01	11	10
a	-/-	c/1	b/-	e/1
b	e/0	-/-	-/-	-/-
c	f/0	f/1	-/-	-/-
d	a/-	-/-	e/-	b/1
e	-/-	f/0	d/1	a/0
f	c/0	-/-	c/1	b/0

(b)

	a	b	c	d	e
b	✓				
c	(c,f) ✓	(f,e) x			
d	(b,e) ✓	(a,e) x	(a,f) x		
e	x	✓	x	x	
f	x	(c,e) x	(c,f) ✓	x	(a,b) (c,d) x

■ **Figure 6.17** State and implication tables for Example 6.8

A set of states where any two states are compatible is called a **compatibility class**. Then, a **maximal compatible** is a compatibility class that will not remain a compatibility class if any state not in the class is added to it. In particular, the compatibility relation is transitive in a compatibility class, but in a maximal compatible, the addition of any other state renders the transitive law invalid on the larger set.

Our next step, therefore, is to establish the set of maximal compatibles. This can be done using the *implication graph* and searching for maximal complete subgraphs, as we did for completely specified tables. However, in contrast with the set of equivalence classes, which consists of disjoint blocks, two distinct maximal compatibles may have *common* states. Hence, we are faced with the problem of determining which subset (if any) of the set of maximal compatibles to choose in order to accomplish the state-reduction process. We will see shortly that, to answer this question, we also need to generate the set of **maximal incompatibles**, obtained by forming the complementary graph of the implication graph. This new graph consists of the same vertices as the implication graph, but now a branch connects two vertices if and only if their corresponding states are incompatible; that is, if and only if there is no branch between them in the implication graph. To generate the set of maximal incompatibles, we search the complementary graph for maximal complete subgraphs. The following examples illustrate the procedure for generating maximal compatibles and maximal incompatibles.

■ **Example 6.9**　Figure 6.18(a) shows the implication graph obtained from the implication table of Figure 6.17(b). As can be seen, the set of maximal compatibles is given by

$$\{a,b\}, \{a,c\}, \{a,d\}, \{b,e\}, \{c,f\} \tag{6.13}$$

and replicates in this case the set (6.12).

The complementary graph is shown in Figure 6.18(b). Two vertices are connected by a branch if and only if there is no branch connecting them in Figure 6.18(a). Hence, the set of maximal incompatibles is given by

$$\{a,e,f\}, \{d,e,f\}, \{c,d,e\}, \{b,c,d\}, \{b,d,f\} \tag{6.14}$$

Note that neither the set (6.13) nor the set (6.14) is pairwise mutually exclusive (disjoint).　　　　　　　　　　　　　　　　　　　　　　■

■ **Example 6.10**　Consider the implication graph shown in Figure 6.19(a). Its complementary graph (the graph of incompatible states) has been generated from it and is shown in Figure 6.19(b). The resulting set of maximal compatibles is

$$\{a,e,g,h\}, \{c,e,g\}, \{b,c,g\}, \{c,d,g\}, \{c,f,g\} \tag{6.15}$$

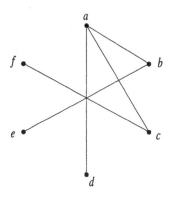

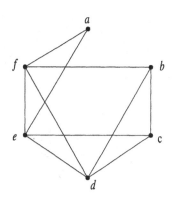

(a) Compatible classes   (b) Incompatible classes

■ **Figure 6.18** Implication graphs for Figure 6.17(b)

and the set of maximal incompatibles is given by

$$\{a,b,d,f\}, \{b,d,e,f\}, \{b,d,f,h\}, \{a,c\}, \{c,h\}, \{g\} \qquad (6.16)$$

■

In general, a subset of maximal compatibles that results in a minimum-row state table must satisfy the following three conditions:

1. *Completeness.* The union of the selected maximal compatibles must contain (cover) all the states of the original state table.
2. *Consistency.* For every input combination, next states of each maximal compatible must be contained within maximal compatibles that belong to the selected subset.
3. *Minimality.* The selected subset should contain the smallest number of maximal compatibles that satisfy conditions 1 and 2.

If the selected subset satisfies these conditions, then each maximal compatible will correspond to a state in the reduced state table.

We can estimate the number of rows in the reduced state table. Let n be the number of states in the original state table and let m be the number of maximal compatibles. Furthermore, let q_i denote the number of states in the ith maximal incompatible. Then, it can be shown that the number of states (k) of the reduced state table is bounded by

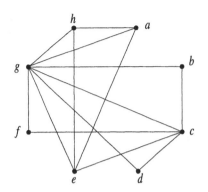

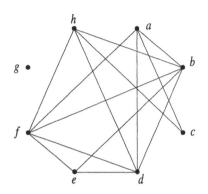

(a) State compatibilities   (b) State incompatibilities

■ **Figure 6.19** Implication graphs for Example 6.10

$$L \le k \le H \tag{6.17}$$

where the upper bound H and the lower bound L are defined by

$$H = \text{minimum}(n, m) \tag{6.18}$$

and

$$L = \text{maximum}(q_1, q_2, \ldots, q_i, \ldots) \tag{6.19}$$

■ **Example 6.11 (a)** In Example 6.9, the number of states in the original state table [Figure 6.17(a)] is $n = 6$ and the number of maximal compatibles [the set (6.13)] is $m = 5$; hence, $H = 5$. Each maximal incompatible in the set (6.14) contains three states; hence, $q_i = 3$ for every i, and $L = 3$. Therefore, the number of states of the reduced state table will be at least 3 and no more than 5.

(b) For Example 6.10, $n = 8$ (see Figure 6.19) and $m = 5$ from the set (6.15); hence, $H = 5$. From the set (6.16) of maximal incompatibles, we obtain

$$L = \text{maximum}(4, 4, 4, 2, 2, 1) = 4$$

Therefore, the reduced state table for this circuit will contain at least four states but no more than five.   ■

Having determined L and H, we begin the search for a minimal number of states with the lower bound. We attempt to establish if there are L maximal compatibles that satisfy the conditions of completeness and consistency. If we fail to find them, we then determine whether there are $L+1$ maximal compatibles that satisfy these conditions. We proceed in this manner by trial and error, knowing that at most H states will be required in the reduced state table.

The selection of a subset of maximal compatibles that satisfies the completeness condition is quite simple and can be done usually by inspecting the set of maximal compatibles. However, to determine whether the selected subset satisfies the consistency requirement, we have to construct a **closure table**. The closure table is similar to a state table except that the maximal compatibles are treated as states, and we find their sets of next states using the original state table. Having constructed the closure table, we then consider the next states of each maximal compatible. If for every input combination the next states are contained within maximal compatibles that belong to the selected subset, then the selected subset is consistent.

The following examples demonstrate this procedure. Example 6.12 is a follow-up on Examples 6.8, 6.9, and 6.11 and shows the derivation of a reduced state table equivalent to the table in Figure 6.17(a). Example 6.13 illustrates the application of the reduction procedure outlined in this section to fundamental-mode flow tables. We will see that a minimal-row flow table can be obtained using a one-step procedure, rather than the two-step procedure (finding equivalent states and then merging rows) outlined in the previous section.

■ **Example 6.12** Recall Example 6.11(a), where we established that $L=3$ for the set (6.13). Hence, the subset $\{a,d\}$, $\{b,e\}$, $\{c,f\}$ is the only one containing three maximal compatibles that satisfies the completeness condition. To test for consistency, we construct the closure table for this subset, shown in Table 6.19. For example, consider the compatible $\{c,f\}$. From Figure 6.17(a), we see that c and f go to f and c under 00, to f under 01, to b under 10, and to c under 11. This information is entered in the closure table in row $\{c,f\}$.

Inspecting the closure table, we conclude that, for every input combination, the next states of each maximal compatible are contained within maximal compatibles that belong to the selected subset. Therefore, the selected subset is consistent. Accordingly, the reduced state table, shown in Table 6.20, contains just three rows. We have renamed the selected maximal compatibles $A = \{a,d\}$, $B = \{b,e\}$, and $C = \{c,f\}$ and obtained the corresponding outputs from the original state table of Figure 6.17(a). Notice that, in general, the reduced state table may still contain unspecified next states and outputs.   ■

■ **Table 6.19** Closure table for Example 6.12

Input Combination $(x_1 x_2)$

PS	00	01	11	10
$\{a,\,d\}$	a	c	b,e	e,b
$\{b,\,e\}$	e	f	d	a
$\{c,f\}$	c,f	f	c	b

■ **Example 6.13** Consider the (primitive) flow table shown in Table 6.21. For the purpose of state reduction, we treat it as we would any incompletely specified state table by disregarding the distinction between circled (stable) and uncircled (unstable) states.

■ **Table 6.20** Reduced state table for Figure 6.17(a)

NS/Output (z)

Input Combination $(x_1 x_2)$

PS	00	01	11	10
A	$A/-$	$C/1$	$B/-$	$B/1$
B	$B/0$	$C/0$	$A/1$	$A/0$
C	$C/0$	$C/1$	$C/1$	$B/0$

■ **Table 6.21** Flow table for Example 6.13

	NS/Output (z)			
	Input Combination $(x_1 x_2)$			
PS	00	01	11	10
a	⟨a/0⟩	c/-	-/-	b/-
b	a/-	-/-	d/-	⟨b/0⟩
c	a/-	⟨c/0⟩	e/-	-/-
d	-/-	f/-	⟨d/0⟩	b/-
e	-/-	g/-	⟨e/1⟩	b/-
f	a/-	⟨f/1⟩	h/-	-/-
g	a/-	⟨g/1⟩	e/-	-/-
h	-/-	f/-	⟨h/1⟩	b/-

The set of compatible pairs derived from the implication table in Figure 6.20 is used to construct the implication graph shown in Figure 6.21(a). Its complementary graph is shown in Figure 6.21(b). The set of maximal compatibles determined from Figure 6.21(a) is given by

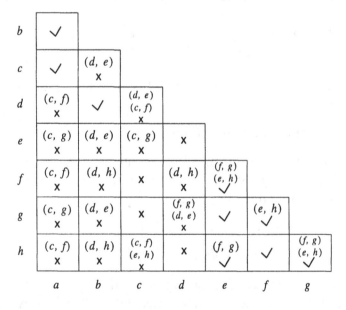

■ **Figure 6.20** Implication table for Example 6.13

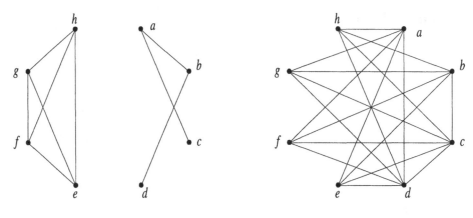

(a) Compatible classes   (b) Incompatible classes

■ **Figure 6.21** Implication graphs for Example 6.13

$$\{a,b\}, \{a,c\}, \{b,d\}, \{e,f,g,h\} \tag{6.20}$$

and the set of maximal incompatibles is obtained from Figure 6.21(b):

$$\{a,d,h\}, \{a,d,g\}, \{a,d,f\}, \{a,d,e\}, \{b,c,h\}, \{b,c,g\}$$
$$\{b,c,f\}, \{b,c,e\}, \{c,d,h\}, \{c,d,g\}, \{c,d,f\}, \{c,d,e\} \tag{6.21}$$

Hence, the upper bound on the number of states [Equation (6.18)] is given by

$$H = \text{minimum}(8, 4) = 4$$

Since each maximal incompatible contains three states, the lower bound [Equation (6.19)] is $L = 3$. Therefore, the reduced flow table will contain either three or four rows.

The subset of three maximal compatibles

$$\{a,c\}, \{b,d\}, \{e,f,g,h\}$$

satisfies the completeness condition, and we can verify that it is consistent by constructing a closure table. The 3-row flow table corresponding to this subset is shown in Table 6.22. ■

■ **Table 6.22**   Minimal row flow table equivalent of Table 6.21

| | NS/Output (z) | | | |
| | Input Combination $(x_1 x_2)$ | | | |
PS	00	01	11	10
A	Ⓐ/0	Ⓐ/0	C/-	B/-
B	A/-	C/-	Ⓑ/0	Ⓑ/0
C	A/-	Ⓒ/1	Ⓒ/1	B/-

We ask the reader to obtain Table 6.22 using the two-step procedure discussed in the previous section. It should be apparent that using the two-step method to reduce fundamental-mode tables requires less work than the one-step technique discussed in this section.

6.5 STATE ASSIGNMENT

The state tables considered thus far have been specified in a symbolic form. However, if we wish to implement an actual circuit, the next logical step must be to encode the state table into binary notation, using a process known as **state assignment**. The result is a *transition table* that combines the *next-state table* and the *output table* (see Section 6.3).

The state-assignment process associates a *unique* binary code with each state (row in the state table). If we have N state variables (memory devices), then the total number of distinct binary codes is 2^N. Therefore, if the state table has n rows, N must be such that $2^N \geq n$ (or $N \geq \lceil \log_2 n \rceil$). It turns out that, for a synchronous state table, this condition is necessary and sufficient. As long as each state is uniquely coded, the state assignment can be made *arbitrarily*. On the other hand, state assignment for an asynchronous flow table often has to satisfy additional constraints in order to avoid critical races, so $\lceil \log_2 n \rceil$ memory devices may not be sufficient.

The number of possible state assignments is quite large, even with a small number of states. With n rows (states) and N state variables ($N \geq \lceil \log_2 n \rceil$), there are

$$P = \frac{2^N!}{(2^N - n)!}$$

possible ways of assigning 2^N binary codes to the n rows. For example, if $n = 4$ and $N = 2$, then $P = 24$; but with $n = 5$ and $N = 3$, $P = 6720$.

The complexity of the combinational part of the circuit usually depends on the particular assignment. Two different assignments may result in hardware implementations that are vastly different. Likewise, two state assignments are considered *equivalent* if they result in circuit implementations of identical complexity. Basically, two state assignments A_1 and A_2 are equivalent if each variable of A_1 is either identical to, or the complement of, a variable of A_2. For example, the three 4-state binary assignments shown in Table 6.23 are equivalent: A_2 is obtained from A_1 by complementing the variables, and A_3 is derived from A_1 by interchanging the variables. Consequently, it can be shown that both assignments determine circuits of identical complexity.

If we discount the equivalent state assignments, then the number of *different* state assignments is given by

$$R = \frac{(2^N - 1)!}{(2^N - n)! N!}$$

For example, if $n = 4$ and $N = 2$, then $R = 3$; but with $n = 5$ and $N = 3$, $R = 140$. This is a considerable reduction in number, but still poses a formidable problem if we want to choose an appropriate state assignment for a given design problem so that the resulting circuit implementation would be the least complex. Unfortunately, there is no simple and efficient method for doing this. A number of guidelines exist, but discussing them is beyond the scope of this book. Instead, our strategy will be to select an arbitrary state assignment for synchronous state tables and a race-free (but otherwise arbitrary) state assignment for asynchronous flow tables. We will consider the former in Chapter 7 and discuss the latter in the following section.

■ **Table 6.23** Example of equivalent 4-state binary assignments

State	A_1		A_2		A_3	
	y_1	y_2	y_1	y_2	y_1	y_2
s_0	0	0	1	1	0	0
s_1	0	1	1	0	1	0
s_2	1	0	0	1	0	1
s_3	1	1	0	0	1	1

Race-Free State Assignment

The analysis example in Section 6.3 demonstrated that a *race* condition can exist in a fundamental-mode asynchronous circuit if two or more state variables must change simultaneously when the circuit makes a transition from one stable state to another. Since the propagation time delays of the memory devices are not the same, a race may cause the circuit to reach one or more incorrect stable states, depending on the order in which the state variables change. In this case, the race is referred to as **critical** and can be detected in the flow table because *only those transitions that take place in columns containing two or more stable states are critical.* In designing a circuit we must always avoid critical races.

If, on the other hand, the circuit always operates properly in the presence of a race, then the race condition is called **noncricital**. Noncritical races can be left unattended, but often can be used to advantage. In fact, sometimes we may introduce noncritical races purposefully to produce a better design.

To avoid critical races, all that we have to do is assign the binary variables so that the state variables will change only one at a time in columns of the flow table that contain two or more stable states. If the flow table consists of only two rows, then the assignment of a single binary variable will always be race-free. However, when the flow table has three or more rows, the state-assignment problem becomes nontrivial.

To illustrate this problem, consider the flow table shown in Figure 6.22(a). We see that a transition is required from row A to row B (under input 01) and to row C (under input 11), from row B to row A (under input 00) and to row C (under input 10), and from row C to row A

x_1x_2

PS	00	01	11	10
A	$\boxed{A/0}$	B/-	C/-	$\boxed{A/0}$
B	A/-	$\boxed{B/0}$	$\boxed{B/1}$	C/-
C	A/-	$\boxed{C/1}$	$\boxed{C/0}$	$\boxed{C/1}$

(a) Flow table

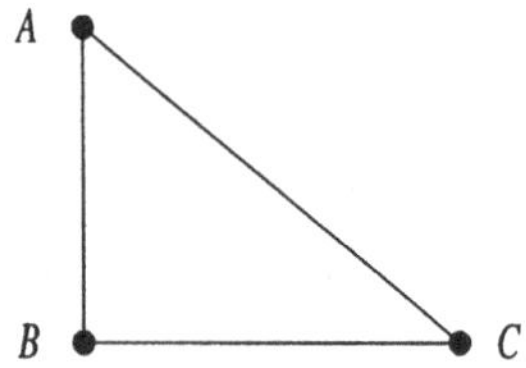

(b) State-assignment graph

■ **Figure 6.22** Example of a 3-row flow table

(under input 00). We can record this information using the **state-assignment graph** shown in Figure 6.22(b). Each vertex in the graph corresponds to a row of the flow table, and a branch connects two vertices if one or more transitions are required between them. Note that the graph is not directed; we only record that a transition is required, between vertices A and C say, and do not care if the actual transition is directed from A to C, or vice versa.

Since there are three states in the flow table of Figure 6.22(a), we need two state variables (y_1 and y_2) to represent them. Let us assume, for example, that the following binary state assignment is made:

State	y_1y_2
A	00
B	01
C	11

This assignment will cause a critical race between rows A and C because it requires that both state variables change simultaneously when the circuit is making a transition from stable state A to stable state C under input 11. Moreover, this assignment will also cause a noncritical race between rows C and A under input 00. In fact, we can verify that *any* permutation of the proposed state assignment for Figure 6.22(a) will result in a critical race.

To avoid critical races, we must assign the binary codes to the vertices in the state-assignment graph so that the codes assigned to any two vertices connected by a branch will differ in only one bit. In other words, the binary codes assigned to any two vertices connected by a branch are **adjacent**. From this perspective, it is now clear why we cannot obtain a critical race-free assignment in Figure 6.22(b): Each pair of vertices is connected by a branch, yet it is impossible to assign adjacent 2-bit binary numbers to all the vertices.

When the vertices of the state-assignment graph are connected so that no assignment can be made to satisfy the required adjacencies, the flow table has to be modified. The structure of the graph will indicate whether such modification is required. For example, in contrast with Figure 6.22(b), the graphs in Figure 6.23(a) illustrate transitions in 3-row flow tables for which critical race-free assignments can be made with only two state variables. A similar case is shown in Figure 6.23(b) for 4-row flow tables. On the other hand, if a 4-row flow table requires the transitions shown in Figure 6.23(c), it must be modified so that critical race-free assignment would be possible.

In general, we modify the flow table by adding extra rows (states) to create cycles in columns where critical races cannot be otherwise resolved. A **cycle** is a sequence of unstable states that terminates in a stable state. Often, the addition of extra rows implies an increase in the number of state variables (memory devices). Sometimes, however, we can obtain a critical race-free assignment either by adding rows *without* increasing the number of state variables or by utilizing some of the unspecified next state entries in the flow table. Let us first illustrate these two cases in Examples 6.14 and 6.15 and then consider, in Example 6.17, a case where adding rows necessitates an increase in the number of state variables.

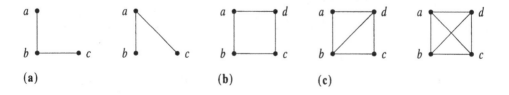

■ **Figure 6.23** Examples of state-assignment graphs

■ **Example 6.14**   Figure 6.24(a) shows the flow table of Figure 6.22(a) with an extra row (D) added to it. The corresponding state-assignment graph with one possible, critical race-free binary coding is shown in Figure 6.24(b). Because two state variables can represent up to four states, additional state variables are not required in this case.

To obtain the race-free assignment, we changed the entry in row A, column 11 to D and added C to the new row D in column 11. Thus, instead of incurring a critical race in going from row A to row C (under input 11) in the original table, we have now created a *cycle* between these rows through row D and, therefore, have eliminated the critical race. The arrows shown in Figure 6.24(a) indicate that all the required transitions between the four rows are now race-free. Note that the additional row is not associated with any stable state but, instead, is used only to convert a critical race into a cycle.   ■

The transition table corresponding to the flow table of Figure 6.24(a) with the selected binary state assignment of Figure 6.24(b) is shown in Table 6.24. The unspecified entries in the flow table can be utilized as don't-cares in the transition table but we should pay special attention to them. In particular, we should not assign 10 to the unspecified next state entries in the fourth row $(D = 10)$ in order to avoid the creation of unwanted stable states in that row.

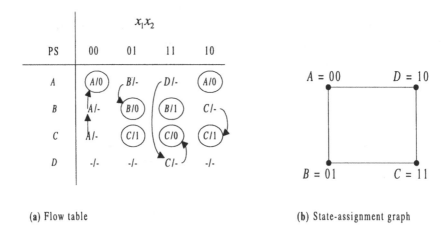

(a) Flow table   (b) State-assignment graph

■ **Figure 6.24**   Adding a fourth row to the flow table of Figure 6.22(a)

■ **Table 6.24** Transition table corresponding to the flow table of Figure 6.24(a)

PS	NS (Y_1Y_2)				Output (z)			
	Input Combination (x_1x_2)				Input Combination (x_1x_2)			
y_1y_2	00	01	11	10	00	01	11	10
00	00	01	10	00	0	×	×	0
01	00	01	01	11	×	0	1	×
11	01	11	11	11	×	1	0	1
10	××	××	11	××	×	×	×	×

Sometimes we can use some of the unspecified next-state entries to obtain a critical race-free assignment, even though at first it may appear that additional rows are necessary. The following example illustrates this point.

■ **Example 6.15** Consider the flow table shown in Figure 6.25(a). For simplicity, the outputs have been omitted from the table. The state-assignment graph in Figure 6.25(b) seems to indicate that additional rows may be required to avoid critical races, since all possible transitions between pairs of rows are needed. Clearly, adding even a single row to a 4-row table will require three state variables and, therefore, will increase the number of memory devices by one.

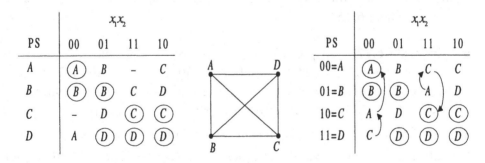

(a) Flow table   (b) State-assignment graph   (c) Critical race-free flow table

■ **Figure 6.25** Four-row flow table for Example 6.15

Although additional rows are indicated in this case, they are not actually needed because cycles can be created to eliminate races by using judiciously the unspecified entries. This is shown in Figure 6.25(c), where the state assignment is selected so that critical races can be eliminated without adding extra rows. The arrows indicate the cycles created to avoid races.   ■

It may not always be possible to create cycles by utilizing unspecified entries without increasing the number of state variables. The following example illustrates this case.

■ **Example 6.16** The 4-row flow table shown in Table 6.25 has the state-assignment graph of Figure 6.25(b). Because there are no unspecified entries, we must use three state variables to avoid critical races. Therefore, four extra rows will be added to the flow table, because three state variables can represent up to eight states.   ■

For cases like the one in Example 6.16, we can obtain the state assignment using the **state-assignment table**. The table is similar in format to a K-map except that each cell corresponds to a row in the modified flow table. The rows are associated with the cells of the state-assignment table such that each can be assigned a unique binary code in terms of the state variables, while maintaining the adjacency requirements dictated by the transitions in the original flow table. The following example illustrates the construction of a state-assignment table for Table 6.25 of Example 6.16.

■ **Example 6.17** To provide a critical race-free assignment for Table 6.25, we add an extra state variable, resulting in a modified flow table with eight rows. The corresponding state-assignment table is shown in Figure 6.26(a). The three state variables are designated by y_1, y_2, and y_3. Cells a, b, c, and d represent the original rows of Table 6.25, and cells e, f, and g (shown in parentheses) designate three of the four additional rows. Since rows associated with adjacent cells differ in only one bit, direct transitions can be made between a and b, b and c, and b and d. On the other hand, a transition between a and d is cycled

■ **Table 6.25** Four-row flow table for Example 6.16

PS	NS			
	Input Combination $(x_1 x_2)$			
	00	01	11	10
a	b	$ⓐ$	$ⓐ$	b
b	$ⓑ$	d	c	$ⓑ$
c	$ⓒ$	a	$ⓒ$	d
d	c	$ⓓ$	a	$ⓓ$

through e, a transition between a and c is cycled through g, and a transition between c and d is cycled through f. The same information can be represented also on a state-assignment graph, as shown in Figure 6.26(b).

Using the assignment given in Figure 6.26, Table 6.25 can be expanded to the 8-row, critical race-free flow table shown in Table 6.26. The arrows indicate the cycles created in order to avoid critical races. Note that, since we only require three rows for the critical race-free assignment, row 110 consists of don't-cares in its entirety. The unspecified entries in the flow table can be utilized as don't-cares as

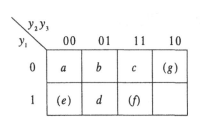

(a) State-assignment table

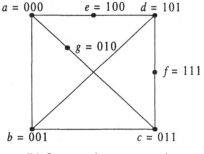

(b) State-assignment graph

■ **Figure 6.26** State assignment for Table 6.25

■ **Table 6.26**   Critical race-free flow table equivalent of Table 6.25

PS	NS			
	Input Combination $(x_1 x_2)$			
$y_1 y_2 y_3$	00	01	11	10
000=a	b	ⓐ	ⓐ	b
001=b	ⓑ	d	c	ⓑ
010=g	—	a	—	—
011=c	ⓒ	g	ⓒ	f
100=e	—	—	a	—
101=d	f	ⓓ	e	ⓓ
110	—	—	—	—
111=f	c	—	—	d

long as we avoid assigning them the binary values corresponding to their rows. In this way, we will not establish unwanted stable states in these rows.   ■

The state-assignment scheme of Figure 6.26 is actually *universal*; it can be used for *any* 4-row table that requires three state variables to avoid critical races. Another example of a universal state-assignment scheme, shown in Figure 6.27, can be used for any 8-row flow table that requires

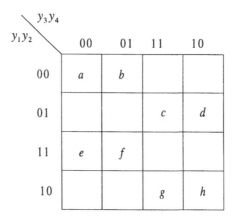

■ **Figure 6.27**   Universal state assignment for 8-row flow tables

four state variables to avoid critical races. The states indicated in the table are the original states, while the empty cells designate the eight additional rows. State transitions must be directed through the empty cells to satisfy the adjacency requirements in order to avoid critical races. For example, a transition between b and c must be cycled through $y_1y_2y_3y_4 = 0101$; between a and c requires a cycle through 0100 and 0101; between a and h requires passage through 0010; and so on. This scheme will also work for 6- or 7-row tables that cannot be made critical race-free with only three state variables. Most 5-row tables can be implemented with only three state variables; but if four variables are necessary, the scheme of Figure 6.27 can be applied to them too.

The main disadvantage of the state-assignment schemes discussed in this section is the introduction of additional unstable states, which, in turn, slow down the speed of operation of the circuit. Other assignment schemes exist in which several state variables are allowed to change simultaneously, but where the state assignment is made such that all races are noncritical. These assignments, referred to as *single transition time assignments*, can be used to achieve faster circuit operation. We do not discuss this topic because it is outside the scope of this book but, instead, turn our attention to the design process.

6.6 DESIGN

The design process for fundamental-mode sequential circuits is schematically illustrated in the flow diagram of Figure 6.28. But since the design and analysis of any system are essentially carried out in opposite directions, Figure 6.28 also describes the steps required to analyze a circuit. (Most of these steps were introduced in Section 6.3.)

We begin with a set of **design specifications** and terminate when the circuit logic diagram is produced. The design specifications can be formulated in terms of (1) a verbal statement, (2) timing diagrams, (3) sequence diagrams, (4) a state diagram, or (5) a flow table. If the design specifications are stated verbally, their translation into a state diagram or table has to be performed manually. No systematic way exists that circumvents the ingenuity of the designer. On the other hand, most of the other design steps can be executed with the aid of computer programs.

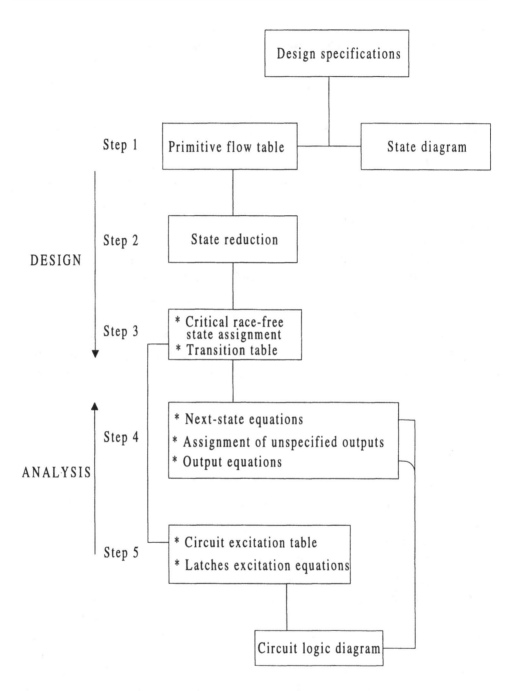

■ **Figure 6.28**   Design/analysis flow diagram for fundamental-mode sequential circuits

We only need to derive a flow table in order to initiate the design process. The state diagram is often helpful in formulating the operation of the circuit, but it is not mandatory. In the process of transforming the verbal description of the circuit into a flow table, the choice between a Mealy or a Moore flow table is usually made ad hoc. Some design problems lend themselves more easily to a Mealy flow table, while others can be more readily formulated as a Moore flow table.

In the following section, we discuss a specific example illustrating the design process. Steps 2 and 3, state reduction and critical race-free state assignment, have already been discussed in Sections 6.4 and 6.5. We discuss the other steps as we proceed.

The Design Process

Design Specifications

Consider the design of an asynchronous sequential circuit that controls the setup and disconnect phases of a communication link between two telephones, A and B. Each telephone is associated with a light bulb, L_A and L_B, respectively, and the control protocol that we wish to implement is defined as follows:

1. Lifting one of the handsets turns the bulb ON at the other telephone. The bulb will remain ON until the handset of this telephone (the called party) is picked up or that of the other (the calling party) is replaced.
2. With both handsets lifted, the communication link is *set up* and both bulbs are turned OFF.
3. Replacing either one of the handsets causes the light bulb associated with the other telephone to turn ON. When the corresponding handset is replaced, the bulb turns OFF and the communication link is *disconnected*.

The design specifications indicate that the circuit should have two inputs and two outputs. The input variables, x_1 and x_2, are associated with the two telephones, A and B, respectively. Let us assume that $x_i = 1$ ($i = 1, 2$) when the corresponding handset is lifted and that $x_i = 0$

otherwise. The two output variables, z_1 and z_2, represent the signals required to turn the light bulbs L_A and L_B, respectively, ON ($z_i = 1$) or OFF ($z_i = 0$).

Note that, since we are designing a fundamental-mode circuit, we must assume that only one input variable is allowed to change at a time. For our example, this condition is satisfied overall since the probability that both handsets will be lifted or replaced simultaneously is almost zero.

Step 1: State Diagram and Flow Table

A Moore state diagram satisfying the design requirements is shown in Figure 6.29. The branches are labeled with bit combinations of $x_1 x_2$. The outputs depend only on the states and are labeled with bit combinations of $z_1 z_2$. To understand how this state diagram represents the operation of the required circuit, assume that all the circuit inputs and outputs are 0 initially. The circuit is in stable state a and will stay there, producing $z_1 z_2 = 00$, as long as the input combination is not changed. Depending on whether the call is initiated by telephone A ($x_1 = 1$) or B ($x_2 = 1$), the circuit will make a transition to stable state b or c, respectively. If the circuit is in state b, L_B is ON ($z_2 = 1$), and if in c, L_A is ON ($z_1 = 1$). If the call initiation is aborted by the calling party at this stage, the circuit returns to state a as shown.

When the circuit is in stable state b or c and the other handset is lifted, the next stable state is d and the communication link is set up. In this state, the light bulbs are both OFF and the circuit will stay in this state as long as the two handsets remain lifted (call in progress).

If either A or B is the first to initiate the disconnect phase, the circuit will make a transition to either stable state e or f, respectively. At this point, however, the terminating party can reconnect the link as long as the handset of the other party is not replaced, and the circuit will make a transition back to state d. But, if the other party has terminated the call, then the circuit goes back to state a and the communication link is disconnected.

Observe that the transitions in Figure 6.29 from e to d and from f to d were not actually specified in the verbal description of the design

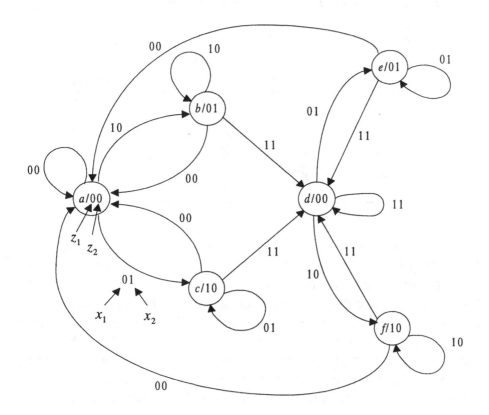

■ **Figure 6.29** Moore state diagram for the telephone circuit

problem. Nevertheless, we must provide these transitions to account for the case where the party initiating the disconnect phase wants to reconnect the link *before* the other party replaces the handset. Without these transitions, the circuit will not function properly, since there will be no outgoing transitions from either e or f when x_1 and x_2 are changed, respectively, from 01 or 10 to 11. This situation illustrates a case where the designer must exercise judgment and supplement the design specifications in order to produce a circuit that operates *reliably*.

The primitive flow table obtained from the state diagram is shown in Table 6.27. We could also have derived Table 6.27 directly from the design specifications without going through the state diagram.

■ **Table 6.27**   Primitive flow table for the telephone circuit

PS	| 00	01	11	10	| Output (z_1z_2)
		Input Combination (x_1x_2)			
a	ⓐ	c	—	b	00
b	a	—	d	ⓑ	01
c	a	ⓒ	d	—	10
d	—	e	ⓓ	f	00
e	a	ⓔ	d	—	01
f	a	—	d	ⓕ	10

In some cases, it is possible to obtain from the design specifications a flow table (or even a minimal-row table), rather than a primitive flow table. Nevertheless, it is *always* a good practice to begin with a primitive flow table and leave the reduction process where it belongs—in Step 2 of the design procedure. If we do so, our design will be less prone to mistakes made in trying to convert the design specifications into a flow table *at the same time* as attempting to eliminate redundant information. In fact, we can add at this point as many extra states as we see fit, since all those redundant states will be eliminated in the reduction process anyway.

Step 2:   State Reduction

As we saw in Section 6.4, state reduction of flow tables is more easily accomplished using the reduction method for completely specified tables, followed by merger. Inspecting Table 6.27, we see no redundant states and, therefore, proceed with the merger process.

At this point, however, we must decide whether we want to implement a Moore or a Mealy circuit. In Step 1 we have specified a Moore circuit; however, we know that a conversion to a Mealy circuit usually results in the least number of rows in the merged table. Indeed, if we now treat Table 6.27 as a Mealy table, then the merger graph shown in Figure 6.30(a) results in the following set of two maximal merger groups,

$$(a,b,c), \ (d,e,f)$$

and the 2-row Mealy flow table of Figure 6.30(b). Hence, we need only one memory device to implement the circuit as a Mealy circuit, and we are assured that the state assignment will be race-free. By contrast, if we continue with the Moore table, then the merged flow table would consist of four rows and we would need *at least* two memory devices to implement a race-free circuit. Based on these observations, we decide to continue the design process a Mealy circuit.

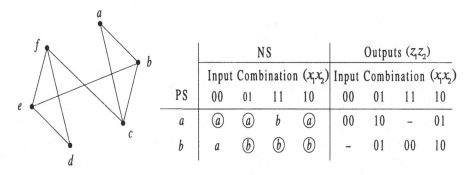

	NS				Outputs $(z_1 z_2)$			
	Input Combination $(x_1 x_2)$				Input Combination $(x_1 x_2)$			
PS	00	01	11	10	00	01	11	10
a	ⓐ	ⓐ	b	ⓐ	00	10	–	01
b	a	ⓑ	ⓑ	ⓑ	–	01	00	10

(a) Merger graph   (b) Minimal-row flow table

■ **Figure 6.30** Row merging for Table 6.27

Step 3: Critical Race-Free State Assignment

The state assignment for our example is trivial. Only one state variable (memory device) is required to implement the Mealy circuit, and there can be no critical races. Therefore, the transition table, shown in Table 6.28, is readily obtained from Figure 6.30(b).

Step 4: Next-State and Output Equations

The next-state equations are obtained from the next-state portion of the transition table, referred to as the *next-state table* (see Section 6.3). The minimal sum (or product) form of these equations is produced by using the K-map or the Quine-McCluskey minimization technique (see Chapter 4).

For our problem, we have only three variables (x_1, x_2, and y), so we can use the next-state K-map, derived from Table 6.28 and shown in Figure 6.31, to obtain the next-state function in minimal form:

$$Y = x_1 x_2 + y x_1 + y x_2 = x_1 x_2 + y(x_1 + x_2) \qquad (6.22)$$

■ **Table 6.28** Transition table for the telephone circuit

PS	NS (Y) Input Combination ($x_1 x_2$)				Outputs ($z_1 z_2$) Input Combination ($x_1 x_2$)			
y	00	01	11	10	00	01	11	10
0	⓪	⓪	1	⓪	00	10	—	01
1	0	①	①	①	—	01	00	10

Similarly, the output functions can be obtained from the output portion of the transition table. However, in deriving the output functions, we may encounter a problem caused by treating the unspecified outputs as don't-cares. Because don't-care outputs are associated with unstable states, we must choose them in such a way that momentary *false* outputs will not occur when the circuit makes transitions between stable states. In the design flow diagram of Figure 6.28, we refer to this procedure as the **assignment of unspecified outputs**.

To understand what happens if we improperly assign the don't-cares, let us reconsider the flow table in Figure 6.30(b). Assume that the circuit is in stable state a under input 01 and is required to make a transition to stable state b under input 11. This transition goes through unstable state b, whose outputs are unspecified. If $z_2 = 1$ is assigned to unstable state b,

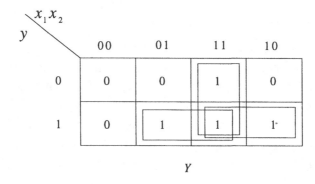

■ **Figure 6.31** Next-state K-map for the telephone circuit

then a *momentary short pulse* will appear on the z_2 output line as the circuit shifts from $z_2 = 0$ in stable state a to $z_2 = 1$ in unstable state b and then back to $z_2 = 0$ when the circuit reaches stable state b. Therefore, to avoid a momentary false output, z_2 *must* be specified as 0 for the unstable state b. On the other hand, for the same transition from stable state a to b, z_1 changes from 1 in a to 0 in b. If $z_1 = 1$ is assigned to unstable state b, then all that will happen is that the change in the z_1 output will take place only at the *end* of the transition. Likewise, if $z_1 = 0$ is assigned to unstable state b, then the change in the z_1 output will occur only at the *beginning* of the transition. Therefore, since it makes no difference when the output changes, z_1 can be assigned a don't-care (×) for unstable state b.

Based on these observations, we can summarize the procedure for assigning unspecified outputs as follows:

1. If the value of an output variable does not change in a transition between two stable states, then it is assigned that *same* value in *each* of the unstable states through which the transition takes place.
2. If the value of any output variable changes in a transition between two stable states, then it can be assigned a don't-care in *each* of the unstable states through which the transition occurs.

Applying this procedure to the telephone circuit, we obtain from Table 6.28 the K-map for the two outputs shown in Figure 6.32. (Of course, we can split the K-map into two maps, one for each output.) The resulting output equations are given by

$$z_1 = x_1'x_2y' + x_1x_2'y \tag{6.23a}$$

and

$$z_2 = x_1'x_2y + x_1x_2'y' \tag{6.23b}$$

Having derived the next-state and output functions, we are already in a position to implement the circuit using delay elements for memory (see the design flow diagram in Figure 6.28). But as already noted in Sections 6.2 and 6.3, we need not insert delay elements in the feedback loops because the input/output propagation time delays that exist in the combinational

$x_1 x_2$

y	00	01	11	10
0	00	10	00	01
1	00	01	00	10

$z_1 z_2$

- **Figure 6.32** Outputs K-map for the telephone circuit

part of the circuit provide the necessary delays. For this reason, the telephone circuit implementation shown in Figure 6.33 depicts the feedback loop without a delay element. We simply designate the "input" to the feedback loop with the variable Y (the next-state output from the combinational part of the circuit) and designate the "output" of the feedback loop with the variable y (the present-state input to the combinational part of the circuit). Viewed alternatively, we can insert a delay element in the feedback loop between Y and y representing the cumulative propagation time delays in the signal path between the input y and the output Y, and then think of the gates between y and Y as having zero propagation time delays.

The propagation time delays inherent in the combinational part of the circuit can be used to provide short-term memory through feedback loops. However, modern asynchronous sequential circuits are usually implemented with latches, as shown in the following section, where we discuss step 5 of the design procedure.

Step 5:   Implementation with Latches

In Chapter 7, we will see that the design of synchronous circuits with flip-flops resembles that of asynchronous circuits with latches. Hence, one of the advantages to latch implementation is in achieving uniformity between the two design procedures. Moreover, when latches are used to implement

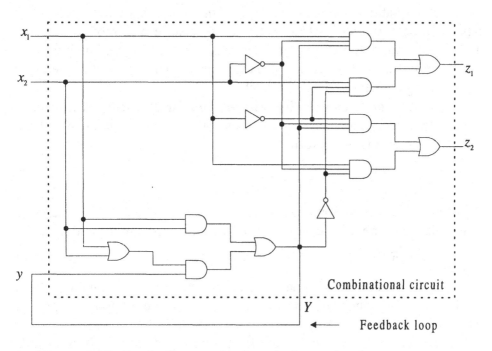

■ **Figure 6.33** Logic diagram for the telephone circuit

asynchronous circuits, the complexity of the combinational part of the circuit is often reduced.

Unfortunately, there is no general procedure to determine which type of latch will produce a circuit implementation with minimal complexity. The selection of a latch type may depend on several factors. In some cases, the choice is restricted by nontechnical considerations; at other times, the designer must choose a latch type to produce the simplest circuit according to some minimization criterion, such as a minimum number of total circuit components or a minimum number of connections between them.

To design a circuit with latches, the next-state information provided by the Y variables has to be converted into the *excitation signals* required by the latch inputs so that state transitions can be executed properly. To do so, we can follow one of two approaches: (1) We can convert the next-state equations into the **excitation requirements** for the latches, or (2) we can embed the excitation requirements into the transition table and derive the **excitation equations** directly from it. In either case, we use the *excitation table* of the selected latch. The former approach requires that we match the next-state equations with the *characteristic equation* of the latch. The latter approach is preferred in most cases because it is more systematic and better suited for computer-aided design. For illustration purposes, however, let us pursue both approaches by implementing the telephone circuit using an active-HIGH SR latch.

Derivation of Excitation Equations from the Next-State Equations: Recall that the characteristic equation of an active-HIGH SR latch [Equation (6.5a)] is of the form

$$Q^+ = R'Q + S \quad (SR = 0)$$

To obtain the excitation equations for the SR latches, we have to manipulate each next-state equation of the designed circuit into the form

$$Y = R'y + S \tag{6.24}$$

Then, by comparing this equation with the characteristic equation of the latch, we will obtain the Boolean expressions for the R and S inputs of each latch.

In our case, there is only one next-state equation, and it is already given in the right format. Hence, by comparing Equations (6.22) and (6.24), we conclude that

$$S = x_1 x_2 \tag{6.25a}$$

and, since $R' = (x_1 + x_2)$,

$$R = (x_1 + x_2)' \tag{6.25b}$$

We then check whether the condition $SR = 0$ is satisfied to ensure proper operation:

$$SR = x_1 x_2 (x_1 + x_2)' = x_1 x_2 x_1' x_2' = 0$$

Derivation of Excitation Equations from the Transition Table: We can also obtain the latch excitation equations by superimposing the excitation requirements on the transition table, as shown in Figure 6.34(a). Since we do not need the output section of the transition table at this point, only the present- and next-state columns are shown. The table is now referred to as the **circuit excitation table**. The excitation requirements under each input combination are filled in using the *excitation table* of the

PS	NS (Y)				$x_1 x_2$			
	$x_1 x_2$				00	01	11	10
y	00	01	11	10	SR	SR	SR	SR
0	0	0	1	0	0x	0x	10	0x
1	0	1	1	1	01	x0	x0	x0

(a) Circuit excitation table

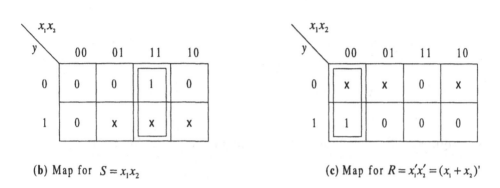

(b) Map for $S = x_1 x_2$ (c) Map for $R = x_1' x_2' = (x_1 + x_2)'$

■ **Figure 6.34** Derivation of latch equations from the transition table

SR latch [Figure 6.6(d)]. To show how this is done, assume for example that $y = 0$. The next state under $x_1 x_2 = 00$ should be $Y = 0$. From the excitation table of Figure 6.6(d), we see that, to effect a transition from $Q = 0$ to $Q^+ = 0$, S should be 0 and R is a don't-care; hence the entries $S = 0$ and $R = \times$ under $x_1 x_2 = 00$. Similarly, the transition from $y = 0$ and $Y = 1$ under input 11 will take place if $S = 1$ and $R = 0$.

Now that we have the latch excitation requirements embedded into the transition table, we can "forget" about the NS columns; the PS to NS transitions have been translated into the inputs of the latch. The Boolean expression for each of these inputs as a function of the present state and the circuit inputs can be obtained from the circuit excitation table, as shown in Figures 6.34(b) and (c). As we see, the excitation equations are the same as those obtained in Equations (6.25a) and (6.25b). Using these equations, we can now produce the logic diagram of the telephone circuit implemented with an SR latch, as shown in Figure 6.35.

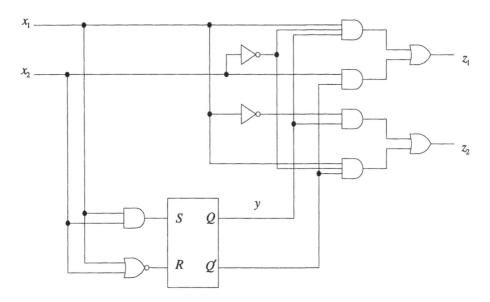

■ **Figure 6.35** Telephone circuit implemented with SR latch

More Design Examples

Edge-Triggered Flip-Flops

Edge-triggered flip-flops are clocked devices that are used to implement synchronous sequential circuits (see Chapter 7). Nevertheless, the internal design of the flip-flop is an asynchronous problem. To illustrate the design process, we consider a *positive* edge-triggered D flip-flop.

Design Specifications: A positive edge-triggered D flip-flop has two inputs, D (data) and C (clock), and a single output, Q. The output $Q = 1$ if $D = 1$ and C changes from 0 to 1 (positive edge triggering); likewise, $Q = 0$ if $D = 0$ and C changes from 0 to 1. Under any other input condition, the output Q remains unchanged.

Figure 6.36(a) shows the D flip-flop in block diagram form. We can approach the design by dividing the circuit into two modules, as shown in Figure 6.36(b). In many cases, this modular approach leads to a reduction in the complexity of the circuit. To illustrate, we design the D flip-flop using active-LOW SR latches.

Step 1: The Moore state diagram for the D flip-flop circuit, shown in

Figure 6.37, starts in the initial state a for the input condition $CD = 00$. The symbol Q assigned to the output in state a designates that the *previous* output value is maintained when the circuit is in state a $(Q^+ = Q)$: if Q was 1, it will remain 1, and if Q was 0, it will remain 0. When CD changes to 01, the circuit goes from state a to state b. Since $C = 0$ in this transition, the output in b is maintained at its previous value. The circuit goes to state d if C changes from 0 to 1, causing the flip-flop to set $(Q^+ = 1)$. If D is changed between 0 to 1 while C is kept at 1, the circuit maintains its set state while making transitions between states d and e. In state e, however, the circuit makes a transition to state a if C changes from 1 to 0. Being in state a, the circuit goes to state c if C changes from 0 to 1, and the output Q is reset $(Q^+ = 0)$. If D is changed between 0 and 1 while C is kept at 1, the circuit maintains its reset state while making transitions between states c and f. In state f, however, the circuit goes to state b if C changes from 1 to 0. The primitive flow table is shown in Table 6.29.

Note that we do not allow simultaneous changes of the input variables because they violate the condition for fundamental-mode operation. This condition is satisfied by requiring that D be maintained at a constant value prior to the application of the clock pulse (*setup time*) and that D be not changed after the application of the positive-going transition of the clock pulse (*hold time*). These requirements were discussed in Section 6.2 and will be reconsidered when the circuit logic diagram is obtained.

To design the circuit following the approach shown in Figure 6.36(b),

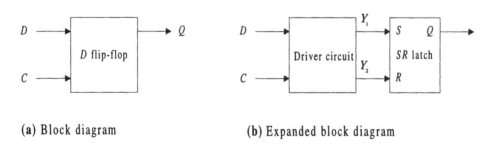

(a) Block diagram   (b) Expanded block diagram

■ **Figure 6.36** D flip-flop

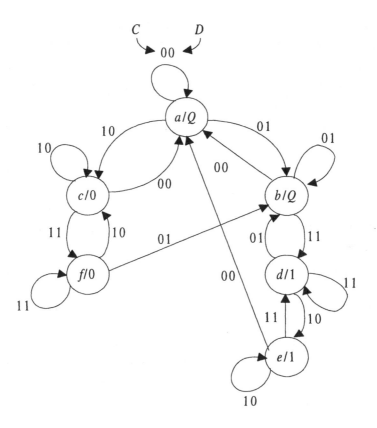

■ **Figure 6.37**   State diagram

■ **Table 6.29**   Primitive flow table

PS	NS Input Combination (CD) 00	01	11	10	Output Q^+
a	ⓐ	b	—	c	Q
b	a	ⓑ	d	—	Q
c	a	—	f	ⓒ	0
d	—	b	ⓓ	e	1
e	a	—	d	ⓔ	1
f	—	b	ⓕ	c	0

■ **Table 6.30** Characteristic tables of *SR* latches

| NAND Latch | | NOR Latch | | |
S	R	S	R	Q+
1	1	0	0	Q
1	0	0	1	0
0	1	1	0	1
0	0	1	1	Indeterminate

we use Table 6.29 to implement the driver circuit; but instead of generating Q^+ directly, we have to produce the appropriate excitation conditions on S and R so that Q^+ can be generated from them. These conditions are obtained directly from the characteristic table of the *SR* latch and are shown for convenience in Table 6.30 for both NAND and NOR latches.

Step 2: There are no equivalent states in Table 6.29, and if we wish to merge rows while keeping the Moore format, then the following maximal merger groups are obtained:

$$(a,b), (c,f), (d,e)$$

Renaming the merger groups A, B, and C, respectively, we show the reduced flow table in Table 6.31.

Step 3: The 3-row Table 6.31 poses no problems as far as critical race-free assignment is concerned. The D flip-flop requires two *SR* latches to implement the driver circuit, in addition to the output *SR* latch that produces Q. If we want the outputs of the driver circuit, Y_1 and Y_2, to feed directly the S and R inputs of the output latch, then the state assignment must correspond to the characteristic table of whichever *SR* latch type is used in the circuit. Hence, choosing to implement the circuit with NAND latches, we must use the following state assignment (see Table 6.30):

$$A = 11$$
$$B = 10$$
$$C = 01$$

■ **Table 6.31** Reduced flow table

	NS				Output
	Input Combination (CD)				
PS	00	01	11	10	Q^+
A	Ⓐ	Ⓐ	C	B	Q
B	A	A	Ⓑ	Ⓑ	0
C	A	A	Ⓒ	Ⓒ	1

The corresponding transition table for the driver circuit is shown in Table 6.32. Note that, due to the selected state assignment, we need not show a separate output table.

Step 4: To obtain the next-state equations, we use the K-maps of Figure 6.38 derived directly from Table 6.32. Hence,

$$Y_1 = C' + y_1 y_2' + D'y_1 = \{C[(Dy_2)'y_1]'\}' \tag{6.26}$$

and

$$Y_2 = C' + y_1' + Dy_2 = [Cy_1(Dy_2)']' \tag{6.27}$$

Step 5: To implement the driver circuit with active-LOW SR latches, recall that the characteristic equation of a NAND SR latch is of the form [Equation (6.5b)]

$$Q^+ = RQ + S' = [S(RQ)']' \tag{6.28}$$

■ **Table 6.32** Transition table for the driver circuit

	NS (Y_1Y_2)			
PS	Input Combination (CD)			
y_1y_2	00	01	11	10
01	11	11	⓪①	⓪①
11	⑪	⑪	01	10
10	11	11	①⓪	①⓪

$y_1 y_2$ \ CD	00	01	11	10
00	x	x	x	x
01	1	1	0	0
11	1	1	0	1
10	1	1	1	1

Y_1

$y_1 y_2$ \ CD	00	01	11	10
00	x	x	x	x
01	1	1	1	1
11	1	1	1	0
10	1	1	0	0

Y_2

■ **Figure 6.38** K-maps for the driver circuit

Comparing Equations (6.26) and (6.27) with Equation (6.28), we obtain

$$S_1 = C \quad \text{and} \quad R_1 = (Dy_2)'$$
$$S_2 = Cy_1 \quad \text{and} \quad R_2 = D$$

The design of the positive edge-triggered D flip-flop is now complete and its logic diagram is shown in Figure 6.39. We can now conclude that, to satisfy the fundamental-mode condition, the setup time must equal the propagation time delay through gates 1 and 4, because their outputs are affected by a change in D. Moreover, the hold time must equal the propagation time delay of gate 2 to ensure proper operation of the lower latch when C changes from 0 to 1. We will discuss the D flip-flop again in Section 7.2.

Timing-Signal Generation

Timing signals are used to control and coordinate various operations in digital systems. Normally, such signals are generated using synchronous circuits (see Chapter 7); however, it is also possible to use a fundamental-mode sequential circuit for this purpose.

Design Specifications: The timing-signal generator, whose block diagram is shown in Figure 6.40(a), consists of a control input x, a clock input C, and a single output z. The input x goes to 1 only after C has gone

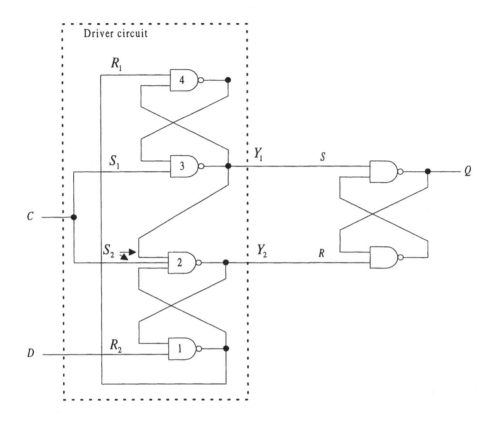

- **Figure 6.39**   Circuit logic diagram of the positive edge-triggered D flip-flop

to 1 and returns to 0 only after C has returned to 0. For each control input pulse, the circuit outputs a pulse on the z line coinciding with the next clock pulse that follows the x pulse. Figure 6.40(b) shows a typical timing diagram to illustrate the operation of the circuit.

The pulse generated on the z output is delayed by one clock period after the pulse on x. The x pulse is not synchronized with the clock, but the z pulse is. Hence, for proper operation, we must ensure that another pulse on x will not occur until at least one clock period after an output pulse has been generated. The train of clock pulses can be *periodic*, as in Figure 6.40(b), or *aperiodic*.

Step 1: The Mealy primitive flow table for the circuit, shown in Table 6.33, starts in state a with the input condition $xC = 00$. Since x can change only after C goes from 0 to 1, the circuit can only go from state a to state b. In state b, if x changes from 0 to 1, the circuit goes to state c; if no x pulse occurs, the circuit goes back to state a when C changes to 0. From state c the circuit goes to state d since C returns to 0 ahead of x, and then on to e when x returns to 0. On the next clock pulse, the leading edge of the output pulse is generated when the circuit makes a transition from state e to f. Since another x input cannot occur until at least one clock period after an output pulse has been generated, the circuit stays in state f for the duration of the clock pulse and then returns to state a on the trailing edge of C.

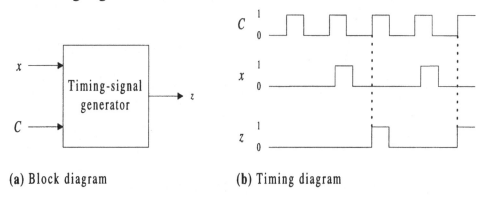

(a) Block diagram **(b)** Timing diagram

■ **Figure 6.40** Timing-signal generator

■ **Table 6.33** Primitive flow table

| | NS/Output (z) | | | |
| | Input Combination (xC) | | | |
PS	00	01	11	10
a	$a/0$	$b/-$	$-/-$	$-/-$
b	$a/-$	$b/0$	$c/-$	$-/-$
c	$-/-$	$-/-$	$c/0$	$d/-$
d	$e/-$	$-/-$	$-/-$	$d/0$
e	$e/0$	$f/-$	$-/-$	$-/-$
f	$a/-$	$f/1$	$-/-$	$-/-$

Step 2:   There are no equivalent states in Table 6.33. From the merger graph shown in Figure 6.41(a), we see that there are two sets of maximal merger groups:

$$(a,b), \ (c,d,e), \ (f)$$

and

$$(a,b,c), \ (d,e), \ (f)$$

Selecting the first set and renaming the groups L, M, and N, respectively, we obtain the reduced flow table shown in Figure 6.41(b).

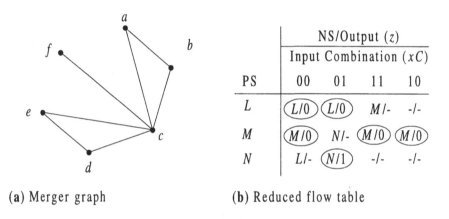

	NS/Output (z)			
	Input Combination (xC)			
PS	00	01	11	10
L	Ⓛ/0	Ⓛ/0	M/-	-/-
M	Ⓜ/0	N/-	Ⓜ/0	Ⓜ/0
N	L/-	Ⓝ/1	-/-	-/-

(a) Merger graph   (b) Reduced flow table

■ **Figure 6.41** State reduction of Table 6.33

Step 3:   To obtain a critical race-free state assignment, we have to utilize an extra row through which the transition from stable state N under $xC = 01$ to stable state L under $xC = 00$ will be cycled. Hence, choosing the assignment $L = 00$, $M = 01$, $N = 11$, and 10 for the extra row, we obtain the transition table shown in Figure 6.42(a).

Step 4:   The next-state equations derived from the K-maps in Figure 6.42(b) are given by

$$Y_1 = x'y_2(y_1 + C)$$
$$Y_2 = x + y_2(y_1' + C)$$

Note our particular selection of 1-cell clusters in the K-map of Y_1. Since the don't-care minterms m_{14} and m_{15} are assigned a 1 in the K-map of Y_2, they should *not* be assigned a 1 for Y_1 to avoid the creation of unwanted stable states.

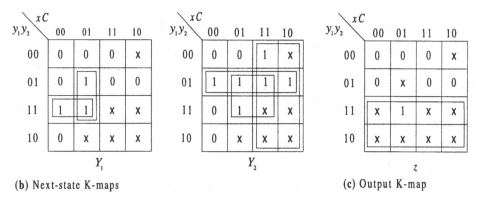

(a) Transition table

(b) Next-state K-maps

(c) Output K-map

■ **Figure 6.42** Implementation information

The output map is shown in Figure 6.42(c). The dashes are assigned values according to the rules established in the previous section. Observe that the output coincides with y_1 :

$$z = y_1$$

We leave it to the reader to implement the circuit with latches.

6.7 HAZARDS IN ASYNCHRONOUS SEQUENTIAL CIRCUITS

Hazards were introduced in Section 4.8 with respect to the design of combinational logic circuits. *Static* and *dynamic* hazards can be present in the combinational part of the asynchronous circuit and should be addressed in the design process. There is yet another type of hazard, called an **essential hazard**, that is specific to fundamental-mode asynchronous circuits.

An essential hazard may result from unequal propagation time delays along two or more signal paths originating from the same circuit input and can cause the circuit to malfunction. To illustrate this, consider the single-input asynchronous circuit of Figure 6.43(a) whose next-state table is shown in Figure 6.43(b). Using the hazard-detection technique of Section 4.8, we can verify that the combinational part of the circuit contains no static or dynamic hazards.

Assume now that the propagation time delay of inverter N_1 is very large in comparison to the delays of the other gates, and let the circuit be in stable state 00 under $x = 0$. If the input is changed to 1, the circuit should make a transition to stable state 01 as indicated in Figure 6.43(b). But, instead, the circuit changes to stable state 10 after some time due to the unequal propagation time delays along the signal paths from x to Y_1 and Y_2 through gates A_1 and A_5 on the one hand, and through gate A_3 on the other hand. This type of behavior constitutes an essential hazard.

To illustrate the sequence of events leading to the incorrect stable state, consider the timing diagram for the circuit shown in Figure 6.44. We have omitted the signal waveforms corresponding to AND gates A_2 and A_4, because they do not contribute any relevant information to our present discussion. As we see, the circuit is in stable state 00 prior to time t_1. If x is changed from 0 to 1 at t_1, the change propagates through the circuit so that, at t_5, $Y_1 Y_2 = 01$ is the correct unstable state. Since N_1 has not changed from 1 to 0 yet, A_3 becomes 1 (rather than stay at 0) and forces Y_1 to 1 at t_6. In turn, this event triggers a succession of changes in the circuit, culminating in the incorrect stable state $y_1 y_2 = 10$ at t_{13}.

Essential hazards cannot be eliminated by adding redundancies to the

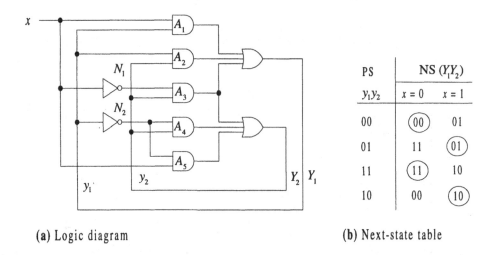

(a) Logic diagram

PS	NS (Y_1Y_2)	
y_1y_2	$x = 0$	$x = 1$
00	(00)	01
01	11	(01)
11	(11)	10
10	00	(10)

(b) Next-state table

■ **Figure 6.43** Circuit with essential hazard

circuit as we did in Section 4.8 to avoid static and dynamic hazards. Nor can they be avoided if latches, rather than delay elements, are used to implement the circuit. But the problem can be solved by inserting delay elements in the feedback paths associated with the state variables that change during the hazardous transitions. This will ensure that the circuit completes its response to the input change before any changes in these state variables occur.

To locate essential hazards, we need not analyze the circuit to the length we did above. The presence of essential hazards can be detected by inspecting the next-state table or the flow table. A flow table has an essential hazard for input variable x_i if and only if, for some stable state, the stable state reached after one change in x_i is different from the stable state reached after three changes in x_i. (Note that, by definition, an essential hazard can exist for *any* input variable.) Applying this detection procedure to Figure 6.43(b), we see that starting in stable state 00, one change in x leads to stable state 01, whereas three changes in x lead to stable state 10. Hence, there is an essential hazard.

Hazards and races are of no concern in synchronous sequential circuits since these circuits operate under clock control. We just make the interval between clock pulses sufficiently large to allow all changes to stabilize.

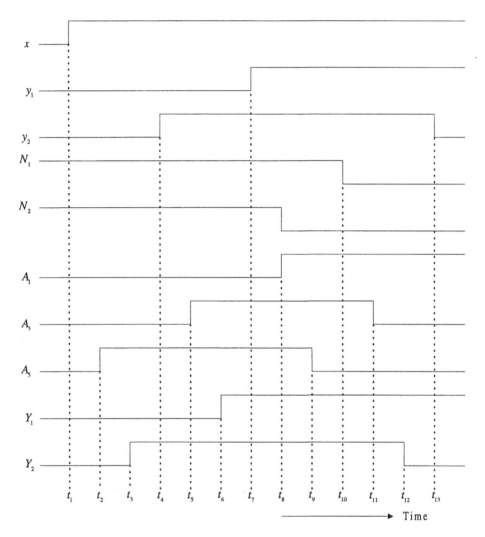

■ **Figure 6.44** Timing diagram for the circuit of Figure 6.43(a)

Hence, we may regard clocking as introducing controlled delays into the circuit so that the timing problems associated with an asynchronous design are eliminated.

PROBLEMS

6.1   A single-input sequential circuit accepts sequences of 0's and 1's
and indicates on two output lines (z_1 and z_2) the remainder for the
number of 1's in the string after division by 4. Obtain a state table
for the circuit. (*Hint*: The circuit has four states and the output
values $z_1 z_2 = 00$, 01, 10, and 11 indicate, respectively, no
remainder, a remainder of 1, of 2, and of 3.)

6.2   Draw a state diagram for each of the following tables:

(a)

	NS		Output	
PS	$x = 0$	$x = 1$	$x = 0$	$x = 1$
A	B	C	a	a
B	B	B	a	b
C	C	A	b	a

(b)

	NS		
	Input Combination $(x_1 x_2)$		
PS	00	01	10
A	A	B	C
B	A	D	C
C	A	C	C
D	A	A	C

6.3   Explain why the sequential circuit represented by the state diagram
in Figure P6.3 may be considered a counter.

6.4   Give a verbal description of the operation of the circuit in Figure
P6.4, assuming that the inputs x_1 and x_2 do not change
simultaneously.

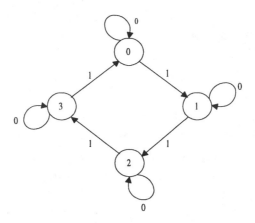

■ **Figure P6.3**

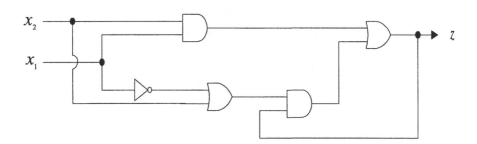

■ **Figure P6.4**

6.5   Each latch in Figure P6.5 is initially reset $(Q = 0)$. Determine the output of each latch in response to the inputs shown.

6.6   **(a)**   Show the Q output waveform of an active-LOW SR latch in relation to the input waveforms given in Figure P6.6(a). Assume that Q is initially at 0

        (b)   Repeat part (a) for the input waveforms shown in Figure P6.6(b).

        (c)   Show the Q and Q' waveforms of an active-HIGH gated SR latch in relation to the input waveforms shown in Figure P6.6(c). Assume that Q is initially 0.

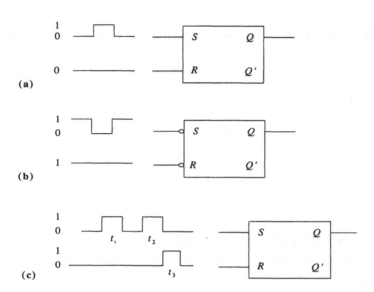

■ **Figure P6.5**

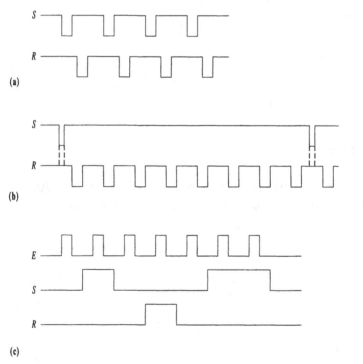

■ **Figure P6.6**

6.7   Analyze the asynchronous sequential circuit shown in Figure P6.7.

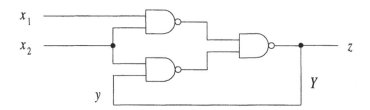

■ **Figure P6.7**

6.8   Analyze the circuit in Figure P6.8. What is the output sequence for the following input sequence: $x_1x_2 = 00,\ 10,\ 11,\ 01,\ 11,\ 10,\ 00,\ 01,\ 00$?

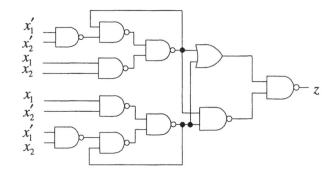

■ **Figure P6.8**

6.9   Analyze the asynchronous circuit in Figure P6.9. Assuming that $z = 0$ initially, determine the output sequence when the input sequence is $x = 0,\ 1,\ 0,\ 1$.

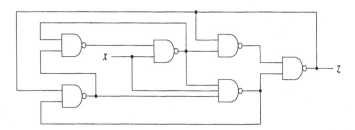

■ **Figure P6.9**

6.10 An asynchronous sequential circuit has two inputs (x_1, x_2) and one output (z). The next-state and output equations describing the circuit are $Y = x_1 x_2' + (x_1 + x_2') y$ and $z = y$.

(a) Draw the circuit logic diagram.
(b) Obtain the transition table.
(c) Obtain a flow table.
(d) Describe in words the behavior of the circuit.

6.11 Repeat Problem 6.10 for the circuit described by the equation

$$Y_1 = x_1 x_2 + x_1 y_2' + x_2' y_1$$
$$Y_2 = x_2 + x_1 y_1' y_2 + x_1' y_1$$
$$z = x_2 + y_1$$

6.12 Find a reduced Moore table for the primitive flow table of Figure P6.12.

	NS				Outputs	
	Input Combination $(x_1 x_2)$				$z_1 z_2$	
PS	00	01	11	10		
1	①	7	–	4	1	1
2	②	5	–	4	0	1
3	–	7	③	11	1	0
4	2	–	3	④	0	0
5	6	⑤	9	–	1	1
6	⑥	7	–	11	0	1
7	1	⑦	14	–	1	0
8	⑧	12	–	4	0	1
9	–	7	⑨	13	0	1
10	–	7	⑩	4	1	0
11	8	–	10	⑪	0	0
12	6	⑫	9	–	1	1
13	8	–	14	⑬	1	1
14	–	12	⑭	11	0	0

■ **Figure P6.12**

6.13   Find a reduced table for the primitive flow table of Figure P6.13.

PS	NS Input Combination (x_1x_2)			
	00	01	11	10
a	-/-	f/-	(a/0)	b/-
b	g/-	-/-	c/-	(b/1)
c	-/-	h/-	(c/1)	d/-
d	e/-	-/-	a/-	(d/0)
e	(e/0)	f/-	-/-	d/-
f	e/-	(f/0)	a/-	-/-
g	(g/1)	h/-	-/-	b/-
h	g/-	(h/1)	c/-	-/-

■ **Figure P6.13**

6.14   Find a minimum-row table equivalent for the table in Figure P6.14.

PS	NS Input (x)		Output Input (x)	
	0	1	0	1
A	E	D	n	n
B	B	F	n	n
C	E	B	n	n
D	B	F	n	n
E	D	A	n	m
F	E	B	n	m

■ **Figure P6.14**

6.15   Obtain a reduced Mealy table for the primitive flow table in Figure P6.15.

| | NS | | | | Outputs | |
| | Input Combination $(x_1 x_2)$ | | | | $z_1 z_2$ | |
PS	00	01	11	10		
1	(1)	11	4	10	0	1
2	5	(2)	–	3	1	1
3	5	2	13	(3)	1	1
4	12	–	(4)	15	0	0
5	(5)	–	8	–	1	0
6	14	(6)	–	10	1	1
7	(7)	6	8	3	0	1
8	7	–	(8)	3	0	0
9	(9)	11	13	10	0	1
10	12	6	13	(10)	1	1
11	5	(11)	–	3	1	1
12	(12)	2	4	15	0	1
13	1	–	(13)	10	0	0
14	(14)	–	8	–	1	0
15	1	6	4	(15)	1	1

■ **Figure P6.15**

6.16 Obtain a race-free state assignment for the flow table in Figure P6.16.

| | NS | | | |
| | Input Combination $(x_1 x_2)$ | | | |
PS	00	01	11	10
a	b	(a)	d	(a)
b	(b)	d	(b)	a
c	(c)	a	b	(c)
d	c	(d)	(d)	c

■ **Figure P6.16**

6.17 Obtain a race-free state assignment for the flow table in Figure P6.17.

PS	NS/Output			
	Input Combination (x_1x_2)			
	00	01	11	10
a	$a/0$	$a/1$	b/-	d/-
b	a/-	$b/0$	$b/0$	c/-
c	a/-	-/-	d/-	$c/0$
d	a/-	a/-	$d/1$	$d/1$

■ **Figure P6.17**

6.18 Consider an asynchronous circuit having two inputs (x_1, x_2) and one output (z). The input sequence $x_1x_2 = 00$, 01, 11 causes the output to become 1. Then, the next input change causes the output to return to 0. No other input sequence will produce a 1 output. Obtain the primitive flow table and the state diagram for the circuit.

6.19 Obtain a primitive flow table for a negative edge-triggered T flip-flop. The circuit has two inputs, T (toggle) and C (clock), and one output, Q. The output is complemented if $T = 1$ and C changes from 1 to 0; otherwise, under any other input condition, the output Q remains unchanged.

6.20 Consider an asynchronous circuit having two inputs, C (clock) and S, and a single output, z. The signal S is an enable signal allowing only complete clock pulses to appear at the output, as shown in Figure P6.20. Obtain a primitive flow table for the circuit. (Note that this is an asynchronous problem because the input S is allowed to change at any time with respect to the clock.)

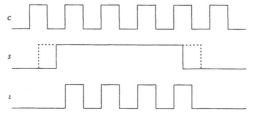

■ **Figure P6.20**

6.21 Obtain a primitive flow table for the circuit in Problem 6.4.

6.22 Obtain a primitive flow table for an asynchronous circuit with two inputs, x_1 and x_2, and two outputs, z_1 and z_2, that operates according to the following specifications:
  (a)   When $x_1 x_2 = 00$, the output is $z_1 z_2 = 00$.
  (b)   If x_2 changes from 0 to 1 while $x_1 = 1$, the output is $z_1 z_2 = 01$.
  (c)   When $x_2 = 1$ and x_1 changes from 0 to 1, the output is $z_1 z_2 = 10$.
  (d)   Otherwise, the output does not change.

6.23 An asynchronous circuit has two inputs, clock (C) and push button (P), and a single output, z. The timing diagram shown in Figure P6.23 illustrates the operation of the circuit. The timing of the push-button signal is random, except that its duration and interval are long compared to the clock duration and interval. The circuit output replicates the first clock pulse after P has been pressed and produces only one such pulse each time the push button is pressed. Find a minimum-row flow table.

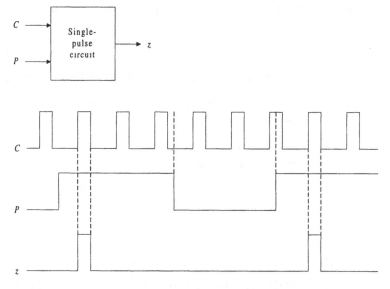

■ **Figure P6.23**

6.24   An asynchronous circuit has two inputs (x_1, x_2) and one output (z). If the input sequence 00, 01, 11 occurs, z becomes 1 and remains 1 until the input sequence 11, 01, 00 occurs. In this case, z becomes 0 and remains 0 until the first sequence occurs again. The last input in one sequence may be the first input in the other sequence. Find a minimum-row flow table.

6.25   Consider the following 2-input (x_1, x_2), single-output (z) asynchronous circuit: The output changes from 0 to 1 only when x_2 changes from 0 to 1 while $x_1 = 1$. The output changes from 1 to 0 only when x_1 changes from 1 to 0 while $x_2 = 1$. Design the circuit to obtain static, hazard-free, next-state and output equations.

6.26   Repeat Problem 6.25 for the following asynchronous circuit with two inputs (x_1, x_2) and two outputs (z_1, z_2): The output z_i (for $i = 1, 2$) is 1 when and only when x_i was the last changing input.

6.27   Implement the circuit specified in Problem 6.11 with SR latches.

6.28   Design an asynchronous circuit whose output z becomes 1 if either of the two inputs (A or B) rises to 1 when the other input is already at 1. The output is 0 if input A drops to 0 regardless of the value of input B. Implement the circuit with **(a)** SR latches, and **(b)** D latches.

6.29   Design an asynchronous circuit having two inputs (x_1 and x_2) and two outputs (z_1 and z_2) according to the following specifications:
   (a)   Initially, both the inputs and outputs are 0.
   (b)   Whenever $z_1 = 0$ and x_1 or x_2 becomes 1, $z_1 = 1$. When the other input becomes 1, $z_2 = 1$.
   (c)   The first input to change from 1 to 0 makes $z_1 = 0$.
   (d) $z_2 = 0$ whenever $z_1 = 0$ and any one of the inputs changes from 1 to 0.

   Implement the circuit with SR latches.

6.30   Design an asynchronous circuit that has two inputs (x_1, x_2) and a single output (z). The circuit is *set* by the input combination

$x_1 x_2 = 10$ so that $z = 1$ after the variable x_2 becomes 1 for the third time without the occurrence of the input combination $x_1 x_2 = 10$. The output remains 1 until the input combination $x_1 x_2 = 10$ *resets* the circuit to the *set* state. The occurrence of $x_1 x_2 = 10$ at any time *resets* the circuit to its *set* state. Implement the circuit with *SR* latches.

6.31 Design the following 2-input (x_1, x_2), single-output (z) asynchronous circuit: The output $z = x_1$ if $x_2 = 1$; but if $x_2 = 0$, the output remains fixed at its last value before x_2 became 0. Obtain the next-state and output equations.

6.32 Design the following asynchronous circuit: The circuit has a single-input push-button switch (x) and a lamp indicating the output (z). The lamp will light whenever the push-button switch is pressed the third time and will stay ON until the switch is pressed again. Obtain the next-state and output equations.

6.33 The asynchronous circuit of Figure P6.33(a) implements the transition table shown in Figure P6.33(b). An analysis of this circuit shows that, if the propagation time delay associated with the inverter is very large, a change of input from $x = 0$ (while $y_1 = y_2 = 0$) to $x = 1$ may result in a transition to stable state 11 instead of to 01. Indicate a point (or points) in the circuit at which the insertion of a delay will prevent the occurrence of the essential hazard.

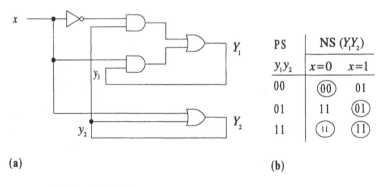

(a)   (b)

■ **Figure P6.33**

7 Synchronous Sequential Circuits

7.1 INTRODUCTION

A **synchronous sequential circuit** is controlled by a timing device, called a **clock generator**, that produces trains of clock pulses. The memory devices in a synchronous circuit change state only in synchronization with a clock pulse. The circuit inputs are allowed to change only when the clock pulses disable the circuit and, thereby, prevent it from changing state. Figure 7.1 is a modification of the basic sequential circuit model of Figure 6.1, showing the clock input to the memory part, and emphasizing that synchronous circuits can be considered as a *special case* of asynchronous circuits.

In Chapter 6, we saw that state transitions in an asynchronous circuit occur only in response to a change in the circuit inputs. Besides that, in making a transition, the asynchronous circuit may go through a succession of unstable states resulting from the propagation delays in the circuit. In contrast, the clock pulses in synchronous circuits *mask* the effects of the delays and, as a result, *all states are stable*. Moreover, the timing in synchronous circuits can be broken down into independent discrete events so that each can be considered separately. When a clock pulse occurs, the circuit is presented with a new input event, regardless of whether the input values have changed or not. To illustrate the distinction between asynchronous and synchronous circuits, assume that we have a single-input (x) sequential circuit. If the circuit is asynchronous and the

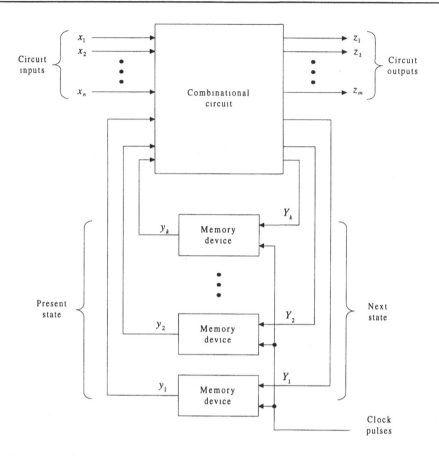

■ **Figure 7.1** Synchronous sequential circuit model

present state is A when $x = 0$, then the circuit must stay in state A as long as $x = 0$. On the other hand, if the circuit is synchronous, then the circuit may make a transition to state B on the next clock pulse even though x remains at 0.

To express the Boolean functions relating the inputs, outputs, and states, let $\mathbf{x} = (x_1, x_2, \ldots, x_n)$ and $\mathbf{y} = (y_1, y_2, \ldots, y_k)$ denote the collections of input and state variables, respectively. Then, the next state variables are given by

$$Y_i(t+1) = G_i(\mathbf{x}(t), \mathbf{y}(t)), \quad i = 1, 2, \ldots, k \tag{7.1}$$

and the output variables for a Mealy circuit are obtained from

$$z_i(t) = F_i(\mathbf{x}(t), \mathbf{y}(t)), \quad i = 1, 2, \ldots, m \tag{7.2}$$

and for a Moore circuit from

$$z_i(t) = F_i(\mathbf{y}(t)), \quad i = 1, 2, \ldots, m \tag{7.3}$$

The analogy between these equations and the corresponding Equations (6.2) through (6.4) for asynchronous circuits is obvious. We have entered the timing element (t) into Equations (7.1) through (7.3) to emphasize the presence of a *controlling* clock input. Nevertheless, we will omit the arguments $(t+1)$ and (t) whenever they are understood.

7.2 FLIP-FLOPS

Latches (see Section 6.2) implement the memory part of asynchronous sequential circuits, while **flip-flops** do the same for synchronous circuits. Latches and flip-flops differ mainly in the way in which they change state. Latches (and gated latches) are *transparent* bistable devices; that is, a change in a latch input affects its output directly. (In a gated latch, the enable control input must also be activated.) The new output value of the latch is delayed only by the propagation time delays of the gates between the inputs and outputs. On the other hand, flip-flops are gated bistable devices without the transparency property. In other words, a change in an input does not affect the flip-flop output directly; rather, the output changes *only* in response to a *transition* in the control (clock) input. Thus, the responses of a flip-flop and a latch are the same except when the inputs change while the control input is activated; in this case, the latch responds to the change, whereas the flip-flop does not. Consequently, a flip-flop can "read in" a new state at the same time that the present state is being "read out." This property, not possible in a latch because of transparency, represents the main advantage of flip-flops over latches.

Basic Flip-Flop Circuits

The most common types of flip-flops are the *SR*, *D*, *JK*, and *T* flip-flops. One way of introducing them is to use the gated *SR* latch discussed in

Section 6.2 and to show how each flip-flop type can be derived from it. However, to do so we must remove the transparency property from the SR latch. Consequently, we will assume in this section that the control input (E) signal [see Figure 6.8(a)] is a very narrow (short duration) pulse so that the latch is enabled (*triggered*) only on the *transition* from 0 to 1 (or from 1 to 0) of this pulse. (Ideally, we can think of the pulses as being impulses of zero width.) We thus "sample" the S and R input signals at the "control pulse times." How this is accomplished in practice will be considered later when we discuss flip-flop triggering techniques.

Using an SR flip-flop to obtain the other flip-flop types does not imply that it is the most important type. In fact, we can hardly find SR flip-flops off the shelf. (Actually, D flip-flops and JK flip-flops are almost the only flip-flops available in integrated-circuit form.) Nevertheless, we use the SR flip-flop because:

1. We are already familiar with the operation of an SR latch.
2. We have already seen in Section 6.6 how to obtain a D latch from an SR latch. Hence, the derivation of a D flip-flop from an SR flip-flop will be straightforward.
3. It is also straightforward to derive the JK and T flip-flops from an SR flip-flop.

SR Flip-Flop

The graphic symbol of an SR flip-flop, shown in Figure 7.2(a), differs from the one for the gated SR latch of Figure 6.8(a) in two aspects: (1) The control input is now referred to as the **clock (C)** input, and (2) a triangle is shown at the C input.

The triangle, called a *dynamic indicator*, indicates that the device responds only to an input clock *transition* from LOW (binary 0) to HIGH (binary 1). Appending a small circle to the C input, as in Figure 7.2(b), indicates that the flip-flop responds to an input clock transition from HIGH to LOW. In either case, the flip-flop is said to be **edge-triggered**, meaning that the flip-flop is sensitive to its S and R input signals either at the *positive edge* or at the *negative edge* of the clock pulse. Hence, Figures 7.2(a) and (b), respectively, depict the graphic symbols of a positive and a negative edge-triggered SR flip-flop.

As we will see later, older flip-flop designs, called *pulse-triggered* (or *master-slave*) flip-flops, operate somewhat differently than the edge-triggered designs. In this type of flip-flop, the information is entered on the leading edge of the clock pulse, but the flip-flop does not change state (the output is *postponed*) until the trailing edge of the clock pulse. Figure 7.2(c) shows the graphic symbol of a pulse-triggered *SR* flip-flop. (The symbol shown by the Q and Q' outputs indicates a postponed output.) Notice that there is no dynamic indicator at the clock input. We will see later that a pulse-triggered flip-flop is implemented with latches.

Figure 7.3 provides the details of a positive edge-triggered *SR* flip-flop by duplicating with minor modifications the corresponding information for the *SR* latch from Figures 6.6(a), (b), and (d). But, remember, the only difference between the *SR* flip-flop and the gated *SR* latch is that *the flip-flop cannot change state except on the triggering edge of a clock pulse*. Hence, while the present and next states in a latch are separated in time by gate delays, they are separated by clock periods in a flip-flop.

In the **characteristic table**, listed in Figure 7.3(a), the Q column denotes the **present state** of the flip-flop, while the $Q(t+1)$ column designates its **next state**. The S, R, and Q values are assumed present *prior* to the application of the next clock pulse, and the next state is attained on the leading edge of the next clock pulse. Since the clock pulse input is implicit to the operation of the flip-flop, there is no need to provide a special column for C in Figure 7.3(a). The operation of the flip-flop is

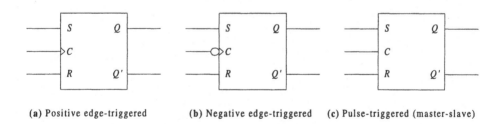

(a) Positive edge-triggered   (b) Negative edge-triggered   (c) Pulse-triggered (master-slave)

■ **Figure 7.2** *SR* flip-flop graphic symbols

triggered (enabled) only when C makes a transition from 0 to 1. Consequently, the S and R inputs are referred to as **synchronous inputs** because data on these inputs can be transferred to the flip-flop's outputs *only* on the triggering edge of the clock pulse. Note, however, that the next state of the *SR* flip-flop is *indeterminate* when $S = R = 1$ and Q is either 0 or 1. Hence, to operate the *SR* flip-flop properly, the S and R inputs must not be 1 simultaneously.

The characteristic table can be reduced to the equivalent table shown in Figure 7.3(b). From either characteristic table we can obtain the **characteristic equation** of the positive edge-triggered *SR* flip-flop:

$$Q(t + 1) = R'Q + S \qquad (SR = 0) \tag{7.4}$$

Note that the condition $SR = 0$ is part of the characteristic equation.

To determine how to achieve a transfer between a given present state and a desired next state, we can use the **excitation table** shown in Figure 7.3(c). Given the present state $Q(t)$ and the desired next state $Q(t + 1)$, the table lists the *excitation values* required at the synchronous inputs *prior* to the application (at time $t + 1$) of the next clock pulse.

S	R	Q	$Q(t+1)$
0	0	0	0
0	0	1	1
0	1	0	0
0	1	1	0
1	0	0	1
1	0	1	1
1	1	0	} Indeterminate
1	1	1	

S	R	$Q(t+1)$
0	0	$Q(t)$
0	1	0
1	0	1
1	1	Indeterminate

$Q(t)$	$Q(t+1)$	S	R
0	0	0	x
0	1	1	0
1	0	0	1
1	1	x	0

  (a) Characteristic table   **(b)** Reduced characteristic table   **(c)** Excitation table

■ **Figure 7.3** *SR* flip-flop operation

D Flip-Flop

The positive edge-triggered D flip-flop, shown in Figure 7.4(a), has a single synchronous input (D) and can be implemented by modifying an SR flip-flop as shown in Figure 7.4(b). The remainder of Figure 7.4 shows the full and reduced characteristic tables and the excitation table of the D flip-flop. These can be easily derived by substituting $S = R'$ into the characteristic table of the SR flip-flop [Figure 7.3(b)]. Correspondingly, the characteristic equation of the D flip-flop is given by

$$Q(t+1) = D \tag{7.5}$$

The D flip-flop is useful when a single bit is to be stored. As seen from Figure 7.4(d), if $D = 0$, the flip-flop resets on the next clock pulse and sets if $D = 1$. Hence, the D flip-flop latches onto the data at the D input, but only in synchronization with the triggering edge of the clock pulse. By contrast, the output of a D latch will follow changes at the D input as long as the enable input is activated.

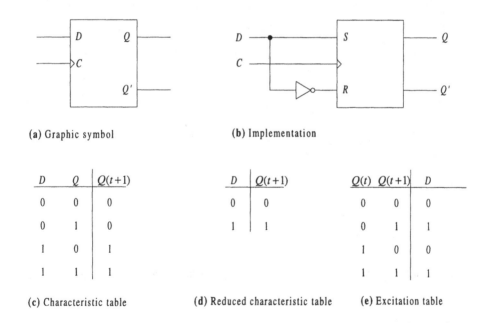

(a) Graphic symbol   **(b)** Implementation

D	Q	$Q(t+1)$
0	0	0
0	1	0
1	0	1
1	1	1

(c) Characteristic table

D	$Q(t+1)$
0	0
1	1

(d) Reduced characteristic table

$Q(t)$	$Q(t+1)$	D
0	0	0
0	1	1
1	0	0
1	1	1

(e) Excitation table

■ **Figure 7.4** D flip-flop

JK Flip-Flop

The positive edge-triggered *JK* flip-flop in Figure 7.5(a) can be implemented by modifying the *SR* flip-flop as shown in Figure 7.5(b). Note that if we restore the transparency property (i.e., remove our assumption that the signal at the *C* input must be of very short duration), then Figure 7.5(b) actually depicts a gated *JK* latch.

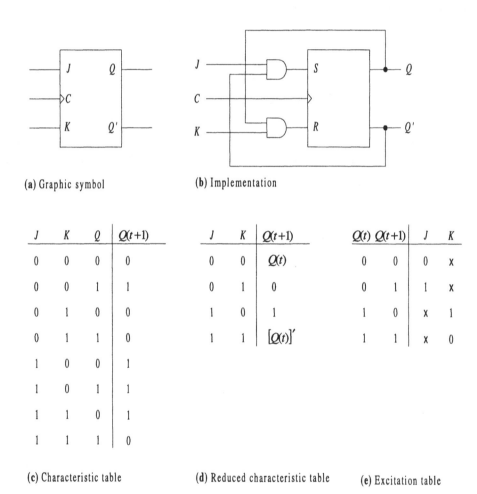

(a) Graphic symbol

(b) Implementation

J	*K*	*Q*	*Q*(*t*+1)
0	0	0	0
0	0	1	1
0	1	0	0
0	1	1	0
1	0	0	1
1	0	1	1
1	1	0	1
1	1	1	0

(c) Characteristic table

J	*K*	*Q*(*t*+1)
0	0	*Q*(*t*)
0	1	0
1	0	1
1	1	[*Q*(*t*)]′

(d) Reduced characteristic table

Q(*t*)	*Q*(*t*+1)	*J*	*K*
0	0	0	x
0	1	1	x
1	0	x	1
1	1	x	0

(e) Excitation table

■ **Figure 7.5** *JK* flip-flop

Observing the characteristic table of the JK flip-flop [Figure 7.5 (c) or (d)], we see that the indeterminate states of the SR flip-flop are now defined. In fact, the JK flip-flop operates exactly like the SR flip-flop as long as $JK = 0$. When $J = K = 1$, the flip-flop *complements* (*toggles*; changes state) on *every* clock pulse. From the characteristic table, we obtain the characteristic equation of the JK flip-flop:

$$Q(t+1) = K'Q + JQ' \tag{7.6}$$

Since the JK flip-flop outperforms the SR flip-flop, the SR flip-flop is no longer manufactured as an off-the-shelf device.

T Flip-Flop

The graphic symbol of a positive edge-triggered T flip-flop is shown in Figure 7.6(a). It can be derived from the SR flip-flop as in Figure 7.6(b). (Alternatively, we can connect together the J and K inputs of a JK flip-flop.) From the characteristic table [Figure 7.6(c) or (d)], we see that a T flip-flop complements on every clock pulse provided that $T = 1$; hence, the designation T (for *toggle*) flip-flop. The characteristic equation of the T flip-flop is given by

$$Q(t+1) = T'Q + TQ' = T \oplus Q \tag{7.7}$$

Flip-Flop Synchronization

We have seen that latches and flip-flops perform the same functions. So why is the removal of the latch transparency property so important to synchronous circuit implementation? To answer this question, consider any of the flip-flops introduced in the previous section and let us drop our assumption about the clock signal by allowing *any* signal at the C input. Being now a gated device, the flip-flop is activated when the control input signal changes *level*. If the synchronous inputs change while C is kept active, the flip-flop output may change more than once. In this case, the output of one flip-flop cannot drive the inputs of another flip-flop when both are activated by the *same* clock pulse. In other words, we cannot *synchronize* their operation.

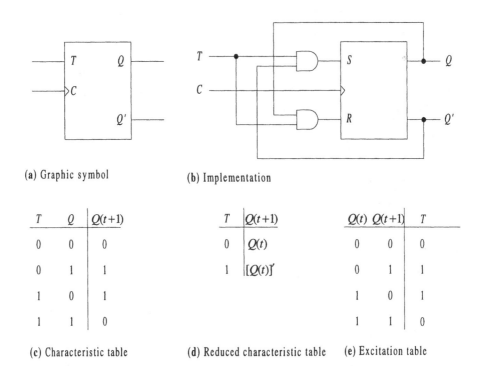

(a) Graphic symbol   (b) Implementation

T	Q	Q(t+1)
0	0	0
0	1	1
1	0	1
1	1	0

T	Q(t+1)
0	Q(t)
1	[Q(t)]'

Q(t)	Q(t+1)	T
0	0	0
0	1	1
1	0	1
1	1	0

(c) Characteristic table   (d) Reduced characteristic table   (e) Excitation table

■ **Figure 7.6** T flip-flop

The two major techniques of synchronizing integrated-circuit flip-flops are *pulse triggering* and *edge triggering*. Recently, however, edge-triggered integrated-circuit flip-flops are manufactured almost exclusively, and pulse-triggered flip-flops are seldom available. Nevertheless, for completeness, we include a discussion of pulse-triggered flip-flops.

Pulse-Triggered Flip-Flops

A **pulse-triggered flip-flop** is a bistable device whose state depends on the values of the synchronous inputs at the leading edge of the clock pulse, but whose state does not change until the trailing edge of the clock pulse. A pulse-triggered flip-flop consists of two latches, where one latch acts as a master and the other as a slave. For this reason, the circuit is also called a **master-slave** flip-flop.

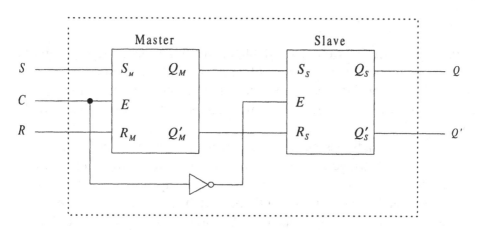

Figure 7.7 Pulse-triggered (master-slave) SR flip-flop

For example, Figure 7.7 shows the block diagram of a master-slave SR flip-flop consisting of two gated SR latches. The graphic symbol for this flip-flop was introduced in Figure 7.2(c). The master is enabled when C changes from 0 to 1. At this point, the slave is inhibited due to the inverter so that the state of the master cannot be transferred to it. But when C returns to 0, the slave is enabled (while the master is disabled), and the information stored in the master is transferred to the slave.

The synchronous inputs must remain stable throughout the period of the clock pulse to prevent the master from changing state more than once. However, since the master is inhibited on the trailing edge of the clock pulse, the synchronous inputs can be changed *at the same time* as the clock pulse goes from 1 to 0. Hence, the input and output information of a master-slave flip-flop can change on the same clock pulse.

Assume now that a number of master-slave flip-flops are connected so that the outputs of some flip-flops drive the inputs of other flip-flops, and all the clock inputs are synchronized (connected to the same clock pulse generator). On the leading edge of each clock pulse, some of the masters will change state, but all the salves will remain at their previous states. On the trailing of the clock pulse, some slaves will change state, but the new states will not affect any of the masters until the leading edge of the next clock pulse. Hence, we can change the states of the flip-flops *simultaneously* during the *same* clock pulse.

Edge-Triggered Flip-Flops

A **edge-triggered flip-flop** is a bistable device whose state depends on the synchronous inputs either at the positive edge or at the negative edge of a clock pulse. Therefore, because input changes do not affect the output except at the triggering edge, the synchronous inputs must remain stable only at that time.

We have already designed an edge-triggered flip-flop in Section 6.6, the positive edge-triggered D flip-flop whose circuit logic diagram is repeated for convenience in Figure 7.8. This flip-flop consists of two "input" SR latches and an "output" SR latch. Assuming that the *setup* time and *hold* time constraints (see Section 6.6) are satisfied, let us ascertain that this design indeed samples the D input signal prior to a positive-going edge in C and then ignores D after the edge (until the next positive-going edge) while the output latch changes state.

We can easily verify that $Y_1 = Y_2 = 1$ when $C = 0$ and D is either 0 or 1. Hence, the output latch is disabled and holds its previous state. If $D = 1$ and C makes a positive-going transition (from 0 to 1), Y_1 changes to 0 while Y_2 remains 1, causing the output latch to set. If D changes while $C = 1$, Y_1 and Y_2 will not change, "locking out" the changes at D. When C returns to 0, Y_1 changes back to 1 and the output latch is disabled.

Similar behavior occurs when $D = 0$ and C makes a positive-going transition. The only difference is that, now, Y_1 remains 1 while Y_2 changes to 0, causing the output latch to reset. Once again, any changes in D while $C = 1$ will be locked out. When C returns to 0, Y_2 goes back to 1 and the output latch is disabled.

Another example of an edge-triggered design, shown in Figure 7.9, depicts a negative edge-triggered JK flip-flop. We leave it to the reader to verify that this device functions as a JK flip-flop when C makes a negative-going transition (from 1 to 0). Changes in J, K, or both while $C = 0$ do not affect Q. Moreover, Q is not affected by a positive-going transition in C, nor is it affected when $C = 1$. To aid in the analysis of this flip-flop, note that edge triggering is accomplished by making the propagation time delay of gates 1 and 2 greater than that of gate 3 and the propagation time delay of gates 4 and 5 greater than that of gate 6.

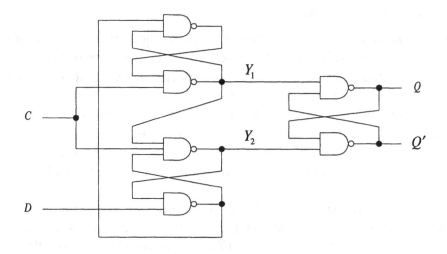

■ **Figure 7.8** Positive edge-triggered D flip-flop

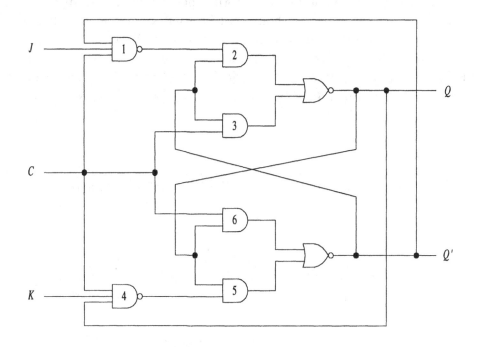

■ **Figure 7.9** Negative edge-triggered JK flip-flop

Asynchronous Inputs

The *synchronous inputs* are so termed because they affect the state of the flip-flop synchronously with the clock pulse. However, most integrated-circuit flip-flops also have one or more **asynchronous inputs** that operate independently of the synchronous inputs and the clock input. The asynchronous inputs are used to set or reset the flip-flop at any time, regardless of the conditions at the other inputs. In other words, they can *override* all the other inputs in order to place the flip-flop in one state or the other. For this reason, the asynchronous inputs are also called *direct inputs*.

Figure 7.10(a) shows a *JK* flip-flop with two asynchronous inputs labeled **present (PR)** and **clear (CL)**. The small circles at the PR and CL inputs indicate that they are active-LOW; that is, the CLEAR and PRESET signals are normally 1, and either will affect the state of the flip-flop when it becomes 0. If the asynchronous inputs are active-HIGH, the small circles would not be shown on the graphic symbol.

The function table of Figure 7.10(b) shows how the asynchronous inputs operate. These inputs should not be used simultaneously because the state of the flip-flop will become indeterminate. When both are 1, the operation of the flip-flop is unaffected by them. If CLEAR = 0 while PRESET = 1, the flip-flop resets provided that Q was 1 or is left in the 0 state if Q was 0. When PRESET = 0 while CLEAR = 1, the flip-flop sets

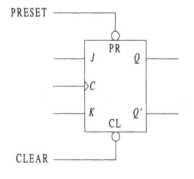

PRESET	CLEAR	Flip-flop Output
0	0	Indeterminate
0	1	Set ($Q = 1$)
1	0	Reset ($Q = 0$)
1	1	Unaffected

(a) Graphic symbol

(b) Function table

■ **Figure 7.10** *JK* flip-flop with asynchronous inputs

if Q was 0 or is left in the 1 state otherwise. As long as CLEAR = 0, the flip-flop will remain reset regardless of the conditions at the other inputs. Similarly, the flip-flop will remain in the set state as long as PRESET = 0. Hence, to resume normal operation, the asynchronous inputs must be returned to 1.

7.3 TIMING CONSIDERATIONS

Clock Rate Determination

When a circuit is clocked, each sequential device must determine its next state on the basis of inputs as they existed during the previous clock period. Consequently, two important timing parameters are associated with sequential devices: the setup time and the hold time. These parameters were already discussed in Section 6.2, so let us just recall their definitions. The *setup time* is the time interval during which the inputs to the sequential device must be held stable *prior* to the occurrence of the triggering edge of the clock input. The *hold time* is the time interval during which the inputs to the sequential device must be held stable *after* the triggering edge of the clock input has occurred.

We must also consider the *propagation time delays* through the combinational part of the sequential circuit. To understand the effect that these delays have, let us refer to Figure 7.1. Observe that, when the flip-flops are clocked, their changing outputs must not be fed through the combinational circuit back to the flip-flops' inputs *faster* than the hold time of the flip-flops. Otherwise, we will get unreliable circuit operation.

Hence, to ensure proper circuit operation, each clock period must allow sufficient time for the following events:

1. The flip-flop outputs must settle to their new states in response to being clocked.
2. The combinational circuit outputs must settle to their new values in response to the new inputs.
3. Before being clocked again, the flip-flops must be allowed enough time to set up to the new output values of the combinational circuit.

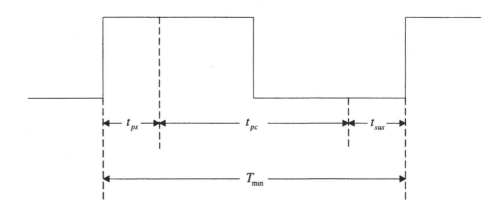

■ **Figure 7.11** Maximum clock rate determination

These timing considerations enable us to determine the *maximum reliable clock rate* for a synchronous design. Since the circuit may be implemented with sequential devices with different timing parameters, worst-case parameters are commonly used to calculate the maximum clock rate. These parameters, shown in Figure 7.11, are defined as follows:

t_{ps}: worst-case, clock-to-output propagation time delay for the slowest flip-flop

t_{pc}: worst-case propagation time delay through the combinational circuit

t_{sus}: worst-case setup time for the slowest flip-flop

Hence, the *minimal* clock period is given by

$$T_{\min} = t_{ps} + t_{pc} + t_{sus}$$

and, therefore, a *conservative* estimate of the maximum clock rate is given by

$$f_{\max} = \frac{1}{T_{\min}}$$

Circuit Inputs Synchronization

In a Mealy synchronous circuit (see Section 6.1), it is crucial to properly synchronize the circuit inputs with the clock; otherwise, the circuit may generate *false* outputs. In fact, there are two situations where this can happen: (1) when the circuit inputs change more than once in between clock pulses, or (2) when the state of the circuit changes, but the inputs are not changed. We can explain the first case by considering Equation (7.2): It shows that the outputs may change momentarily even though the present state is still present.

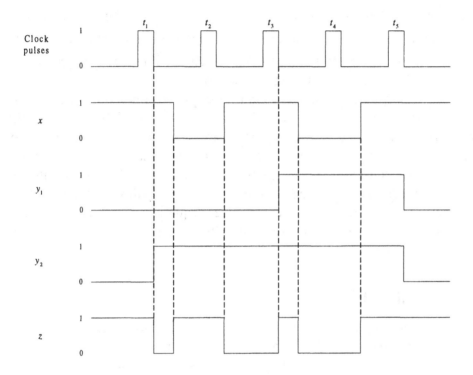

■ **Figure 7.12** Timing diagram of a Mealy synchronous circuit

On the other hand, in the second case, the outputs may change momentarily because the circuit has assumed a new state while the old inputs associated with the previous state are still present. To understand this, consider the timing diagram shown in Figure 7.12. It illustrates the operation of a single-input (x), single-output (z) Mealy circuit implemented with two negative edge-triggered flip-flops (y_1 and y_2), where the output function is given by

$$z = xy_2' + xy_1 + x'y_1'y_2$$

Note that, if x is changed from 1 to 0 *after* the trailing edge of the clock pulses at t_1 and t_3, the circuit has already assumed its next state while x remains at the value associated with the previous state. As a result, rather than maintain an output $z = 1$ through t_2 (as implied by the output function), z becomes 0 momentarily after t_1. Similarly, rather than maintain an output $z = 0$ through t_4, z becomes 1 momentarily after t_3. Hence, to eliminate the false outputs, the circuit input x must change *at the same time* as the trailing edge of the clock pulses.

We can conclude that, to ensure the proper operation of a Mealy circuit, all input changes must be between clock pulses (up to the triggering edge of the next clock pulse) and occur only once. In practice, either we must generate the input sequences in synchronization with the clock, or we must use special circuits to synchronize the inputs with the clock. By contrast, in a Moore circuit, the outputs will not change as long as the present state does not change [see Equation (7.3)], so the circuit inputs do not have to be synchronized with the clock pulses.

7.4 ANALYSIS

In previous chapters, we emphasized time and again the necessity for circuit analysis and the duality between the analysis and design procedures. Moreover, for sequential circuits, we have already established that synchronous circuits are basically a special case of asynchronous circuits. Therefore, it is not surprising that Figure 7.13, depicting the steps involved in analyzing and designing synchronous circuits, is almost identical to the corresponding Figure 6.28 for asynchronous circuits. In fact, the major difference between them is only in the state assignment phase.

For an asynchronous circuit, $N \geq \lceil \log_2 n \rceil$ state variables (memory devices) may not be sufficient to represent n states since additional constraints have to be satisfied to avoid critical races. But in synchronous circuits, $N \geq \lceil \log_2 n \rceil$ state variables are necessary and sufficient to represent n states because the clock enables simultaneous changes in the state-variable values and, thereby, prevents the occurrence of races. Hence, for a synchronous circuit we can choose an *arbitrary* state assignment as long as we are not concerned with the complexity of the combinational part of the circuit.

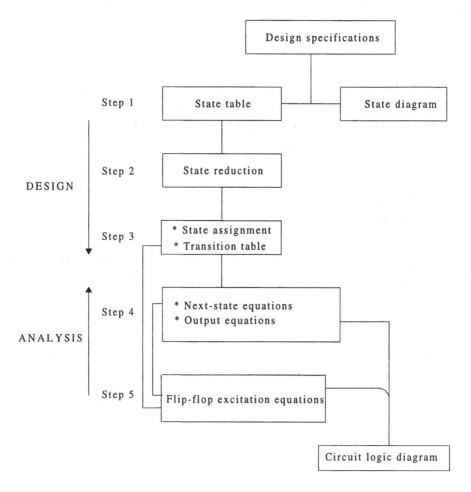

■ **Figure 7.13** Design/analysis flow diagram for synchronous sequential circuits

Analysis Example

To illustrate the analysis procedure for synchronous circuits, consider in Figure 7.14 the single-input (x), single-output (z) circuit consisting of two negative edge-triggered flip-flops connected to the same clock input. The Q outputs of the flip-flops, labeled y_1 and y_2, collectively determine the present state (PS) of the circuit. With two flip-flops, the circuit can be in any one of the following four states: $y_1 y_2 = 00$, 01, 10, or 11.

We analyze the circuit of Figure 7.14 by proceeding "bottom-up" in Figure 7.13.

Excitation Equations

The **excitation equations** express each synchronous input of each flip-flop as a function of the present state and the inputs of the circuit. These Boolean functions are derived directly from the combinational part of the circuit. Hence, we obtain from Figure 7.14

$$J = xy_2 \tag{7.8a}$$
$$K = x + y_2' \tag{7.8b}$$
$$D = x'y_2' + y_1'y_2 \tag{7.9}$$

This information can be transformed into the **circuit excitation table** shown in Table 7.1. The table lists the synchronous input conditions for each (present) state under all possible combinations of the circuit inputs.

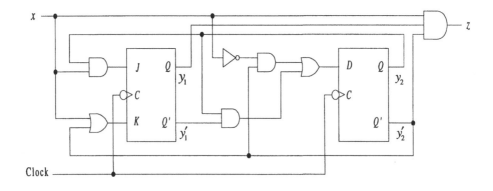

■ **Figure 7.14** Logic diagram of analyzed circuit

■ **Table 7.1**   Circuit excitation table for Figure 7.14

PS	$x=0$			$x=1$		
$y_1 y_2$	J	K	D	J	K	D
00	0	1	1	0	1	0
01	0	0	1	1	1	1
10	0	1	1	0	1	0
11	0	0	0	1	1	0

Output Equations

The **output equations** are also obtained directly from the combinational part of the circuit. These Boolean functions express each of the circuit outputs as a function of the present state and the inputs of the circuit. The same information can also be represented by an **output table** listing the outputs for each (present) state under all possible combinations of the circuit inputs. Thus, from Figure 7.14 we obtain the output equation

$$z = xy_1 y_2' \tag{7.10}$$

and Table 7.2 is the corresponding output table.

■ **Table 7.2**   Output table for Figure 7.14

PS	Output (z)	
$y_1 y_2$	$x=0$	$x=1$
00	0	0
01	0	0
10	0	1
11	0	0

Next-State Equations

To determine the **next-state equations** as a function of the present state and the circuit inputs, we substitute the excitation equations of each flip-flop into its characteristic equation. The characteristic equation of the JK flip-flop [see Equation (7.6)] with $Q(t+1) = Y_1$ and $Q = y_1$ is given by

$$Y_1(t+1) = K'y_1 + Jy_1'$$

Substituting Equations (7.8a) and (7.8b) into this expression, we obtain

$$Y_1(t+1) = (x + y_2')'y_1 + xy_2y_1' = x'y_1y_2 + xy_1'y_2 \qquad (7.11)$$

Similarly, the characteristic equation of the D flip-flop [see Equation (7.5)] with $Q(t+1) = Y_2$ and $Q = y_2$ is given by

$$Y_2(t+1) = D$$

Substituting Equation (7.9) into this expression yields

$$Y_2(t+1) = x'y_2' + y_1'y_2 \qquad (7.12)$$

The next-state equations can be represented collectively in the **next-state table** by listing the transitions between the present and next states of the circuit under all possible circuit input combinations. Table 7.3 is the next-state table for the circuit of Figure 7.14.

State Table

The next-state and output tables can be readily combined to form the **transition table**. Then, the **state table** is obtained from the transition table by using some shorthand notation to designate the states, rather than carry along their binary values. Using the letter assignments

$$a = 00$$
$$b = 01$$
$$c = 10$$
$$d = 11$$

for our circuit, we obtain the state table shown in Table 7.4. Since the circuit is a Mealy circuit, we can merge the next-state and output parts of the table (see Section 6.1) as shown in Table 7.5. Remember, *all the entries in the state table of a synchronous sequential circuit represent stable states.*

■ **Table 7.3** Next-state table for Figure 7.14

PS	NS (Y_1Y_2)	
y_1y_2	$x=0$	$x=1$
00	01	00
01	01	11
10	01	00
11	10	00

■ **Table 7.4** State table for Figure 7.14

PS	NS (Y_1Y_2)		Output (z)	
y_1y_2	$x=0$	$x=1$	$x=0$	$x=1$
a	b	a	0	0
b	b	d	0	0
c	b	a	0	1
d	c	a	0	0

State Diagram

The state diagram provides a graphical representation of the circuit. Derived directly from Table 7.4, the state diagram for our circuit is shown in Figure 7.15.

To find what the circuit is supposed to do, let us assume that it always starts in state a, called the **initial state**. As long as $x = 1$, the next state on successive clock pulses will always be a and the output z will be maintained at 0. If the input is the sequence $x = 0, 1, 0, 1$, the circuit will follow the state transitions $b - d - c - a$, returning to a on the fourth clock pulse and emitting 1 at the output. In fact, we can verify that, for *any* input sequence of *any* length, the output is 1 *only* when the binary pattern 0101 is detected. Therefore, we can call the circuit a **0101 sequence detector**.

We will return to sequence detectors in Section 7.5, where we will actually design one.

■ **Table 7.5** Alternative representation of Table 7.4

PS	NS/Output (Y_1Y_2 / z)	
y_1y_2	$x = 0$	$x = 1$
a	$b/0$	$a/0$
b	$b/0$	$d/0$
c	$b/0$	$a/1$
d	$c/0$	$a/0$

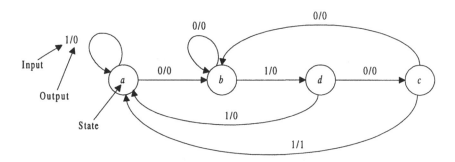

■ **Figure 7.15** State diagram for the circuit of Figure 7.14

Timing and Sequence Diagrams

Figure 7.16(a) shows an example of a **timing diagram** for the 0101 sequence detector. This timing diagram is *ideal* in the sense that we have not considered any of the propagation time delays in the circuit and have assumed that all the signals have zero rise time and fall time. The clock is shown as a periodic signal but could be aperiodic just as well. The values of the input, present state variables (y_1 and y_2), and output are shown between the dashed lines so that they can be related to the clock pulses. To read the values of the next state variables (Y_1 and Y_2) from the diagram, look for the values of y_1 and y_2 immediately following the trailing edge of each clock pulse. Figure 7.16(a) shows that the input sequence 001010101 produces the output sequence 000010001; indeed, the circuit outputs a 1 each time the pattern 0101 is detected.

We can also plot the same information in the form of a **sequence diagram**. Figure 7.16(b) shows the sequence diagram corresponding to the input sequence used in Figure 7.16(a). We leave its verification as an exercise for the reader.

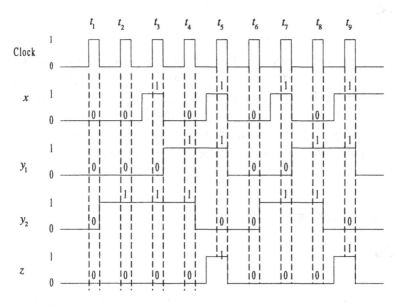

(a) Timing diagram

Clock	t_1	t_2	t_3	t_4	t_5	t_6	t_7	t_8	t_9
x	0	0	1	0	1	0	1	0	1
PS	a	b	b	d	c	a	b	d	c
NS	b	b	d	c	a	b	d	c	a
z	0	0	0	0	1	0	0	0	1

(b) Sequence diagram

■ **Figure 7.16** Time-domain behavior of the 0101 sequence detector

We can ask, what time frame should be allocated to describe the operation of the circuit along the time axis? There is no single answer to this question, but, because any sequential circuit exhibits eventually some kind of *cyclic* behavior, we should allow sufficient time so that some cyclic behavior can be traced. In Figures 7.16(a) and (b) we have allowed enough time to show two such cycles.

7.5 DESIGN

The analysis procedure of digital circuits begins with a logic diagram and culminates in a description of the output functions of the circuit. The design process, on the other hand, starts with a set of specifications and terminates when the logic diagram of the required circuit is produced.

The design of synchronous circuits is shown "top-down" in the flow diagram of Figure 7.13. To illustrate the various steps involved in design, we consider a specific example in the next section. Other design examples will be outlined later.

The Design Process

Design Specifications

We wish to design a synchronous circuit that accepts serially transmitted messages of arbitrary length. The circuit is controlled by a periodic clock. When the circuit is turned ON, the output assumes the value of the first input signal. The output will then change value only if three consecutive input signals have opposite values to the output value.

Let us interpret these specifications. Since the message is transmitted serially, only one input (x) is required. Also, a single output (z) is required so that, initially, $z = 0$ if $x = 0$, or $z = 1$ if $x = 1$. To interpret the last statement of the specifications, consider the following example of an input/output sequence:

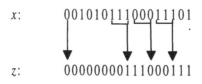

Because $x = 0$ initially, z starts at 0. Only after receiving three consecutive 1's (input signals of opposite value to $z = 0$) does z change to 1. After receiving three consecutive 0's, z changes to 0, and so on. In between these changes, z maintains its last value. The following input/output sequence indicates a similar behavior except that, now, $x = 1$ initially:

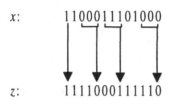

$$x: \qquad 1100011101000$$

$$z: \qquad 1111000111110$$

Observing these sequences, we see that the circuit functions as a *filter*; it filters out input changes that are less than three clock periods. We will, therefore, refer to this example as the *filter example*.

Step 1: State Diagram and State Table

A state table is always the starting point from which the design proceeds. However, the derivation of a state diagram, while not mandatory, often helps to describe the operation of the circuit.

A Mealy state diagram for the filter example is shown in Figure 7.17. Let us briefly explain how it is derived. To begin with, we specify an *initial state* and label it a. Since the circuit has a single output, there are only two possible ways to proceed if a is the present state (PS). When $x = 0$, the output must be 0 too, and the circuit should make a transition to another state labeled b. Similarly, if $x = 1$, $z = 1$ and the next state (NS) is c. If PS $= b$ and x continues to be 0, the output should not change, and the circuit may just as well stay in b (i.e., NS is the same as PS) and continue to produce a 0 output. Hence, b designates the condition that $z = 0$ unless three consecutive inputs of opposite value (1) were received. Similarly, if PS $= c$ and x continues to be 1, NS $= c$ and $z = 1$. Therefore, c indicates the condition that $z = 1$ unless three consecutive inputs of opposite value (0) were received. When PS $= b$, three consecutive 1 inputs would change z to 1 on the occurrence of the third 1 input. Therefore, the circuit makes a transition to state c through states d

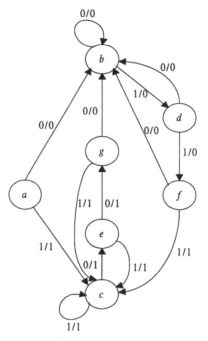

■　**Figure 7.17** Filter state diagram

and f. Similarly, if PS $= c$, three consecutive 0 inputs would change z to 0 on the occurrence of the third 0 input, and the circuit makes a transition to state b through states e and g. Finally, if the present state is either d or f and $x = 0$, then the assumed sequence of three consecutive 1's is broken after one 1 input or two 1 inputs, respectively, and the circuit must

■　**Table 7.6** Filter state table

	NS		Output (z)	
PS	$x = 0$	$x = 1$	$x = 0$	$x = 1$
a	b	c	0	1
b	b	d	0	0
c	e	c	1	1
d	b	f	0	0
e	g	c	1	1
f	b	c	0	1
g	b	c	0	1

go back to b with z unchanged. Exactly the opposite happens if the present state is either e or g.

From the filter state diagram, we readily obtain the state table shown in Table 7.6.

Step 2: State Reduction

Since the filter state table is *completely specified*, we can apply the state-reduction algorithm for completely specified sequential circuits outlined in Section 6.4. We can readily verify that the set of *equivalence classes* is

$$\{a,f,g\},\{b\},\{c\},\{d\},\{e\}$$

resulting in the reduced state table shown in Table 7.7. Note that, although

■ **Table 7.7** Filter reduced state table

	NS		Output (z)	
PS	$x = 0$	$x = 1$	$x = 0$	$x = 1$
a	b	c	0	1
b	b	d	0	0
c	e	c	1	1
d	b	a	0	0
e	a	c	1	1

■ **Table 7.8** Filter transition table

PS	NS $(Y_1Y_2Y_3)$		Output (z)	
$y_1y_2y_3$	$x = 0$	$x = 1$	$x = 0$	$x = 1$
$a = 000$	001	010	0	1
$b = 001$	001	011	0	0
$c = 010$	100	010	1	1
$d = 011$	001	000	0	0
$e = 100$	000	010	1	1

we have reduced the state table from seven to five states, the number of flip-flops required to implement the filter circuit remains the same. Hence, only the complexity of the combinational part of the circuit is simplified in this case.

Step 3: State Assignment

Three flip-flops are required to implement the filter circuit. Since they can represent $2^3 = 8$ distinct states, the three unused state assignments can be utilized as don't-care conditions. We have arbitrarily selected a straight binary assignment for the five states of the reduced state table, resulting in the transition table shown in Table 7.8. The present state variables (the Q outputs of the flip-flops) are labeled y_1, y_2, and y_3, and the corresponding next-state variables are labeled, respectively, as Y_1, Y_2, and Y_3.

Step 4: Next-State and Output Equations

From the transition table, we obtain the K-map for the filter output shown in Figure 7.18(a). The don't-cares correspond to the three unused state assignments. Because x can be 1 or 0 with either one of them, there are six don't-care conditions. The resulting output equation is therefore

$$z = y_1 + y_3'x + y_2 y_3' \tag{7.13}$$

Similarly, we obtain from the transition table the K-maps for the next-state variables shown in Figure 7.18(b). From these maps, we get the next-state equations:

$$Y_1(t+1) = y_2 y_3' x' \tag{7.14}$$
$$Y_2(t+1) = y_2'x + y_3'x \tag{7.15}$$
$$Y_3(t+1) = y_1'y_2'x' + y_3x' + y_2'y_3 \tag{7.16}$$

Step 5: Flip-Flop Excitation Equations

We can convert the next-state equations into the excitation requirements for the flip-flops. Alternatively, rather than derive the next-state equations, we can skip this step altogether by *embedding* the flip-flop

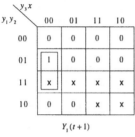

(a) Output map

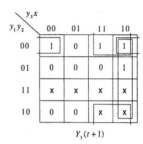

(b) Next-state maps

■ **Figure 7.18** Filter K-maps

excitation requirements into the transition table and deriving the excitation
equations directly from it (see Figure 7.13). We will pursue both
approaches to illustrate them, but first we must select a flip-flop type.

Unfortunately, there is no general algorithm to determine which flip-
flop type will produce a circuit implementation of minimal complexity.
The selection of a flip-flop type may depend on several factors. In some
cases, the choice is restricted by nontechnical considerations, but at other
times the designer must choose a flip-flop type to produce the simplest
circuit according to some minimization criterion, such as a minimum
number of total circuit components or a minimum number of connections
between them.

In our case we are not bound by any consideration other than
demonstrating the design process. Therefore, we choose arbitrarily to
implement the filter using positive edge-triggered *JK* flip-flops.

Derivation of Excitation Equations from Next-State Equations:
Recall that the characteristic equation of a *JK* flip-flop [Equation (7.6)] is
of the form

$$Q(t+1) = K'Q + JQ'$$

Hence, to use the next-state equations to derive the Boolean expressions for the J and K inputs of the flip-flops, we have to rewrite Equations (7.14) through (7.16) in the form

$$Y_i(t+1) = K'y_i + Jy_i', \qquad i = 1, 2, 3$$

Starting with Equation (7.14), we note that its form is not suitable for matching against the characteristic equation because both y_1 and y_1' are missing. To rectify this, we can AND to it $(y_1 + y_1')$ to obtain

$$Y_1(t+1) = y_2 y_3' x'(y_1 + y_1') = (y_2 y_3' x')y_1 + (y_2 y_3' x')y_1'$$

Hence, by comparison with the characteristic equation, we obtain $K_1' = y_2 y_3' x'$, so that

$$K_1 = y_2' + y_3 + x \tag{7.17a}$$

and

$$J_1 = y_2 y_3' x' \tag{7.17b}$$

Similarly, from Equation (7.15) we have

$$Y_2(t+1) = y_2' x + x(y_2 + y_2')y_3' = (xy_3')y_2 + (x)y_2'$$

so that

$$K_2 = x' + y_3 \tag{7.18a}$$

and

$$J_2 = x \tag{7.18b}$$

Finally, Equation (7.16) yields

$$Y_3(t+1) = y_1' y_2' x'(y_3 + y_3') + y_3 x' + y_2' y_3 = (x' + y_2')y_3 + (y_1' y_2' x')y_3'$$

so that

$$K_3 = xy_2 \tag{7.19a}$$

and

$$J_3 = y_1' y_2' x' \tag{7.19b}$$

Derivation of Excitation Equations from Transition Table: Rather than derive the next-state equations and then manipulate them to obtain the excitation equations, we can obtain the excitation equations directly by superimposing the flip-flops' excitation requirements on the transition table, as shown in Table 7.9. Because we do not need the output section of the transition table at this point, only the present- and next-state columns are shown in Table 7.9, which is now referred to as the **circuit excitation table**. Using the *excitation table* of the *JK* flip-flop [Figure 7.5(e)], we fill in the excitation requirements under each input combination. To illustrate how this is done, let us assume, for example, that the present state is $y_1 y_2 y_3 = 011$. From Table 7.9, we see that the next state under $x = 0$ should be $Y_1 Y_2 Y_3 = 001$, implying the following:

1. $y_1(t) = 0$ should remain unchanged at $t+1$ [since $Y_1(t+1) = 0$, too]. From Figure 7.5(e), we see that, to effect a transition from $Q(t) = 0$ to $Q(t+1) = 0$, J should be 0 and K is a don't-care; hence the entries $J_1 = 0$ and $K_1 = \times$ under the heading $x = 0$.
2. $y_2(t) = 1$ should change to 0 at $t+1$ [$Y_2(t+1) = 0$]. From Figure 7.5(e), we find that, to effect a transition from $Q(t) = 1$ to $Q(t+1) = 0$, $J = \times$ and $K = 1$; hence the entries $J_2 = \times$ and $K_2 = 1$ under the heading $x = 0$.
3. $y_3(t) = 1$ should remain unchanged at $t+1$ [$Y_3(t+1) = 1$]. Using the *JK* flip-flop excitation table, we obtain the entries $J_3 = \times$ and $K_3 = 0$ under $x = 0$.

■   **Table 7.9**   Circuit excitation table for the filter example

PS	NS($Y_1 Y_2 Y_3$)		$x = 0$						$x = 1$					
$y_1 y_2 y_3$	$x{=}0$	$x{=}1$	J_1	K_1	J_2	K_2	J_3	K_3	J_1	K_1	J_2	K_2	J_3	K_3
000	001	010	0	×	0	×	1	×	0	×	1	×	0	×
001	001	011	0	×	0	×	×	0	0	×	1	×	×	0
010	100	010	1	×	×	1	0	×	0	×	×	0	0	×
011	001	000	0	×	×	1	×	0	0	×	×	1	×	1
100	000	010	×	1	0	×	0	×	×	1	1	×	0	×

We can now "forget" the next-state (NS) columns of the circuit excitation table because the present-state to next-state transitions have been translated into the synchronous inputs of the flip-flops. We can use K-maps to obtain the Boolean expression for *each* of these inputs as a function of the present state and the circuit input, as shown in Figure 7.19. Note that here, too, we utilize the unused state assignments as don't-care conditions in addition to the don't-cares inherent in the operation of the *JK* flip-flop. The excitation equations shown in Figure 7.19 are the same as those derived in Equations (7.17) through (7.19), except that $K_1 = 1$ in Figure 7.19, while K_1 equals $y_2' + y_3 + x$ according to Equation (7.17a). Both are equivalent, but the first is a minimal expression. (To verify that they are equivalent, use the dashed clusters indicated on the K-map to obtain K_1.)

Circuit Logic Diagram

Figure 7.20 shows the circuit logic diagram for the filter example. A good practice is always to *analyze* the circuit obtained to ensure that it does, in fact, satisfy the design specifications. Analysis is particularly important if unused state assignments were utilized as don't-cares, because they can cause undesirable operation of the circuit.

Unless a sequential circuit is reset to the initial state at the same time it is switched on, the circuit may be in *any* state; in particular, it may be in any one of the unused states. To see what happens in this case, we analyze the circuit of Figure 7.20 and obtain the state diagram shown in Figure 7.21. As we can see, if the circuit is initially started in the *invalid* state 110, it will go to one of the valid states ($a = 000$ or $c = 010$) within one clock pulse and resume normal operation. On the other hand, if the circuit enters one of the other two invalid states (101 or 111), then it may or may not resume normal operation, depending on the input sequence. For example, if x starts with 0 and then changes back and forth between 0 and 1, the circuit will alternate between these two invalid states and will not resume the normal operation of filtering the changes.

A circuit that resumes normal operation within one or two clock pulses by allowing invalid states to go to valid states is called **self-starting**. If the analysis reveals that the circuit is not self-starting, then it must be redesigned. For example, we can redesign the circuit by providing for an automatic reset to an initial state [utilizing the asynchronous clear (CL)

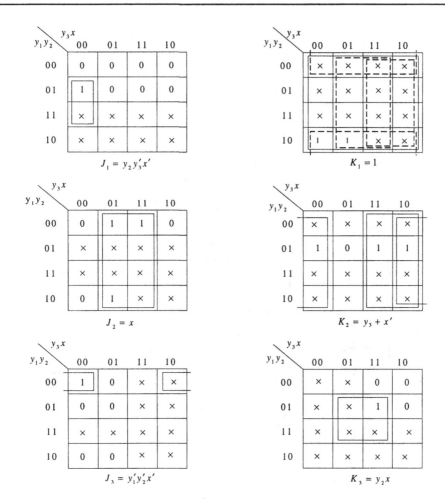

■ **Figure 7.19** Maps for deriving the flip-flops' excitation equations for the filter

and preset (PR) inputs of the flip-flops] or by specifying a valid next state for any unused state. The latter approach means that we would not be able to utilize the unused states as don't-cares (see Problem 7.13).

More Design Examples

Sequence Detectors

To illustrate the design of sequence detectors, let us reconsider the 0101 detector used to demonstrate the analysis process in Section 7.4.

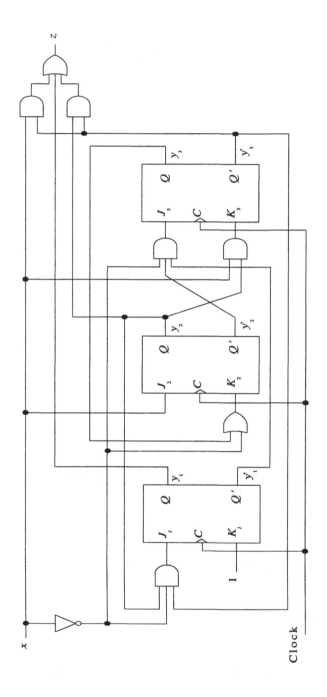

■ **Figure 7.20** Circuit logic diagram of the filter

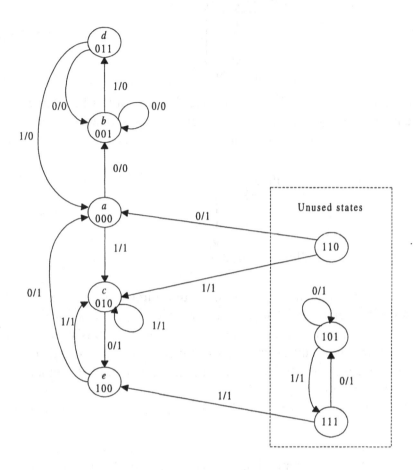

- **Figure 7.21** State diagram for the circuit of Figure 7.20

- **Example 7.1 Design Specifications:** Design a synchronous circuit that will detect every occurrence of 0101 in a serially transmitted message of any length.

Since the message is transmitted serially, only one input (x) is required. To detect the occurrence of the sequence 0101, we need a single output (z); let $z = 1$ designate that the sequence has been detected. To illustrate the operation of the required circuit, consider the following input/output sequence:

x: 0100010101010

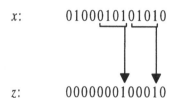

z: 0000000100010

The output sequence shown indicates that the circuit has detected *two* occurrences of 0101. But, for the same input sequence, there is also another possible output sequence:

x: 0100010101010

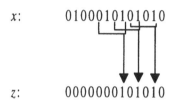

z: 0000000101010

This output sequence shows that the circuit has detected *three* occurrences of 0101. The reason for this is that now we allow the detection of *overlapping sequences*, whereas, previously, we assumed *tacitly* that the circuit returns to its initial state after *every* detection of 0101. Because the design specifications do not indicate which of the two possibilities is required, let us go along with the second. (The 0101 detector in Section 7.4 can detect only nonoverlapping sequences.)

Step 1: A Mealy state diagram is shown in Figure 7.22(a). For illustration purposes, we also show a Moore state diagram in Figure 7.22(b). A Mealy circuit is probably more advantageous to implement since it requires only two flip-flops as opposed to the three required by the Moore circuit. Table 7.10 shows the state table derived from the Mealy state diagram.

Step 2: It can be easily verified that Table 7.10 is irreducible. Hence, two flip-flops are required to implement the circuit.

Step 3: Table 7.11 shows the transition table resulting from the chosen state assignment. The minimal expression for the output function

$$z = xy_1y_2$$

can be derived directly from Table 7.11.

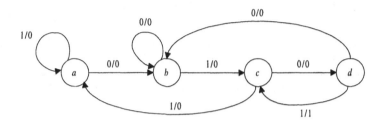

(a) Mealy circuit

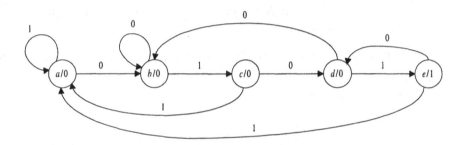

(b) Moore circuit

■ **Figure 7.22** State diagram for the 0101 sequence detector

■ **Table 7.10** State table for the 0101 sequence detector of Figure 7.22(a)

PS	NS		Output (z)	
	$x = 0$	$x = 1$	$x = 0$	$x = 1$
a	b	a	0	0
b	b	c	0	0
c	d	a	0	0
d	b	c	0	1

■ **Table 7.11** Transition table for the 0101 sequence detector

PS	NS (Y_1Y_2)		Output (z)	
y_1y_2	$x=0$	$x=1$	$x=0$	$x=1$
$a=00$	01	00	0	0
$b=01$	01	10	0	0
$c=10$	11	00	0	0
$d=11$	01	10	0	1

Step 4: Table 7.12 shows the excitation requirements for two implementations of the 0101 sequence detector: one using SR flip-flops and the other with T flip-flops. To derive this table, we used the excitation tables for the SR and T flip-flops shown in Figures 7.3(c) and 7.6(e), respectively.

Using four 3-variable K-maps for the SR flip-flops and two 3-variable K-maps for the T flip-flops, the reader can verify the following excitation equations:

$$S_1 = xy_2, \quad R_1 = x \oplus y_2, \quad S_2 = x', \quad R_2 = x$$
$$T_1 = x'y_1y_2 + x(y_1 \oplus y_2), \quad T_2 = (x \oplus y_2)'$$

and derive the corresponding circuit logic diagrams. We can see from the excitation equations that, for the chosen state assignment, a circuit using SR flip-flops would have a less complex combinational part than the circuit obtained if T flip-flops are used.   ■

Counters

Counters are important building blocks in digital systems and can be configured in a variety of forms depending on their application. A counter is a sequential circuit that follows a prescribed sequence of distinct states under the control of input clock pulses. Counters are classified according to the way in which they are clocked as either *ripple counters* or

■ **Table 7.12** Circuit excitation table for the 0101 sequence detector

PS	NS (Y_1Y_2)							$x = 0$					$x = 1$		
y_1y_2	$x = 0$	$x = 1$	S_1	R_1	S_2	R_2	T_1	T_2	S_1	R_1	S_2	R_2	T_1	T_2
00	01	00	0	×	1	0	0	1	0	×	0	×	0	0
01	01	10	0	×	×	0	0	0	1	0	0	1	1	1
10	11	00	×	0	1	0	0	1	0	1	0	×	1	0
11	01	10	0	1	×	0	1	0	×	0	0	1	0	1

synchronous counters and are extensively discussed in Section 7.7. The purpose of the present section is to introduce the fundamentals of synchronous counter design.

The counter counts externally applied pulses (*count pulses*) by cycling through a sequence of states. If the count is incremented for each input pulse, the counter is called an **up counter**. If the count is decremented for each input pulse, the counter is called a **down counter**. We will see in Section 7.7 that a down counter can be reconfigured as an up counter, and vice versa. For this reason, we only consider up counters in this section.

The number of distinct states in the counting sequence is called the **modulus** of the counter. A Modulo-M (abbreviated mod-M) counter is a counter that can count up to M input pulses. For example, a mod-6 counter follows a sequence of six distinct states; it starts in an initial state and returns to it on the application of the 6*th* count pulse. If the modulus is an *integral* power of 2, the counter is said to have a **natural binary modulus**. For example, mod-4, mod-8, and mod-16 counters all have a natural binary modulus. A natural modulus counter requires $N = \log_2 M$ flip-flops to implement and is also referred to as an *N-bit counter*.

The following examples illustrate the design of synchronous counters. Example 7.2 details the design of a mod-8 counter that follows a straight binary count sequence.

■ **Example 7.2** Consider the design of a mod-8 synchronous counter. Since $8 = 2^3$, the counter has a natural binary modulus and is a 3-bit counter. It follows a count sequence of eight distinct states, as shown in the state diagram of Figure 7.23. If we let state s_0 be the initial state, the counter will go to s_1 on the occurrence of the first count pulse, then to s_2 on the occurrence of the second count pulse, and so on. (The transitions are understood to occur whenever a count pulse is present.) On the occurrence of the 8*th* count pulse, s_7 goes to s_0 and the count sequence repeats itself. As we can see, the chosen labels are quite helpful: For every i, s_i simply indicates the number of count pulses that have been applied thus far.

The associated state table is given in Table 7.13. If we want the counter to follow a straight binary code, all that we have to do is assign to s_i the binary equivalent of decimal i, for every i. If we do so and also select negative edge-triggered T flip-flops for implementation, the circuit excitation table for this mod-8 counter can be derived as shown in Table 7.14. (The next-state columns have been omitted from the table.) For example, if the present state is $y_1 y_2 y_3 = 011\,(s_3)$, indicating that three count pulses have been obtained (counted) thus

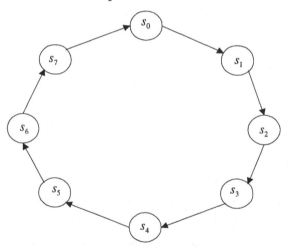

■ **Figure 7.23** State diagram for a mod-8 counter

far, then on application of the 4*th* count pulse the next state will be 100 (s_4). Therefore, because all the flip-flops must complement to accomplish this transition, we must make $T_2 = 1$, $T_1 = 1$, and $T_0 = 1$.

We obtain the excitation equations from Table 7.14,

$$T_2 = y_1 y_0, \quad T_1 = y_0, \quad T_0 = 1$$

and show the circuit logic diagram of the mod-8 binary counter in Figure 7.24. (You may find it instructive to consider how this counter would operate if implemented with positive edge-triggered T flip-flops.)   ■

■ **Table 7.13** State table of a mod-8 counter

PS	NS
s_0	s_1
s_1	s_2
s_2	s_3
s_3	s_4
s_4	s_5
s_5	s_6
s_6	s_7
s_7	s_0

■ **Table 7.14** Circuit excitation table for the counter of Example 7.2

Count Sequence			Flip-flop Inputs		
y_2	y_1	y_0	T_2	T_1	T_0
0	0	0	0	0	1
0	0	1	0	1	1
0	1	0	0	0	1
0	1	1	1	1	1
1	0	0	0	0	1
1	0	1	0	1	1
1	1	0	0	0	1
1	1	1	1	1	1

Note that Figure 7.24 also shows a circuit output $z = y_2 y_1 y_0$, designated the **count output carry**. This output becomes 1 whenever the counter completes its count sequence cycle, that is, on every *8th* pulse. Such an output is very useful when we want to connect counters in cascade. In general, the cascade connection of a mod-*M* counter with a mod-*N* counter results in a mod-*MN* counter. For example, Figure 7.25 shows the cascade connection of two mod-8 synchronous counters. To cascade the counters, we use the count output carry of the least significant digit to drive the count pulse input of the following digit.

We can see that the circuit diagram of the mod-8 counter possesses a well-defined regularity. Moreover, we can exploit this regularity to expand the counter without having to go through the design process again. For example, we can expand the counter to a mod-16 counter simply by adding another T flip-flop and connecting its T input to z and its C input to the count pulses line. If the Q output of the added flip-flop is designated as y_3, then we can easily generate the new count output carry by using a 2-input AND gate; that is,

$$z_{\text{mod}-16} = y_3 y_2 y_1 y_0 = y_3 z_{\text{mod}-8}$$

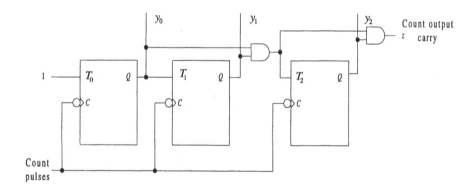

■ **Figure 7.24** Circuit logic diagram for the counter of Example 7.2

Clearly, by adding more flip-flops in this fashion, we can configure a synchronous counter of *arbitrary* natural modulus.

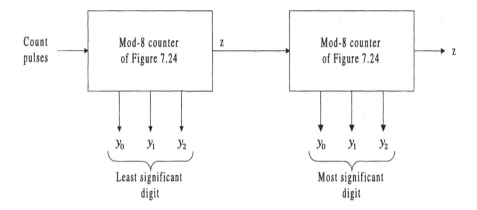

■ **Figure 7.25** Cascade connection of two mod-8 synchronous counters

From Example 7.2 we observe that the state assignment determines the specific code sequence that the counter follows. We illustrate this observation in the following example.

■ **Example 7.3** Consider the design of a 3-bit Gray code synchronous counter. The 3-bit Gray code provides the binary coding for the count sequence and can be obtained from Figure 2.1(d). Any flip-flop type can be used to implement this counter, so let us choose T flip-flops for simplicity.

The circuit excitation table for the 3-bit Gray code counter, shown in Table 7.15, is self-explanatory and provides the following excitation equations and count output carry equation:

$$T_0 = y_0', \qquad T_1 = y_0(y_1 \oplus y_2)', \qquad T_2 = y_0'(y_1 \oplus y_2)$$

and

$$z = y_2 y_1' y_0'$$

We leave it to the reader to draw the circuit logic diagram for the Gray code counter.   ■

The counters designed in Examples 7.2 and 7.3 have a natural binary modulus. In Example 7.4 we illustrate the design of a counter whose modulus is not an integral power of 2.

■ **Table 7.15** Circuit excitation table for the Gray code counter

Count Sequence			Flip-flop Inputs		
y_2	y_1	y_0	T_2	T_1	T_0
0	0	0	0	0	1
0	0	1	0	1	0
0	1	1	0	0	1
0	1	0	1	0	0
1	1	0	0	0	1
1	1	1	0	1	0
1	0	1	0	0	1
1	0	0	1	0	0

■ **Example 7.4**   A mod-10 counter has a count sequence of ten states and requires, therefore, four flip-flops to implement. Such a counter is also called a **decimal counter** or a **decade counter**. If the binary-coded decimal (BCD) format is used for state assignment, the counter is called a **BCD counter** and counts from $(0000)_2$ to $(1001)_2$ and back to $(0000)_2$ on the 10*th* count pulse.

Let us implement a BCD counter using *JK* flip-flops. The circuit excitation table is shown in Table 7.16. Utilizing the unused states that correspond to minterms 10 through 15 as don't-cares, the reader can derive the following excitation equations and count output carry equation:

$$J_0 = K_0 = 1 \tag{7.20a}$$
$$J_1 = y_3'y_0, \qquad K_1 = y_0 \tag{7.20b}$$
$$J_2 = K_2 = y_1y_0 \tag{7.20c}$$
$$J_3 = y_2y_1y_0, \qquad K_3 = y_0 \tag{7.20d}$$

and

$$z = y_3y_0 \tag{7.21}$$

We obtain Equation (7.21) by noting that $z = 1$ only on the 9*th* count pulse.

To see the effect of the unused states on the operation of the counter, we must analyze the BCD counter circuit obtained from the excitation equations. From this analysis we obtain the state diagram shown in Figure 7.26 and observe that the designed counter is *self-starting*. ■

■ **Table 7.16** Circuit excitation table for the BCD counter

Count Sequence				Flip-flop Inputs							
y_3	y_2	y_1	y_0	J_3	K_3	J_2	K_2	J_1	K_1	J_0	K_0
0	0	0	0	0	×	0	×	0	×	1	×
0	0	0	1	0	×	0	×	1	×	×	1
0	0	1	0	0	×	0	×	×	0	1	×
0	0	1	1	0	×	1	×	×	1	×	1
0	1	0	0	0	×	×	0	0	×	1	×
0	1	0	1	0	×	×	0	1	×	×	1
0	1	1	0	0	×	×	0	×	0	1	×
0	1	1	1	1	×	×	1	×	1	×	1
1	0	0	0	×	0	0	×	0	×	1	×
1	0	0	1	×	1	0	×	0	×	×	1

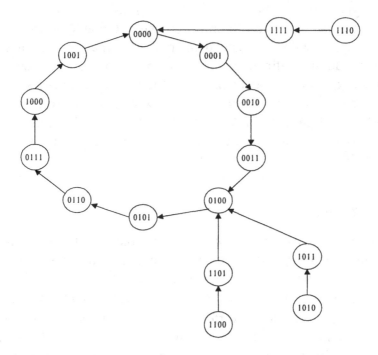

■ **Figure 7.26** State diagram for the BCD counter of Example 7.4

7.6 MSI AND LSI SEQUENTIAL CIRCUITS

Various types of sequential circuits are available in MSI and LSI integrated-circuit (IC) chips and are used extensively in the design of digital systems. The design of digital systems cannot be completely understood without knowing how these devices operate and where they can be applied. The purpose of this and the next section is to introduce the operating principles of some of these components and to list a variety of their applications.

We introduce two classes of sequential circuits: registers and counters. *Registers* are digital circuits commonly used to store binary information. They can be classified into two types: *parallel* registers used solely for storing binary information and *shift* registers used to store information and also to process data. The operating principles of both types are considered in this section. *Counters*, introduced in Section 7.7, are classified as ripple counters or synchronous counters. In a *ripple* counter, the flip-flops are not clocked simultaneously, whereas in a *synchronous* counter all the flip-flops are synchronized (the clock input is common to all the flip-flops).

When we use MSI and LSI chips to design digital systems, we are not really concerned with their actual internal construction but, rather, with their input/output relations. There is no need to design a register or a counter if one is already available on an IC chip. Nevertheless, since we are concerned with design principles, we will find it instructive to study the internal configuration of some available devices. For this reason, the various devices shown are quite similar to those IC chips that are available off-the-shelf.

Parallel Registers

A **register** consists of a collection of binary storage **cells**, each implemented by a flip-flop. The number of cells determines the **length** of the register and, therefore, the length of the binary word stored within the register. Thus, an n-bit register contains n flip-flops and can store n bits. Since n bits give rise to 2^n different binary combinations (values), each of the distinct n-tuples stored in an n-bit register corresponds to a *distinct state* of the register. Therefore, each register state can be associated with a unique item of information, and the system designer is free to define (encode) the meaning of each state.

Two operations, write and read, are associated with a register. To **write** (or **load**) information into the register, each information bit is input to one of the flip-flops. To **read** the contents of a register, we simply access (sample) the Q outputs of all the flip-flops. Since the read and write operations are applied to all the cells simultaneously, the register is commonly referred to as a **parallel register**.

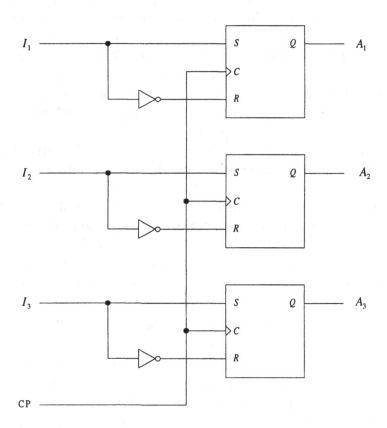

- **Figure 7.27** Three-bit parallel register

Figure 7.27 shows an example of a 3-bit parallel register implemented with three positive edge-triggered SR flip-flops. The clock pulse (CP) input, common to all the flip-flops, controls the loading of new data into the register. Therefore, when the CP input goes from 0 to 1, the information present on the input lines transfers to the flip-flops. For example, if $I_2 = 1$, then A_2 will be 1 on the next positive-going transition of CP; on the other hand, if $I_2 = 0$, then A_2 will be 0 on the next transition of CP from 0 to 1. If CP is kept 0, then the content of the register cannot be changed.

The register shown in Figure 7.27 is perhaps the simplest one, consisting only of flip-flops with no external gates (except those required for proper excitation of the flip-flops). Its simplicity, however, might be a disadvantage if the register is incorporated into a digital system. The digital system is controlled by a central clock pulse generator that produces a continous train of clock pulses. Because the register of Figure 7.27 uses the CP input to control the transfer of information, a continuous train of pulses on this input would force the register to change state on *every* clock pulse, rendering its operation uncontrollable.

This problem is commonly solved by adding a **load control** input to the register. Figure 7.28 shows the 3-bit register of Figure 7.27 modified to incorporate the load control input (at the expense of additional 2-input AND gates). Although the clock pulses are continuously present, the LOAD input, rather than CP, now controls the operation of the register. If LOAD = 0, the AND gates are disabled, the S and R inputs of each flip-flop are 0, and the register cannot change state on any clock pulse. On the other hand, when LOAD = 1, the AND gates are enabled and the input information can be transferred into the register on the next triggering transition of CP. Once new information is transferred into the register, the previously stored information will be destroyed.

Many applications require that the register be cleared (its content made 0) *prior* to its synchronous (clocked) operation. To accomplish this, the flip-flops shown in Figure 7.28 also include an asynchronous clear (CL) input. The small circle at the CL input indicates that the flip-flop is cleared when the CLEAR signal becomes 0. As long as CLEAR = 0, the content of the register will be 0 independent of the values of CP and LOAD. To resume normal clocked operation, the CLEAR input must be brought to 1 and kept at 1. The register is usually cleared when LOAD = 0 and in between clock pulses (i.e., when CP = 0 in our case) by placing a momentary negative-going pulse on the CLEAR line.

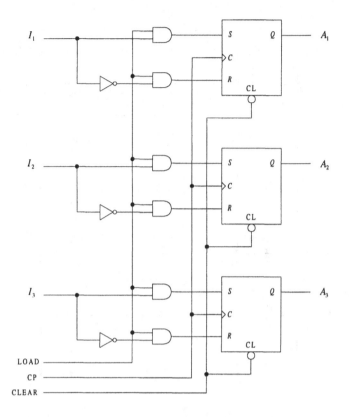

■ **Figure 7.28** Three-bit parallel register with LOAD and CLEAR

Examples of Parallel Register Applications

Parallel Transfer

As we mentioned before, parallel registers are used to store information. Information transfers can take place between registers or between an external source and a register. The block diagram of Figure 7.29 shows a parallel transfer of information between two n-bit registers constructed with SR flip-flops. To explain this transfer, assume that register A has already been loaded with some information. Prior to transferring this information into register B, we clear register B by making CLEAR $= 0$. The CLEAR input is then returned to 1 and, at the appropriate *transfer time t*, the LOAD line of register B is raised to 1. Hence, at the next

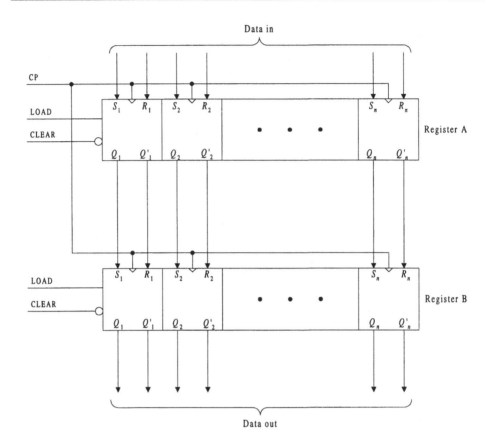

■ **Figure 7.29** Parallel information transfer between registers

clock pulse, the entire content of register A will be loaded into register B. Notice in passing that we disposed of the inverters at the R inputs in register B by using both the Q and Q' outputs of the flip-flops in register A.

Parallel Binary Addition with Register Storage

The parallel binary adder introduced in Section 5.5 accepts two binary numbers and outputs their sum. Usually, the two numbers are stored in registers and are transferred one after the other to the parallel adder unit. The circuit of Figure 7.30, consisting of two 4-bit registers constructed

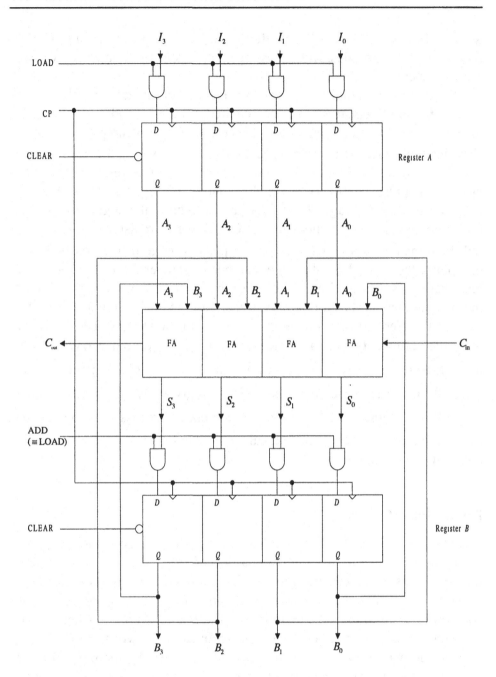

■ **Figure 7.30** Four-bit parallel adder with register storage

with D flip-flops and a 4-bit parallel binary adder unit consisting of four full-adder (FA) modules, illustrates register transfer in the addition of two 4-bit numbers.

To see how the addition of two binary numbers is accomplished, first assume that both registers have been cleared. Since we add 4-bit numbers, the carry-in (C_{in}) is kept equal to 0 because there is no carry into the least significant bit position of the parallel adder. By activating the LOAD line of register A, we load the first number (the augend) into register A on the next clock pulse (CP). Because the content of register B is 0 while register A holds the augend, the adder unit produces the augend at its sum (S) outputs. Activating the ADD (LOAD) line of register B, the augend will be transferred to register B on the positive edge of the next CP. At this point, the second number (the addend) is transferred to register A by reactivating its LOAD line. The adder unit will now produce the sum of the two numbers at its sum outputs. Reactivating the ADD line of register B will transfer the sum into register B on the next CP. The *overflow* from the addition process is indicated by C_{out}, the output carry bit from the adder unit. If the sum exceeds binary 1111, $C_{out} = 1$; otherwise, $C_{out} = 0$. In either case, the final result is given by the 5-bit number $C_{out}B_3B_2B_1B_0$. Figure 7.31 summarizes the entire process by showing the timing diagram of the control signals associated with the parallel binary addition of 4-bit numbers.

Read/Write Memory

A memory unit in a digital system is a collection of registers used to store information relevant to the operation of the system. The binary information is organized into groups called **words**. Each word is stored in a memory register and is treated as an entity. The word length is sometimes referred to as the memory **width**. Each memory register (word) is assigned a unique number, called the **address** of the word, starting from 0 up to the maximum number of words available in the memory. Figure 7.32 shows a pictorial view of a memory unit containing n m-bit words. The **capacity** (size) is the total number of bits that a memory can store. Thus, the storage capacity of the memory unit shown in Figure 7.32 is nm

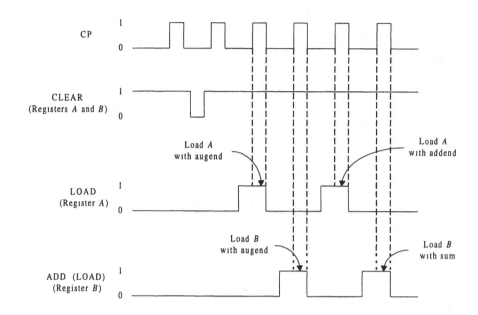

■ **Figure 7.31** Control timing diagram for parallel binary addition

bits. For large memory capacities, we use the abbreviation K to designate multiples of $1024 (= 2^{10})$. (Note that although K is called *kilo*, it does not stand for 1000.) For example, the capacity of a 32K memory is 32,768 bits. Frequently, the smallest addressable memory unit is a byte (8 bits) and the capacity is then expressed in *kilobytes* (KB). Hence, a 64KB memory can store 65,536 bytes (or 524,288 bits). For very large memories, we use the abbreviation M (*mega*) to designate multiples of 2^{20}. Thus, a 1 MB memory can store 1,048,576 bytes.

We have already encountered three types of memory in Section 5.9: the Read-Only Memory (ROM), the Programmable Logic Array (PLA), and the Programmable Array Logic (PAL). These are storage devices in which a *fixed* set of binary information is embedded into the unit and cannot be changed in real time during the system operation. They are logically organized as shown in Figure 7.32; however, due to the process of implanting the information into them, words are not stored in registers per se. Rather, programmable arrays are strictly combinational devices from which data can only be read.

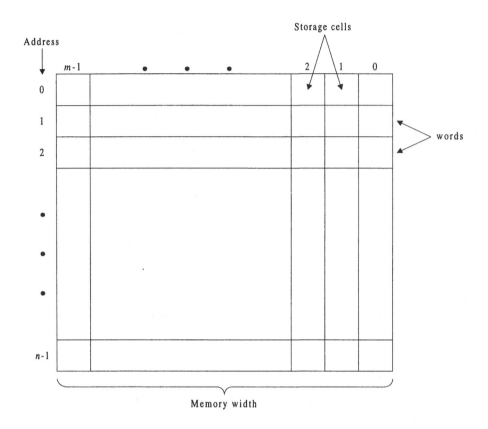

■ **Figure 7.32** Logical organization of a memory unit

A different type of memory, called a **read/write memory (RWM)**, is required if the contents of the registers need to be changed during system operation. The process of storing data in a RWM is called *writing* into memory. Writing into a RWM is a *destructive* process because the data stored in the location are destroyed when the new data are written in. The process of retrieving data from a RWM is called *reading* from memory and is usually a nondestructive process. Writing into or reading from a memory location is done by accessing the location through its address.

The operating speed of a memory is measured in terms of its **access time,** the amount of time required to perform a read operation (or a write operation). If the access time is the same for any address in memory, the memory is called a **random-access memory (RAM)**. (In fact, the term RAM is commonly used, albeit incorrectly, to denote a random-access RWM.) In other words, the actual physical location of a memory word has no effect on how long it takes to read from or write into that location. For example, programmable arrays are (*read only*) random-access memory devices. Most RWMs are implemented with semiconductor integrated circuits and are random access. One disadvantage of semiconductor memories is that they lose the information stored in them when power is interrupted or removed. Such a memory is called a **volatile** memory. In contrast, programmable arrays are *nonvolatile* because of the way in which they are constructed. Other examples of nonvolatile memory devices are magnetic tapes, disks, and diskettes.

Figure 7.33 shows a block diagram of a random-access RWM unit. To access a memory word, we specify its address in binary form. If the number of words is N and if we were to access each memory register *directly*, then the number of address lines required would be very large

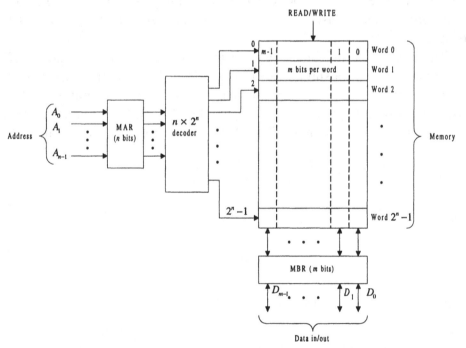

■ **Figure 7.33** Organization of a random-access RWM

even for a relatively small N. Therefore, to keep the number of address lines manageable, we use a register-decoder combination. The register, called the **memory address register (MAR)**, is used to specify the address of the memory word selected. Hence, a combination of an n-bit MAR and an $n \times 2^n$ decoder can specify up to 2^n memory words. For example, a memory whose capacity is 32,768 (32K) bits requires 12 address lines if the memory width is 8 bits (because $32768/8 = 2^{12}$) or 13 address lines if the memory width is 4 bits (because $32768/4 = 2^{13}$).

To transfer information to and from memory locations, we use a special-purpose register called the **memory buffer register (MBR)**. The length of the MBR is equal to the memory width. Reading from or writing into a memory register is controlled by the READ/WRITE line. To *read* the memory, the address of the memory word is placed in the MAR, the READ/WRITE signal is set to 1, and a copy of the content of the addressed word is then brought into the MBR by the memory logic circuit. On the other hand, to *write* a word into memory, the data are placed in the MBR, the address of the location into which the data are to be written is loaded into the MAR, and the READ/WRITE line is set to 0. The memory logic circuit then transfers the content of the MBR into the addressed location.

The basic building block of a memory register is the **memory cell (MC)** that can store a single bit. The circuit logic diagram of such a cell, shown in Figure 7.34(a), is implemented with a D flip-flop, but other flip-flop types would have been equally suitable. The cell has one input, two control lines, and one output. The ENABLE control line selects the cell

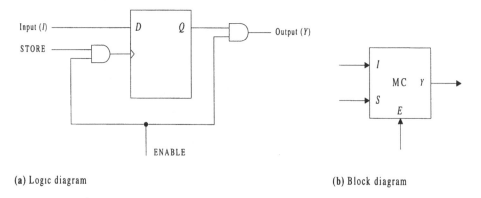

(a) Logic diagram   (b) Block diagram

■ **Figure 7.34** Memory cell

for reading or writing, and the STORE control line enables the information bit present at the input (I) to be written into the cell. Hence, if ENABLE$=1$, then raising STORE to 1 will *set* the flip-flop ($Q=1$) if $I=1$, or *reset* it ($Q=0$) if $I=0$. To read the cell, STORE is kept at 0, ENABLE is raised to 1, and the information in Q is presented to the output Y. A block diagram representation of the memory cell is shown in Figure 7.34(b).

We can use memory cells to construct a random-access RWM unit as shown in Figure 7.35. The capacity of this memory unit is $4\times3=12$ bits, implying a 3-bit MBR and an addressable space of 4 words. To select these words, we need a 2×4 decoder and, therefore, a 2-bit MAR. The operation of the memory unit is controlled through the MEMORY ENABLE input to the decoder. When this line is 0, all the decoder outputs are 0 and none of the memory words are selected; if MEMORY ENABLE $=1$, one of the four words is selected depending on the value of the two address lines A_1 and A_0.

The 3-bit MBR in Figure 7.35 is comprised of three D flip-flops, each associated with two AND gates and one OR gate. The right-hand AND gates are enabled by the read/write (R/W) control and the left-hand AND gates are enabled by (R/W)$'$. The second input to each of the right-hand AND gates comes from the Y outputs of the MCs in the column above it. These outputs are *wire-ORed* (see Section 3.7) to the AND gate input. The second input to each of the left-hand AND gates is connected to an input (I) line. In a write mode, the information on these input lines is written into the memory via the MBR. The outputs of the two AND gates are ORed to the corresponding D input of the MBR flip-flops so that

$$D_i = (R/W)Y_i + (R/W)'I_i, \qquad i=0, 1, 2$$

Finally, the Q outputs of the MBR flip-flops are each connected to the I inputs of the MCs in the column above it.

To understand the operation of this memory unit, consider for example the process of accessing memory word 2, first for write and then for read. In each case, we must go through the following steps:

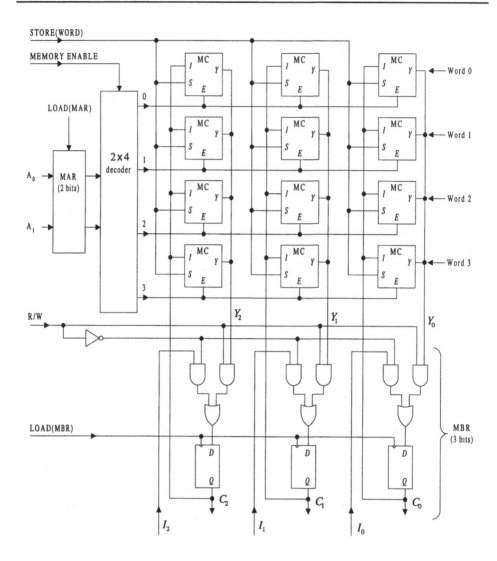

■ **Figure 7.35** Organization of a 4×3 random-access RWM unit

Writing into memory:

  1. Place the address of the selected memory register in the MAR:
     (a) $A_1 A_0 = 10$
     (b) $\text{LOAD(MAR)} = 1$

2. Raise MEMORY ENABLE to 1. The output labeled 2 of the decoder is now 1, enabling all the MCs in the corresponding (word 2) row.
3. Set R/W to 0, indicating that information is to be written into memory.
4. Set LOAD(MBR) to 1 to place the information present on the input lines I_2, I_1, and I_0 into the MBR. The input information is now available at the Q outputs of the MBR flip-flops.
5. Raise STORE(WORD) to 1 to transfer the information from the MBR into the MCs of word 2.

Reading from memory:

1. Place the address of the selected memory register into the MAR:
   (a) $A_1A_0 = 10$
   (b) $LOAD(MAR) = 1$
2. Raise MEMORY ENABLE to 1. The output labeled 2 of the decoder is now 1 and the MCs of word 2 are enabled.
3. Set R/W to 1, indicating that information is to be read from memory.
4. Set LOAD(MBR) to 1 to transfer the information present at the Y outputs of the MCs corresponding to word 2 into the MBR.

Figures 7.36(a) and (b) show the timing diagrams of the control signals associated with the write and read operations, respectively. Notice that the occurrences of some control signals must be staggered to allow for various propagation time delays through the memory circuits.

Shift Registers

A **shift register** consists of flip-flops connected such that the output of one flip-flop feeds the input of the next flip-flop. The clock inputs of all the flip-flops are connected to a common clock pulse source. The number of flip-flops in the chain determines the **length** of the shift register.

Figure 7.37 shows an example of a 4-bit shift register implemented with SR flip-flops. The Q and Q' outputs of each flip-flop are

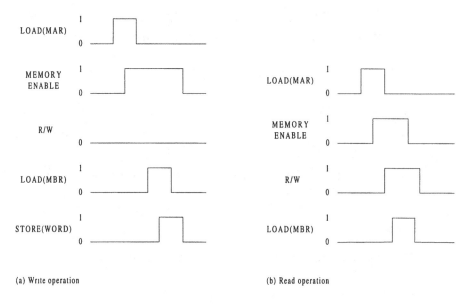

(a) Write operation   (b) Read operation

■ **Figure 7.36** Memory timing diagram

connected, respectively, to the S and R inputs of the flip-flop to its right. (Had we used other types of flip-flops, the interconnections would have to match their excitation requirements.) All the flip-flops are connected to a common clock and are cleared asynchronously when the CLEAR input changes from 1 to 0. The **serial input (SI)**, x, determines the input information bit to the leftmost flip-flop on the next clock pulse. The **serial output (SO)**, z, is taken from the Q output (or the Q' output, or both) of the rightmost flip-flop and is sampled *prior* to the next clock pulse.

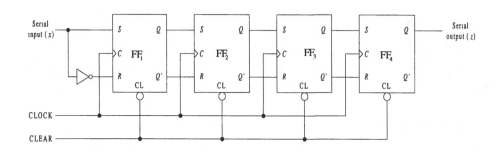

■ **Figure 7.37** Four-bit shift register

To understand how this shift register operates, consider the timing diagram of Figure 7.38. Assume that the register is cleared first and that x is a single 1 pulse. At the next clock pulse, the serial input bit will be stored in FF_1. On subsequent clock pulses, the bit will be *shifted to the right* through the chain FF_2, FF_3, and FF_4 and will become available at the serial output z after four clock pulses. The amount of time (or, equivalently, the number of clock pulses) necessary to shift the information from the input to the output is called the **shift period**. The shift period of the 4-bit shift register in Figure 7.38 is four clock pulses. In general, the shift period for an n-bit shift register is n clock pulses.

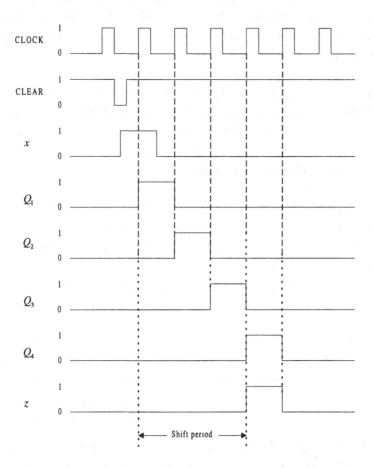

■ **Figure 7.38** Timing diagram for the shift register of Figure 7.37

A shift register not only stores but also processes the data stored in it. To illustrate this point, recall the discussion of binary multiplication and division in Section 2.4. We saw that multiplication (division) of a binary number by 2^k is accomplished by shifting the number k positions to the left (right). These processes can be implemented using a shift-left (shift-right) register. We will consider other applications of shift registers in the next section.

Although the register of Figure 7.37 shifts its content to the right, we can easily reconstruct the register to shift its content to the left. In general, a shift register that can shift in only one direction is called a **unidirectional** shift register. But, with some additional logic gates, we can construct a shift register that can shift in *both* directions (not simultaneously of course). A right/left shift register is referred to as **bidirectional**.

Shift registers are also classified by their loading and output capabilities. The register of Figure 7.37 is called a **serial-in/serial-out (SISO)** shift register because information bits are both loaded and read out serially. If all the Q outputs are accessible and can be read at the same time, the shift register is said to have a parallel output capability and is called a **serial-in/parallel-out (SIPO)** shift register. Note that a parallel-out shift register is also capable of serial-out.

A shift register can have a parallel load capability. In this mode, data are loaded into all the flip-flops simultaneously. A serial-out shift register with parallel load is called a **parallel-in/serial-out (PISO)** shift register, but if the register is also capable of parallel outputs, it is called a **parallel-in/parallel-out (PIPO)** shift register. Any parallel-in shift register is also capable of serial-in.

We can configure a **universal** shift register that combines all these features, as shown in Figure 7.39. The register consists of four SR flip-flops and four 4×1 multiplexers (MUX). The mode of operation is selected using two control variables, s_1 and s_0, common to the four multiplexers. Table 7.17 relates the state of the control variables to the mode of operation. When $s_1 s_0 = 00$, input 0 of each multiplexer is selected so that the Q output of the corresponding flip-flop is connected to its S and R inputs. Hence, the next clock pulse transfers into each flip-flop the value it held previously, so the content of the register remains

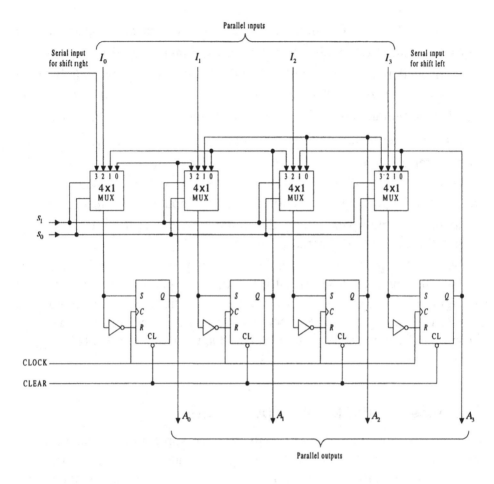

■ **Figure 7.39** Four-bit universal shift register

■ **Table 7.17** Function table for the shift register of Figure 7.39

| Control | | Mode |
s_1	s_0	
0	0	No change
0	1	Shift left
1	0	Parallel load
1	1	Shift right

unchanged. If $s_1 s_0 = 01$, input 1 of each multiplexer is selected and the register is configured to shift left. The serial input enters flip-flop A_3 via its associated multiplexer, and the content of the register can be read serially through A_0 or in parallel by accessing the Q outputs of all the flip-flops simultaneously. When $s_1 s_0 = 10$, input 2 of each multiplexer is selected and the binary data on the parallel input lines I_0, I_1, I_2, and I_3 are transferred simultaneously into the register on the next clock pulse. If $s_1 s_0 = 11$, input 3 of each multiplexer is selected and the register is configured for shift-right operation. The serial input enters flip-flop A_0 via its associated multiplexer, and the content of the register can be read serially through A_3 or in parallel.

MSI shift registers are available in a variety of forms: serial-in/serial-out, parallel-in/parallel-out, serial-in/parallel-out, parallel-in/serial-out, and configurations capable of any of these combinations. Also available are left-shift, right-shift, and bidirectional registers. Besides that, most of these registers come with several bit lengths, with 4, 5, and 8 bits being the most common.

Examples of Shift Register Applications

We present three examples of shift register applications: data format conversions, serial transfer between registers, and serial binary addition. A fourth application, that of using a shift register to implement a ring counter, is discussed in Section 7.7.

Data Format Conversions

In a variety of applications, data are transferred in parallel within the digital system, but have to be serially transferred in and out of the system for communication with external devices. For example, data are handled in parallel inside a digital computer, but transferred serially between the computer and a remote terminal. To implement this communication process, we need to convert the data formats. For information sent from the computer to the terminal, a *parallel-to-serial* data converter is required, but a *serial-to-parallel* data converter is needed when the information is transmitted from the terminal to the computer.

To implement the process of converting the data from a parallel transfer mode to a serial transfer mode, we can use a PISO shift register. With a PISO register, data are loaded in parallel and then clocked out of the serial output one bit at a time. To implement the reverse process, that of serial-to-parallel format conversion, we can use a SIPO shift register. Now data are clocked in through the serial input and are available in parallel once the register has been loaded.

The communication process just described is actually more complex than it appears at first sight. Because the registers are of finite length, the data streams must be divided into *frames*. The frame length is equal to the length of the register, and each frame is considered as a separate entity. To delineate the frame boundaries, the frame contains, in addition to the data, a *header* and a *trailer*. In a parallel-to-serial transfer, each frame is loaded into the PISO shift register and then clocked out serially. In a serial-to-parallel transfer, the transmitter organizes the serial bit stream into frames and clocks in each frame serially into the SIPO shift register. In each case, the receiver recognizes that a frame has been received by looking for the header and trailer information.

Serial Transfer Between Registers

Previously, we saw that the transfer of information into an n-bit parallel register (or between n-bit parallel registers) requires n input lines. However, if the data are transmitted over a long distance, it may be impractical to use n communication lines to transfer the n bits simultaneously. In situations like this, the information is transmitted serially one bit at a time, rather than in parallel.

Serial data transfers can be accomplished using shift registers, as shown in Figure 7.40(a). The two shift-right SISO registers are connected such that the serial output (SO) of the *source* register X goes to the serial input (SI) of the *destination* register Y. Their operation is synchronized by having their clock (CP) inputs connected together. The *shift control* input determines when the registers are shifted and the number of shifts allowed. This is accomplished by using an AND gate that allows clock pulses (on the CLOCK line) into the CP inputs only when the shift control is 1.

For example, assume that the registers are three bits each and that we want to transfer the entire content of register X into register Y. The timing diagram associated with this transfer is shown in Figure 7.40(b).

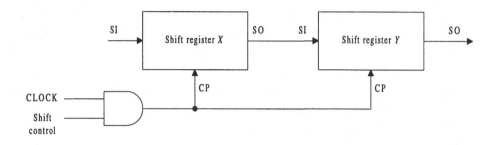

(a) Block diagram

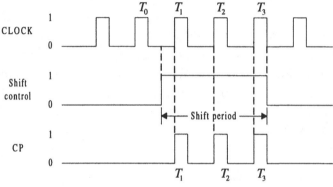

(b) Timing diagram

■ **Figure 7.40** Serial data transfer

Because the number of shifts required for this transfer is three, the shift period is equal to three clock pulses, and the shift control line is set to 1 for the length of this period. Note that the shift control signal in Figure 7.40(b) is actually set to 1 prior to the commencement of the transfer operation in order to take care of existing propagation time delays. In fact, the shift control signal can be activated asynchronously between clock pulses, or it can be synchronized with the trailing edge of a clock pulse. In either case, when the shift control signal is set to 1, the next three clock pulses, T_1, T_2, and T_3, are fed into the CP terminals of the two registers. On the trailing edge of the third clock pulse, the shift control is changed to 0 and the registers are inhibited.

To see how the serial transfer is actually accomplished, assume that the initial content of register X is 101 and that of Y is 010. Assume also that the serial input (SI) of register X is connected to 0. As shown in Table 7.18, the serial transfer from X to Y occurs in three steps. Note that the initial content of Y is lost and that the final content of X is zero (since we assumed that its serial input is permanently connected to 0). In order not to lose the data stored in the source register, the serial output (SO) of X must be also connected to its serial input. In this way, the content of register X will be fed back to itself at the same time as it is transferred to register Y.

We can now compare the serial data transfer discussed above to the parallel data transfer. A serial transfer of N bits requires N clock pulses and a single communication line, whereas a parallel transfer of N bits requires a single clock pulse, but N communication lines. A parallel transfer, then, is much faster but less economical than serial transfer. The choice between serial or parallel data transfer depends on the particular system application and is basically a trade-off between the *economy over distance* of the serial transfer and the *speed* of the parallel transfer. Often, a combination of the two types of transfer is used to realize at least some of the advantages offered by both.

■ **Table 7.18** Example of serial transfer

	Shift Register X			Shift Register Y			SO of Y
Initial value	1	0	1	0	1	0	→ 0
After T_1	0	1	0	1	0	0	→ 1
After T_2	0	0	1	0	1	0	→ 0
After T_3	0	0	0	1	0	1	→ 1

Serial Binary Addition

A parallel binary adder performs the addition of two binary numbers at a relatively fast rate because all the bit positions are operated on simultaneously. The speed of a binary parallel adder is limited by the carry propagation time, but this can be minimized using a *look-ahead carry* technique (see Section 5.5). Nevertheless, the major disadvantage of the parallel adder is that the complexity of the logic circuit increases in direct proportion to the number of bits being added. By contrast, serial binary addition results in a much simpler logic circuit, but one that operates at a much slower speed. Once again we encounter a trade-off between speed and economy. In a digital computer, for example, speed is of the essence, so parallel processing of data is used despite the complexity of the logic circuits. On the other hand, if speed is of secondary consideration, serial processing of data can be used, resulting in logic circuits of reduced complexity.

The 4-bit serial binary adder, shown in the block diagram of Figure 7.41, uses two 4-bit, shift-right, SISO registers to store the numbers to be added. The loading time of the registers (prior to the commencement of the addition operation) can be decreased by using registers with parallel load capability. The least significant bits (*lsb*) of the two registers, in cells A_0 and B_0, are fed through the serial outputs (SO) to a single full-adder (FA) unit, together with the Q output of the carry flip-flop. The carry flip-flop stores the carry output (C_0) of the FA so that the carry can be added to the next significant bit position of the numbers in the registers. The operation of the carry flip-flop is synchronized with the occurrence of the shift-right (SR) control signal and successive clock pulses by using an AND gate that feeds into the flip-flop's clock (C) input.

Assume that the augend and addend have been loaded into registers A and B, respectively. Assume also that the carry flip-flop has been cleared so that the carry input (C_i) to the FA is 0 initially. The contents of the two least significant cells, A_0 and B_0, at the x and y inputs of the FA are added together with C_i (which is 0 at present), and the FA produces their sum on S and the resulting carry on C_0. Note that the S output of the FA is fed back to the most significant cell (A_3) of register A through its serial input (SI) and that the serial output (SO) of register B is circulated into its SI. If we now raise the shift-right (SR) control line to 1, the following events will take place *simultaneously* on the next clock pulse:

1. The carry flip-flop will set ($Q = 1$) if C_0 was 1; otherwise, Q will remain unchanged.

2. The contents of A and B will be shifted to the right by 1 bit position.

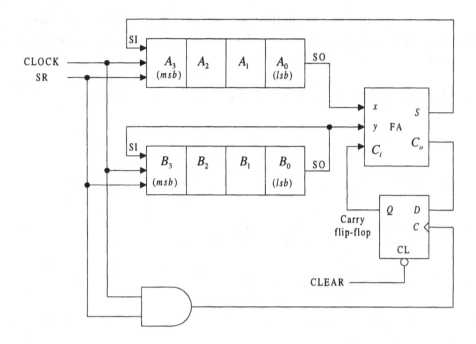

■ **Figure 7.41** Serial binary adder

3. S will be stored in A_3.
4. B_0 will be stored in B_3.

Between this clock pulse (the first after raising SR to 1) and the next clock pulse, FA adds the former contents of cells A_1 and B_1 (now shifted to A_0 and B_0) together with the new C_i and produces the new sum and carry. On the second (next) clock pulse, the events described above will be repeated. Continuing in this fashion, the serial adder completes the addition of two 4-bit binary numbers after four clock pulses. On the trailing edge of the fourth clock pulse, SR is changed to 0 and the serial adder is disabled.

The entire process is tabulated in Table 7.19, assuming that the augend is 0111 and the addend is 0010. The arrows show what happens on the first clock pulse (CP_1) after raising SR to 1. Similar events occur on subsequent clock pulses, but the arrows are not shown to prevent crowding the table. At the termination of the addition operation, register A stores the binary sum of the two numbers, the Q output of the carry flip-flop indicates any carry resulting from the addition, and register B retains the addend.

■ **Table 7.19** Example of serial binary addition

	A_3	A_2	A_1	A_0	B_3	B_2	B_1	B_0	C_i	S	C_o
Initial value	0	1	1	1	0	0	1	0	0	1	0
After CP_1	1	0	1	1	0	0	0	1	0	0	1
After CP_2	0	1	0	1	1	0	0	0	1	0	1
After CP_3	0	0	1	0	0	1	0	0	1	1	0
After CP_4	1	0	0	1	0	0	1	0	0	1	0
		Final sum in A									Last Carry

7.7 COUNTERS

Ripple Counters

A **binary ripple counter** consists of a cascade connection of flip-flops, each operating in a complementing mode. For example, Figure 7.42(a) shows a 4-bit binary ripple counter implemented with negative edge-triggered JK flip-flops. The external count pulses are applied to the clock input of the left-hand flip-flop (A_0), and the outputs of this and subsequent flip-flops are each connected to the clock input (C) of the following flip-flop. Since the J and K inputs of all the flip-flops are connected to 1, each flip-flop will toggle (complement; change state) on the transition from 1 to 0 at its clock input. (Other flip-flop types can be used; all that is required is that they operate in a complementary mode.)

To understand how the binary ripple counter operates, refer to the timing diagram of Figure 7.42(b). The diagram does not consider the propagation time delays inherent in the operation of a ripple counter, but does explain how the counter works. The effects of delays will be discussed later in this section. The count pulses are shown as a periodic train of pulses, but this is done for simplicity only; the following analysis applies equally to aperiodic pulse sequences.

Assume that the counter is cleared (each flip-flop is reset to 0), prior to the application of the count pulses, by applying a negative-going pulse to the CLEAR input. When the count sequence commences, A_0 complements on the trailing edge of the first input pulse as indicated by the arrow in Figure 7.42(b). Since A_0 makes a transition from 0 to 1, flip-flop A_1 will not change state. Hence, the present content of the counter is $A_3 A_2 A_1 A_0 = (0001)_2 = (1)_{10}$, indicating a count of one pulse. Flip-flop A_0 will complement again on the trailing edge of the second clock pulse. The transition of A_0 from 1 to 0 causes A_1 to change state, as indicated by the arrow in Figure 7.42(b), and the content of the counter after the second pulse becomes $A_3 A_2 A_1 A_0 = (0010)_2 = (2)_{10}$. This process continues in a similar fashion, with the count being incremented by 1 on the application of every input pulse. When 15 pulses have been received, the count is $A_3 A_2 A_1 A_0 = (1111)_2 = (15)_{10}$. On the application of the $16th$ pulse, A_0 is reset, sequentially causing A_1, A_2, and A_3 to reset. Hence, the counter is cleared and ready to receive the next train of 16 pulses.

Because each flip-flop stage following the least significant one is triggered by the output of the previous stage, the input signal propagates, or **ripples**, through the counter from the least significant to the most significant flip-flop. Hence the name **ripple counter**. Furthermore, since the count pulses input is not connected directly to the clock input of each flip-flop, the flip-flops do not change states at exactly the same time and operate, therefore, asynchronously. Consequently, the ripple counter is also called an **asynchronous counter**.

The counter in Figure 7.42 has 16 distinct states and is, therefore, a mod-16 counter. If we "delete" the most significant flip-flop (i.e., configure a counter with only three stages), the counting sequence would have 8 states, resulting in a mod-8 counter. On the other hand, if we add a fifth (JK) flip-flop by connecting its clock input to the Q output of A_3, the counting sequence would have 32 distinct states, resulting in a mod-32 counter. In general, a mod-M binary ripple counter, where M is a *natural* modulus, requires $\log_2 M$ flip-flops and can be very easily configured.

From the timing diagram in Figure 7.42(b), we see that the least significant flip-flop A_0 changes state on each count pulse and goes from 1

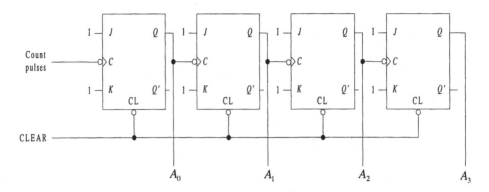

(a) Logic diagram

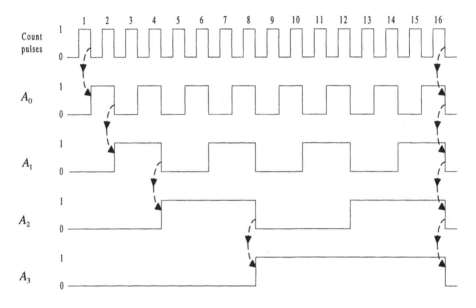

(b) Timing diagram

■ **Figure 7.42** Four-bit binary ripple counter

to 0 on the triggering edge of every second pulse. Therefore, the frequency of the output signal of the least significant flip-flop is exactly *half* the frequency of the count pulses. Whenever A_0 goes from 1 to 0, A_1 changes state so that A_1 goes from 1 to 0 on the triggering edge of every fourth pulse. The frequency of the output of this flip-flop is, therefore, exactly one-fourth the frequency of the count pulses. Similarly, A_2 goes from 1 to 0 on the triggering edge of every eighth pulse, or at one-eighth the frequency of the count pulses, and A_3 goes from 1 to 0 on the triggering edge of every sixteenth pulse, or at one-sixteenth the frequency of the count pulses. Hence, starting with the least significant flip-flop, *each* stage of the counter *divides* the input frequency by 2. The output of the last flip-flop in Figure 7.42 divides the frequency of the count pulses by 16. In general, the output from the most significant flip-flop in a mod-M counter divides the input frequency by (the modulo number) M. Therefore, a mod-M counter is also called a **divide-by-M counter**.

The counter of Figure 7.42 counts *upward* from zero and is, therefore, called an **up counter**. But a ripple **down counter**, counting *downward* from some maximum count to zero, is also simple to construct. One approach is to indicate the count by using the Q' outputs of the flip-flops, rather than their Q outputs. Hence, while $A_3 A_2 A_1 A_0$ indicates the count in the upward direction, $A_3' A_2' A_1' A_0'$ provides the count in the downward direction *at the same time*. Alternatively, we can reconfigure the counter as shown in Figure 7.43. We only need to connect the Q' (rather than the Q) output of each flip-flop to the clock input of the one following it; $A_3 A_2 A_1 A_0$ still indicates the count, but now the counter goes from 1111 to 0000. We ask the reader to show the timing diagram and count sequence table for this counter (see Problem 7.39).

All along we have been using negative edge-triggered flip-flops to implement the counters. We could have used positive edge-triggered flip-flops instead (see Problems 7.40 and 7.41). We ask the reader what would the counter outputs indicate if we change to positive edge-triggered flip-flops in Figure 7.42 or Figure 7.43.

In Figure 7.44 we combine the up and down counters of Figures 7.42 and 7.43 into an **up/down counter** by adding some logic gates and a single control line labeled U/D (up/down). If U/D = 1, the counter functions as an up counter, and when U/D = 0, the counter operates as a down counter. To see this, let us write the logic function at the output of any of the OR gates. Calling this function $f_i (i = 1, 2, 3)$, we obtain

$$f_i = (\text{U/D}) A_{i-1} + (\text{U/D})' A_{i-1}' = (\text{U/D}) \odot A_{i-1}$$

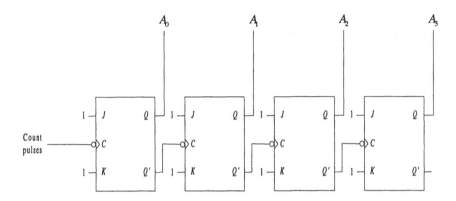

■ **Figure 7.43** Four-bit binary ripple down counter

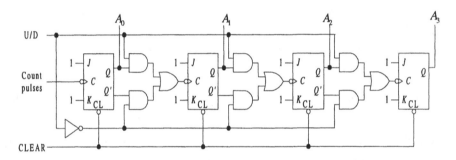

■ **Figure 7.44** Four-bit binary ripple up/down counter

Therefore, if U/D = 1, the upper-level AND gates are enabled (while the lower-level AND gates are inhibited), the Q output of each flip-flop is transferred through the OR gate to the clock input of the following flip-flop, and the counter operates as the up counter of Figure 7.42. When U/D = 0, the reverse happens, and the counter functions as the down counter of Figure 7.43.

Ripple Counters with Arbitrary Modulus

The ripple counters discussed so far are characterized by modulus numbers that are *integral* powers of 2. In Section 7.5 we referred to such counters as having a *natural* binary modulus. For various applications, however, we may be required to construct a counter whose modulus is *not* natural.

We can easily modify a binary ripple counter to produce *any* modulus by *resetting* it to 0 at the desired count. In other words, rather than let the counter follow its natural sequence of states, we make it *skip* those states whose count is greater than or equal to the modulus and then return to 0. Hence, to construct a mod-*M* counter, where *M* is *any* integer, find the smallest number of flip-flops *N* such that $2^N \geq M$, connect them as a natural binary ripple counter, and make them reset to 0 on the *Mth* count pulse input. This technique is referred to as **premature resetting**.

To illustrate the premature resetting technique, consider the design of a mod-10 (*decimal*) ripple counter. This counter follows a sequence of 10 states and resets to 0 on the 10*th* count pulse input. Since $M = 10$, we require $N = 4$ flip-flops to implement this counter. Figure 7.45(a) shows the four flip-flops connected as a natural binary ripple counter. To reset the counter on the 10*th* count pulse input, we use a NAND gate. Because decimal 10 is binary 1010, the NAND gate inputs are derived from A_3 and A_1. Hence, the NAND gate output is 1 for the count sequence 0 through 9. On the 10*th* count pulse, the NAND gate output changes to 0, causing the counter to reset (clear) and, therefore, forcing the gate's output back to 1.

The timing diagram of Figure 7.45(b) shows that this counter follows the natural binary sequence up to the trailing edge of count pulse 10. But then at this point, A_0 makes a transition from 1 to 0, forcing A_1 to change to 1. Since A_3 is also 1, RESET (the NAND gate output) makes a transition from 1 to 0 and clears the counter. Because A_0 and A_2 are 0 anyway at this time, only A_1 and A_3 are reset. As a result, a **spike** (a pulse of short duration, also called a **glitch**) occurs in A_1, caused by the momentary appearance of state 1010 before the counter is cleared. Figure 7.45(c) shows an expanded view of the operation of the counter on the 10*th* count pulse. On the trailing edge of this pulse, A_1 changes to 1 while A_3 remains at 1. With $A_3 A_1 = 11$, the output of the NAND gate will change from 1 to 0 *after* a time t_g, the gate's propagation time delay. As a result, the two flip-flops (A_1 and A_3) will reset after a time t_{FF}, the propagation time delay between the CL input and the Q output of the flip-flop. With $A_3 A_1 = 00$, the output of the NAND gate will change back to 1 once t_g has elapsed.

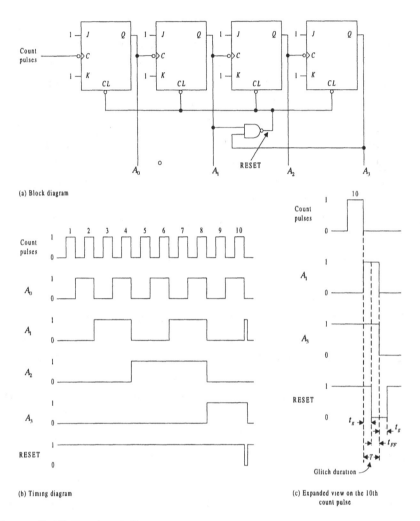

(a) Block diagram

(b) Timing diagram

(c) Expanded view on the 10th count pulse

■   **Figure 7.45** Decimal Counter

With the premature resetting technique, we can use a mod-M natural binary counter to implement *any* mod-N counter where $N \leq M$. In fact, we can easily construct a counter with *selectable* modulus by using a $(\log_2 M) \times M$ decoder followed by an $M \times 1$ multiplexer (MUX). Figure 7.46 shows an example of such a counter in which any modulus up to 16 (0 through 15) can be selected. It uses a mod-16 natural counter to drive a 4×16 decoder followed by a 16×1 MUX. To understand how this counter operates, assume that a mod-6 counter is required. We select the modulus by setting the selection lines of the multiplexer to 6 (i.e.,

$s_3 s_2 s_1 s_0 = 0110$), thus routing I_6 to the multiplexer output. As long as $I_6 = 0$, CLEAR $= 1$ (due to the inverter at the output of the multiplexer) and the counter follows the count sequence 0, 1, 2, 3, 4, and 5. When the decoder detects the *6th* count (i.e., when $A_3 A_2 A_1 A_0 = 0110$), I_6 becomes 1, the CLEAR signal goes to 0, and the counter is reset.

The glitch phenomenon is inherent to the premature resetting technique. We can reduce the glitch duration T [see Figure 7.45(c)] if we use components with small propagation time delays. Nevertheless, the presence of even a narrow glitch may still cause problems in some applications. The glitch limits the occurrence of the trailing edge of the next count pulse to a time that is greater than T. Furthermore, if the counter outputs are used to drive another circuit and if that circuit is fast enough, the glitch would be recognized as an input and might cause the circuit to malfunction. In other applications, however, if the time constant of the driven circuit is larger than T (a seven segment display is a good example of such a circuit), the glitch would not affect the operation of the circuit at all.

The glitch can be eliminated altogether if, instead of using premature resetting, we approach the design of the counter differently. Let us illustrate this approach in terms of the decimal counter introduced previously. Table 7.20 shows the count sequence of the decimal counter. (Count 10 is the same as count 0 but is included for clarity.) By closely observing the table, we can draw the following conclusions:

1. A_0 complements on each count pulse. The count pulses input should, therefore, be connected to the clock input of the least significant flip-flop. The J and K inputs of this flip-flops should be connected to 1.

2. (a) A_1 is complemented when A_0 changes from 1 to 0 as long as $A_3 = 0$.

   (b) When A_3 changes to 1 and A_0 goes from 1 to 0 (count 8), A_1 changes to 0 and stays 0 through the remainder of the count sequence [count 9 and count 10 (= count 0)].

To satisfy (a), A_0 should be connected to the clock input of A_1, the K input of flip-flop A_1 should be connected to 1, and its J input to A_3'. Thus, as long as $A_3 = 0$, $A_3' = 1$, the J and K inputs of A_1 are 1, and A_1 complements on the negative-going transition of A_0 (counts 0 through 7). But on the 8*th* count pulse, A_0 goes from 1 to 0 while A_3 changes to 1; hence, $A_3' = 0$, causing J of A_1 to be 0. With $J = 0$ and $K = 1$, A_1 will reset on the 1-to-0 transition of A_0 and remain so for counts 8 and 9. In counts 0 and 1, A_1 will remain at 0 (even though A_3 changes to 0) because A_0 makes no transition from 1 to 0. Hence, (b) is also satisfied.

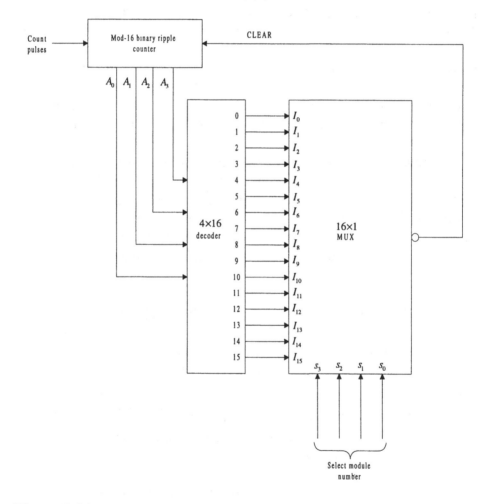

■ **Figure 7.46** Counter with selectable modulo number

■ **Table 7.20** Count sequence table of a decimal counter

Count	A_3	A_2	A_1	A_0
0	0	0	0	0
1	0	0	0	1
2	0	0	1	0
3	0	0	1	1
4	0	1	0	0
5	0	1	0	1
6	0	1	1	0
7	0	1	1	1
8	1	0	0	0
9	1	0	0	1
10	0	0	0	0

3. A_2 complements whenever A_1 changes from 1 to 0. Connect, therefore, A_1 to the clock input of A_2 and the J and K inputs of this flip-flop to 1.

4. (a) A_3 changes from 0 to 1 if A_2 and A_1 are both 1 and A_0 changes from 1 to 0 (count 7 to count 8).

(b) A_3 changes from 1 to 0 if either A_2 or A_1 or both are 0 and A_0 changes from 1 to 0.

The common denominator for these conditions is "A_0 changes from 1 to 0." Hence, connect A_0 to the clock input of A_3. Connect the K input of A_3 to 1 and its J input to the output of an AND gate whose inputs are A_2 and A_1. If condition (a) is present, both J and K of A_3 are 1, and A_3 will change from 0 to 1 as A_0 changes from 1 to 0. If $A_3 = 1$ and condition (b) exists, $K = 1$ and $J = 0$, and A_3 will change to 0 and stay at 0 on the first occurrence of a 1-to-0 transition of A_0 (count 9 to count 0).

The block diagram of the decimal counter resulting from this process is shown in Figure 7.47. The reader is encouraged to verify the operation of the counter by obtaining its timing diagram and to note that there is no glitch.

Cascade Connection of Ripple Counters

We have already considered the cascade connection of synchronous counters in Section 7.5. In contrast, cascade connection of binary ripple counters is more easily accomplished, because all we need do is connect the most significant output of one counter to the count pulses input of the following counter. Figure 7.48 shows the cascade connection of two decimal counters resulting in a 2-decade (mod-100) counter capable of counting from decimal 0 to decimal 99.

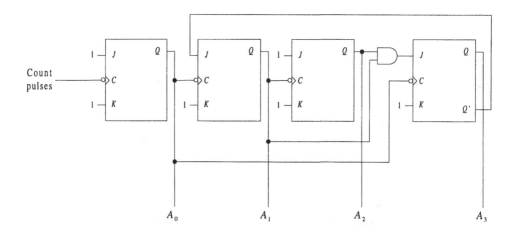

■ **Figure 7.47** Alternative decimal counter

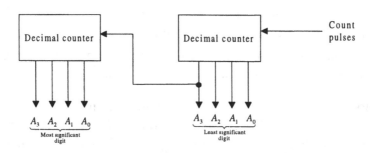

■ **Figure 7.48** Two-decade ripple counter

Presettable Counters

The counters described so far count either up from 0 or down toward 0. In many applications, however, the count sequence must start from some initial number and then proceed up or down from it. In other words, we must **preset** the counter by loading a given preset count number.

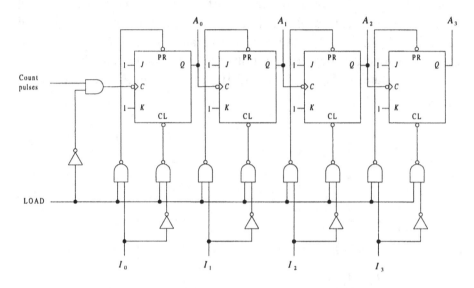

■ **Figure 7.49** Four-bit presetttable binary ripple counter

The counter in Figure 7.49 is a mod-16 binary ripple up counter implemented with *JK* flip-flops. To preset this counter, we incorporate a parallel load capability by using the asynchronous inputs, PR (preset) and CL (clear), of the flip-flops. When the LOAD control line is set to 1, the count pulses input is inhibited and the preset data present on $I_0 - I_3$ are transferred in parallel into the counter. Hence, if $I_i = 1$, only the output of the left-hand NAND gate will change from 1 to 0, presetting A_i to 1 if it was 0 or leaving A_i unchanged if it was 1. On the other hand, if $I_i = 0$, only the output of the right-hand NAND gate will change from 1 to 0, clearing A_i to 0 if it was 1 or leaving A_i unchanged if it was 0. Once the preset count has been loaded into the counter, LOAD is brought back to 0, the count pulses input is enabled, and the counter can resume counting up from the preset number.

We can use a presettable mod-M counter to implement any mod-N counter such that $N \leq M$. Figure 7.50 shows two examples of implementing a mod-7 counter using the mod-16 presettable counter of Figure 7.49. When the counter in Figure 7.50(a) reaches the $7th$ count (state 0111), the AND gate output becomes 1 and the 0 inputs are loaded into the counter. This causes the output of the AND gate to change back to 0 and the counter can resume counting. Thus, the counter goes through states 0, 1, 2, 3, 4, 5, and 6 and returns to state 0 on the $7th$ count pulse. Similarly, by loading the preset number 4, the counter can be made to count through states 4, 5, 6, 7, 8, 9, and 10, as shown in Figure 7.50(b).

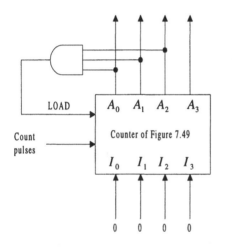 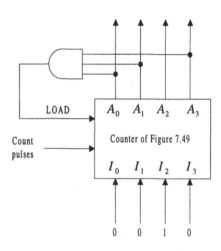

(a) Counting through states 0,1,2,3,4,5,6　　　(b) Counting through states 4,5,6,7,8,9,10

■　**Figure 7.50**　Implementing a mod-7 counter using the presettable counter of Figure 7.49

Propagation Delay in Ripple Counters

Ripple counters are very easy to construct, but they have one major drawback due to their operating principle. The count ripples through the counter, where each flip-flop is triggered by the transition at the output of the preceding flip-flop. Due to the propagation time delay (t_p) between the clock input and the Q (or Q') output of the flip-flop, the second flip-flop will not respond until a time t_p *after* the first flip-flop receives a triggering input pulse; the third flip-flop will not respond until a time $2t_p$ after the count pulses input occurs; and so on. Hence, the propagation time delays of the flip-flops accumulate as the counter follows the counting sequence and, therefore, limit the *speed* at which the counter can operate.

The speed of operation of the counter determines the *maximum* frequency allowed for the count pulses. To see this, assume periodic count pulses and let T_C designate the time interval between successive count pulses. If the ripple counter contains N flip-flops, then, to ensure proper operation, we must have

$$T_C \geq N t_p$$

Hence, the maximum count pulse frequency is given by

$$f_{max} = \frac{1}{T_C}$$

and is inversely proportional to the number of flip-flops in the ripple counter: if N is increased, then f_{max} will decrease. For example, with $t_p = 60$ nanoseconds, the maximum allowed clock frequency for a 4-bit ripple counter is

$$f_{max} = \frac{1}{4 \times 60} = 4.16 \text{ MHz}$$

while that for a 5-bit counter is

$$f_{max} = \frac{1}{5 \times 60} = 3.33 \text{ MHz}$$

Synchronous Counters

Synchronous counters differ from ripple counters in that the count pulses input is connected to the clock inputs of *all* the flip-flops. Ripple counters are simpler than synchronous counters, but they have two major disadvantages: (1) speed of operation, and (2) a forced straight binary count sequence. The first disadvantage is due to the rippling of the count pulses signal, which, in turn, limits the speed of the ripple counter. In contrast, a synchronous counter with the same number and type of flip-flops can operate at much higher clock frequencies because all the flip-flops are clocked simultaneously.

The second disadvantage of the ripple counter is due to its construction. In Example 7.3, we demonstrated that the design of a synchronous counter that follows *any* count sequence is just as simple as the design of a counter that follows a straight binary code. On the other hand, a ripple counter can follow only a straight binary code and requires a code converter if a different count sequence code is desired.

While synchronous and ripple counters have distinguishing characteristics, they perform, nevertheless, identical functions. Similar to ripple counters, MSI synchronous counter chips are commercially available in a variety of modulus numbers, with up, down, and up/down configurations, and with parallel load capabilities. We already know how to design synchronous counters and are familiar with the various configurations of ripple counters. Therefore, rather than repeat the discussion here, we assign the design of some of these counters to problems at the end of this chapter. However, one special type of synchronous counters, the ring counter, requires an introduction.

Ring Counters

Shift registers can be arranged to form several types of counters, referred to as **ring counters**. In a ring counter, the serial output of the shift register is connected in some way to its serial input. Strange as it may sound, in most applications the ring counter is not really used for a counting function, but rather as a timing sequence generator. We will consider the design of a timing sequence generator when we discuss counter applications in a subsequent section.

There are basically two types of ring counters: standard-ring and twisted-ring. A mod-N **standard-ring counter** requires N flip-flops and is implemented using a serial-in, serial-out (SISO) shift register whose serial output (SO) is fed back to its serial input (SI). Moreover, one of the flip-flops is initially set while all the others are cleared. On subsequent clock pulses, the single 1 bit is shifted from one flip-flop to another.

Because the SO is fed back to the SI, the bit will keep on *circulating* through the shift register as long as clock pulses are applied. Since the shift register contains N flip-flops, circulating a single bit among them provides N distinct states. Figure 7.51(a) shows an example of a 4-bit standard-ring counter implemented with JK flip-flops. This counter is organized so that the least significant flip-flop (A_0) is initially loaded with a 1, while all other flip-flops are cleared. The counting sequence, shown in Table 7.21, implies that this is a mod-4 counter. The timing diagram for this ring counter is shown in Figure 7.51(b).

With four flip-flops, the register can assume up to 16 states, but only four are valid in a standard-ring counter configuration. In general, a standard-ring counter with N flip-flops has N *valid* states and $2^N - N$ *invalid* states. Figure 7.52 illustrates the state diagrams for the valid and invalid states of the counter in Figure 7.51. As we can see, the 4-bit ring counter has five invalid count sequences using the $2^4 - 4 = 12$ invalid states. Note that each invalid state is characterized by having more than one set flip-flop (i.e., more than a single 1 bit circulating through the shift register). In contrast, each of the valid states is characterized by exactly one set flip-flop. The problem with a standard-ring counter is that, once the counter is in any one of the invalid states, it will never resume the proper count sequence unless we force the counter to get to a state where *one and only one* flip-flop is set.

There are a number of ways to modify the standard-ring counter so that it will be *self-starting*. One such procedure is illustrated in Figure 7.53. Rather than connect the SO to the SI as in Figure 7.51, the serial input is now obtained from $(A_0 + A_1 + A_2)'$. This feedback connection will *automatically* correct the counter when it gets locked in any of the invalid count sequences. Note, however, that the counter may take up to four count pulses to make the correction.

The number of valid states of a ring counter can be *doubled* if the shift register is connected as a **twisted-ring counter**. Also called a *switch-tail* or *Johnson counter*, a twisted-ring counter is a circular shift register in which the *complement* of the serial output is connected to the serial input. Figure 7.54(a) shows such a counter obtained by modifying the standard-ring counter of Figure 7.51 (a). The counter starts from a cleared state (obtained by using the CLEAR line) and follows the sequence of eight distinct states shown in Figure 7.54(b).

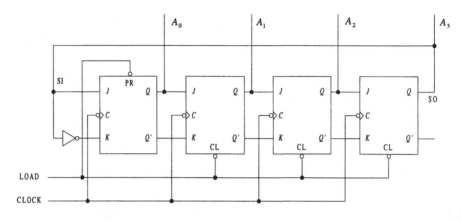

(a) Logic diagram

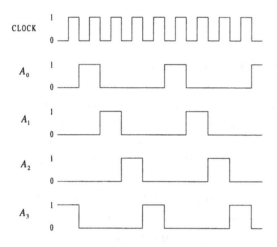

(b) Timing diagram

■ **Figure 7.51** Standard-ring counter

■ **Table 7.21** Count sequence for the ring counter of Figure 7.51

A_3	A_2	A_1	A_0
0	0	0	1
0	0	1	0
0	1	0	0
1	0	0	0

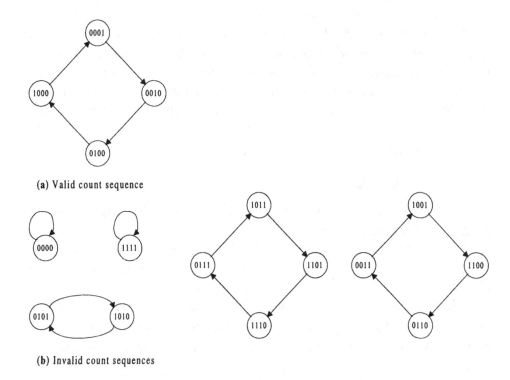

(a) Valid count sequence

(b) Invalid count sequences

■ **Figure 7.52** State diagrams for the ring counter of Figure 7.51

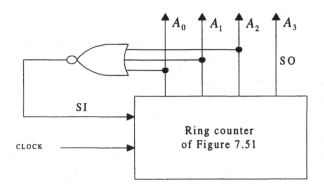

■ **Figure 7.53** Self-starting standard-ring counter

In general, a twisted-ring counter with N flip-flops has a count sequence of $2N$ *valid* states and $2^N - 2N$ *invalid* states. Like the standard-ring counter, a twisted-ring counter can get locked into a sequence of invalid states. To avoid this condition, the circuit can be modified so that it will be self-starting. For example, one modification of Figure 7.54(a) disconnects the inverter into J_0 (the J input of flip-flop A_0) and, instead, enables it by the function

$$J_0 = (Q_2 + Q_3)'$$

We as the reader to verify that, with this modification, the counter of Figure 7.54(a) is self-starting.

Since a twisted-ring counter with N flip-flops goes through a sequence of $2N$ states, the length of the count sequence is always *even*. Nevertheless, we can configure an *odd-length* counter by modifying an even-length counter. For example, if we modify the counter in Figure 7.54(a) so that K_0 (the K input of flip-flop A_0) is enabled by A_2 rather than by A_3, the result will be the mod-7 counter shown in Figure 7.55. (Note that J_0 is now fed from A_3', so we can dispose of the inverter in Figure 7.54.) In general, we can modify an even-length twisted-ring counter implemented with N JK flip-flops by connecting K_0 (the K input of the least significant flip-flop A_0) to A_{N-2} (the Q output of the next-to-last flip-flop in the shift register chain). Alternatively, we can leave K_0 connected to A_{N-1} and connect J_0 to A_{N-2}' (rather than to A_{N-1}').

To summarize, standard-ring counters can be constructed for any modulus with a mod-N counter requiring N flip-flops. In general, a standard-ring counter requires more flip-flops than a binary counter for the same modulus. For example, a mod-8 standard-ring counter uses eight flip-flops, while a mod-8 binary counter requires only three. A twisted-ring counter requires half as many flip-flops as a standard-ring counter, but usually more than a binary counter. For example, a mod-8 twisted-ring counter requires four flip-flops as opposed to three by its binary counterpart. On the other hand, both counter types require three flip-flops to implement a mod-5 counter. Other distinguishing factors between ring counters and binary counters are considered in the following section.

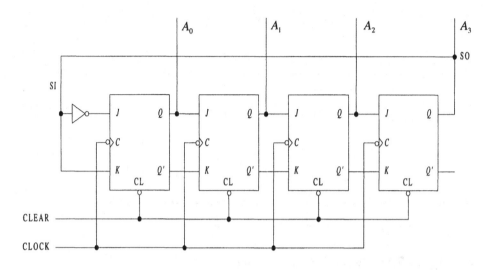

(a) Logic diagram

State	Count Sequence			
	A_3	A_2	A_1	A_0
S_0	0	0	0	0
S_1	0	0	0	1
S_2	0	0	1	1
S_3	0	1	1	1
S_4	1	1	1	1
S_5	1	1	1	0
S_6	1	1	0	0
S_7	1	0	0	0

(b) Count sequence

■ **Figure 7.54** Twisted-ring counter

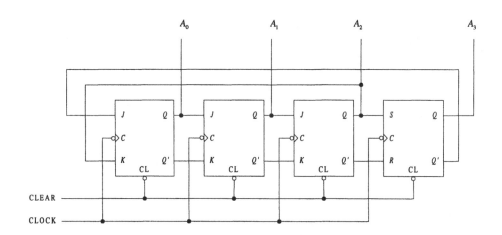

- **Figure 7.55** Mod-7 twisted-ring counter

Decoding a Counter

In various applications of counters, the count must be determined or displayed. For example, a certain operation might have to be initiated when a counter reaches the 101110 state [count of $(46)_{10}$]. To obtain a signal indicating that the counter has reached this state, the counter outputs must be decoded. To decode a mod-M binary counter, we need an $N \times M$ decoder, where $N \geq \lceil \log_2 M \rceil$. Usually, the complements of the counter outputs are also available, so the decoder will contain M N-input decoding gates (AND gates for example; see Chapter 5). Hence, to decode a mod-16 binary counter, we require a 4×16 decoder having 16 AND gates of four inputs each.

On the other hand, the standard-ring counter, while less efficient than a binary counter in the use of flip-flops, can, nevertheless, be decoded *without* the use of decoding gates. The decoded signal for each of the counter states is simply obtained from the output of its corresponding flip-flop. For example, in the mod-4 standard-ring counter in Figure 7.51, the four signals corresponding to the four valid states are each obtained from the output of the corresponding flip-flop. We can verify easily that the same signals can be generated by using a mod-2 binary counter combined with a 2×4 decoder, a circuit that requires two flip-flops and four 2-input AND gates. In some cases, a standard-ring counter might be a better choice because a binary counter requires a decoder.

For a given modulus, a twisted-ring counter requires half as many flip-flops as a standard-ring counter, but it must use gates to decode each count. However, unlike a binary counter, each decoding gate in a twisted-ring counter requires only two inputs, *regardless* of the number of flip-flops in the counter. To illustrate this, the decoding of the mod-8 twisted-ring counter of Figure 7.54 is shown in Figure 7.56. The decoding follows a regular pattern: The all 0's (all 1's) state is decoded by taking the Q' outputs (Q outputs) of the two extreme flip-flops; all the other states are decoded from an adjacent $\{0, 1\}$ or $\{1, 0\}$ pattern in the sequence. To produce the same signals using a binary counter, we would require a mod-3 counter and a 3×8 decoder (i.e., three flip-flops and eight 3-input AND gates).

Thus, twisted-ring counters represent a middle ground between standard-ring counters and binary counters. A twisted-ring counter requires less flip-flops than a standard-ring counter, but usually more than a binary counter. The decoding circuit of a twisted-ring counter is less complex than that of a binary counter, whereas a standard-ring counter requires no decoding at all.

Examples of Counter Applications

Counting and Frequency Division

The two most obvious applications of counters are counting and frequency division. Counting is performed whenever the count pulses represent certain events whose number of occurrences we want to accumulate. For example, using a magnetic card to access an automated bank teller, we are only allowed three mistakes in keying our (secret) identity number (ID). After three mistakes, the card will be kept by the machine and can only be claimed in person. Each time we key in the ID number, a counter is activated. If the keyed ID is the one expected, the counter remains in the zero state. On the other hand, with each erroneous ID keyed in, the counter advances so that, when it reaches count (state) 3, its decoded outputs disable the mechanism that returns the card.

Counting can also be used to control a sequence of operations. For example, consider a machining process that requires five steps to complete

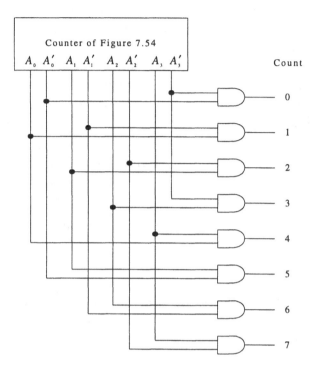

■ **Figure 7.56** Decoding a mod-8 twisted-ring counter

and has three possible operations that can be performed at each step. Using a mod-5 counter, we can associate each counter state with one of the steps of the machining process. The three state variables (the Q outputs of the flip-flops) can be encoded such that a 1 value indicates that an operation is performed, while a 0 value indicates that no operation is performed. Hence by following a coded count sequence, the counter can be used to control the process.

To illustrate the function of frequency division, consider a digital clock that displays the time of the day in hours, minutes, and seconds. To drive the clock, we need a source that will generate one count pulse per second. One common approach is to generate the count pulses from a 60-Hz power line. The power-line signal is first reshaped into a square wave and then divided down to a frequency of 1 Hz (one pulse per second) using a mod-60 counter. This counter requires six flip-flops, and the 1-Hz signal is taken from the most significant flip-flop.

Frequency Measurement

One of the most straightforward methods to measure the unknown frequency of a periodic signal is to use a counter as shown schematically in Figure 7.57. The signal is represented by a train of pulses; if the frequency of other types of periodic signals is to be measured, the signals have to be reshaped into pulses first. The GATE signal enables the AND gate for a preselected (known) time interval, called the *sampling interval*. Prior to the application of the GATE signal, the counter is cleared to zero. During the sampling interval, the pulses of unknown frequency are applied to the count pulses input of the counter and counted. At the termination of the sampling interval, the counter stops counting, and its content is a direct measure of the frequency of the signal.

Note that the accuracy of this method depends on the duration of the sampling interval. For example, if the unknown frequency is 5487 Hz and the sampling interval is 1 second, then 5487 pulses will enter the counter so that its content will be 5487 at the end of the sampling interval. On the other hand, if the sampling interval is 0.1 second, then 548.7 pulses will pass through the AND gate and the content of the counter will be either 548 or 549. Various methods are used in actual frequency counters to control the accuracy of the sampling interval and to adapt its width to the frequency range in which the unknown frequency lies.

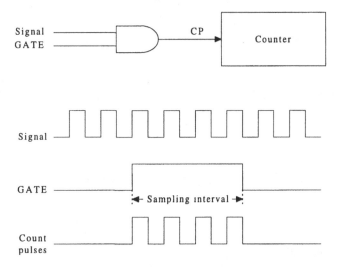

■ **Figure 7.57** Frequency measurement with a counter

Generation of Timing Signals

The control of any sequence of operations in a digital system requires the generation of various timing sequences and signals. We have already encountered several such situations in this chapter: the control of read and write operations in a random access RWM (Figure 7.36), the control of serial data transfers using a shift control signal (Figure 7.40), and the control of frequency measurement using a GATE signal (Figure 7.57).

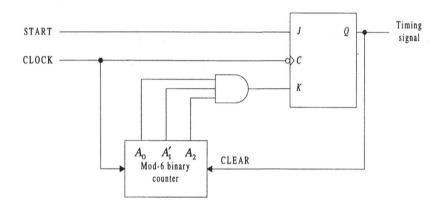

(a) Logic diagram

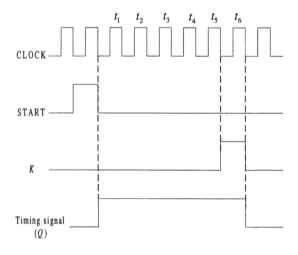

(b) Timing diagram

■ **Figure 7.58** Generation of timing signals

As we already know, the generation of timing sequences can be easily accomplished by means of a counter/decoder combination or a circulating shift register (ring counter). To illustrate the generation of timing signals, assume that some control signal is required to stay ON for a period of six clock pulses. To generate this timing signal, we can use the circuit shown in Figure 7.58(a). With the counter and flip-flop initially reset to 0, a START = 1 signal will set the flip-flop on the next clock pulse and thus enable the counter. After reaching a count of five clock pulses (binary 101), the output of the decoding AND gate becomes 1. Since $K = 1$ and $J = 0$ at this point, the next clock pulse (the 6th) will reset the flip-flop and clear the counter. As long as $Q = 0$, the counter is disabled. As seen in Figure 7.58(b), the timing signal generated at the Q output of the flip-flop stays ON for the required duration of six clock pulses.

Problems

7.1   The Q output of a positive edge-triggered SR flip-flop is shown in Figure P7.1 in relation to the clock signal. Determine the input signals on the S and R inputs that are required to produce this output.

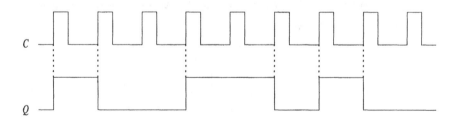

■ **Figure P7.1**

7.2   Draw the Q output of a positive edge-triggered D flip-flop for the
inputs shown in Figure P7.2. Assume that Q is initially 0.

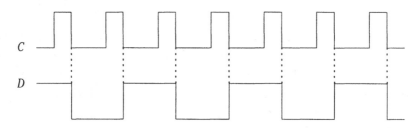

■ **Figure P7.2**

7.3   Draw the Q output of a pulse-triggered (master-slave) JK flip-flop
for the inputs shown in Figure P7.3. Assume that Q is initially 0.

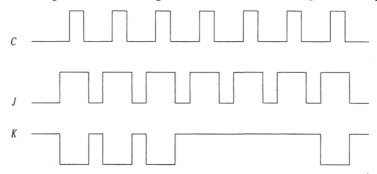

■ **Figure P7.3**

7.4   An SR flip-flop is connected as shown in Figure P7.4. Determine
the Q output in relation to the clock. What specific function does
this device perform?

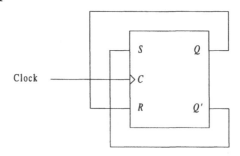

■ **Figure P7.4**

7.5   Use a T flip-flop and additional gates to implement an SR flip-flop, a D flip-flop, and a JK flip-flop.

7.6   Use a JK flip-flop and additional gates to implement an SR flip-flop, a D flip-flop, and a T flip-flop.

7.7   Use a D flip-flop and additional gates to implement an SR flip-flop, a JK flip-flop, and a T flip-flop.

7.8   Consider the flip-flop shown in Figure P7.8.

(a) Obtain the characteristic table.
(b) Obtain the characteristic equation.
(c) Derive the excitation table.

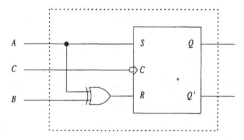

■ **Figure P7.8**

7.9   Suppose that the D input of a flip-flop changes from 0 to 1 in the middle of a clock pulse.

(a) What happens if the flip-flop is positive edge-triggered?
(b) What happens if the flip-flop is pulse-triggered (master-slave)?

7.10   Analyze the circuit shown in Figure P7.10. Obtain the state table and the state diagram, and determine the function of the circuit.

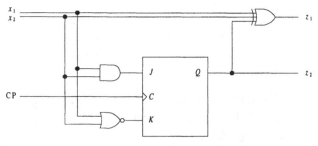

■ **Figure P7.10**

7.11   Derive the state table and state diagram of the sequential circuit of Figure P7.11. What is the function of the circuit?

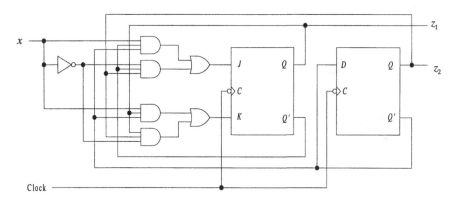

■   **Figure P7.11**

7.12   A synchronous sequential circuit is implemented with three negative edge-triggered JK flip-flops (A_1, A_2, and A_3) and has a single input (x) and single output (z). The flip-flop excitation equations and the circuit output equation are as follows:

$$J_1 = K_1 = A_2$$
$$J_2 = K_2 = 1$$
$$J_3 = x'(A_1 \oplus A_2) + xA_1'A_2', \qquad K_3 = A_2$$

and

$$z = (x \oplus A_3)'$$

Obtain the state table and state diagram. What is the function of the circuit?

7.13   Redesign the filter circuit of Figure 7.20 so that it will be self-starting by making all the invalid states go to the initial state under all input combinations.

7.14   The state diagram of a single-input, single-output synchronous sequential circuit is shown in Figure P7.14. Design a self-starting circuit with T flip-flops.

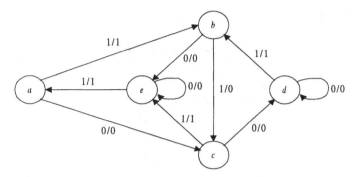

■ **Figure P7.14**

7.15 Design a synchronous sequential circuit that accepts a serially transmitted message and produces its 2's complement. The message enters the circuit starting from the least significant bit. Implement with *SR* flip-flops.

7.16 Reduce the state table in Table P7.16 and tabulate the reduced state table.

■ **Table P7.16**

	NS/Output (z)	
PS	$x = 0$	$x = 1$
a	$d / 1$	$e / 0$
b	$d / 0$	$f / 0$
c	$b / 1$	$e / 0$
d	$b / 0$	$f / 0$
e	$f / 1$	$c / 0$
f	$c / 0$	$b / 0$

7.17 Repeat Problem 7.16 for the state table given in Table P7.17.

7.18 Design the logic circuit shown in Figure P7.18. The circuit will set the *JK* flip-flop if the binary number $x_1 x_2 x_3$ is odd and will reset it otherwise.

■ **Table P7.17**

PS	NS/Output (z)	
	$x = 0$	$x = 1$
a	$f/1$	$d/1$
b	$g/1$	$h/1$
c	$d/1$	$a/0$
d	$a/0$	$h/1$
e	$c/1$	$d/1$
f	$d/1$	$e/0$
g	$h/1$	$b/0$
h	$b/0$	$d/1$
i	$f/0$	$a/0$

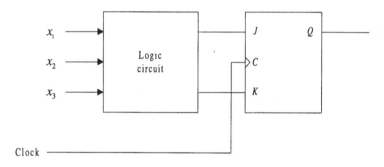

■ **Figure P7.18**

7.19   Design a synchronous sequential circuit to control the operation of an automatic coffee machine. A cup of coffee costs 15¢ and the machine has two input slots. In one slot, only 10¢ coins can be inserted; in the other, only 5¢ coins. The machine will give change in 5¢ coins only, and only one such coin per transaction. Implement the circuit with T flip-flops.

7.20   Design a single-input, single-output synchronous sequential circuit whose output will be 1 if the input sequence 101 or 110 is detected. Overlapping sequences are allowed. Use T flip-flops to implement the circuit.

7.21 Design a synchronous sequential circuit with two inputs (x_1 and x_2) and a single output (z) so that $z = 1$ if and only if the following rule is not satisfied: *i before e except after c*. Use the following encoding: $c = 01$; $e = 11$; $i = 10$; and all other letters = 00. Implement with D flip-flops.

7.22 Design a synchronous sequential circuit with two inputs (x_1 and x_2) and a single output (z) in which the input pair represents, in coded form, the first four letters of the English alphabet. That is, $A = 00$, $B = 01$, $C = 11$, and $D = 10$. The output is equal to 1 if and only if the most recent two inputs are in alphabetical order (that is, AB, BC, and CD). Obtain the next-state and output equations.

7.23 Design a 1111 sequence detector. Overlapping sequences are allowed. Implement the circuit with SR flip-flops.

7.24 Implement the 0101 sequence detector of Example 7.1 using the modified SR flip-flop shown in Figure P7.8.

7.25 Redesign the mod-8 counter of Example 7.2 using negative edge-triggered SR flip-flops.

7.26 Design a mod-12 synchronous binary counter and implement it with negative edge-triggered JK flip-flops.

7.27 The period of the count pulses used to clock the counter in Problem 7.26 is 10 seconds. The Q outputs of the flip-flops are used as inputs to a logic circuit that controls the *sequence* of traffic lights in the following way:

> Green is ON for 40 seconds,
> Yellow is ON for 20 seconds,
> Red is ON for 40 seconds,
> Red and yellow are ON for 20 seconds,
> Green is ON for 40 seconds,

and so on. Design the circuit assuming that 1 represents a light that is ON.

7.28 Design a synchronous sequential circuit that will count through the sequence 0, 2, 4, 6 when its *control input D* is 0 and through the sequence 6, 4, 2, 0 when $D = 1$. The circuit should be self-starting so that, whenever it finds itself in an invalid state, it will always return to the 0 state. Implement the circuit with *JK* flip-flops.

7.29   A serial parity-bit generator is a circuit that receives *n*-bit messages, each spaced by a single clock pulse, and regenerates them so that the parity bit for each message is inserted in the space. Hence, the circuit output is a continuous string of bits without spaces. Let $n = 3$ and assume an even-parity method; that is, the parity bit is set to 1 so that the total number of 1's in the message (including the parity bit) is an even number (see Section 2.5). Design the circuit and implement it with *T* flip-flops.

7.30   Implement the 3-bit parallel register of Figure 7.28 with *D* flip-flops.

7.31   The memory unit of Figure 7.33 has a capacity of 8192 words of 32 bits per word.
    (a) How many flip-flops are needed for the memory address register (MAR) and the memory buffer register (MBR)?
    (b) How many words will the memory unit contain if the MAR has 15 bits?

7.32   Using an *SR* flip-flop, redesign the memory cell of Figure 7.34 so that it will also include as input the read/write (R/W) control line.

7.33   **(a)** What is the bit capacity of a memory unit that has 512 addresses and can store 8 bits at each address?
    (b) How many address bits are required for a 512×4 memory?
    (c) What is the total capacity of a memory unit with 10 address lines and 4 output lines?
    (d) How many bits can actually be stored in a 64K memory? In a 100MB memory?

7.34   A random-access RWM containing 256 words has an access time of 300 nanoseconds (from address to data output). How long does it take to read the data from *all* memory addresses?

7.35   What is the state of the shift register in Figure P7.35 after each clock pulse if it starts initially in 101001111000?

7.36   The content of a 4-bit shift register is initially 1101. The serial input is 101101, and the register is shifted six times to the right.

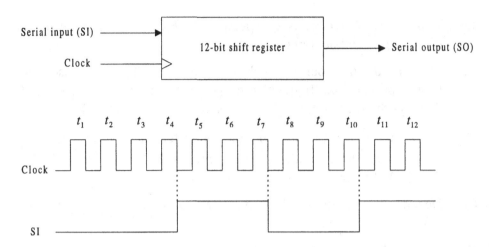

- **Figure P7.35**

What is the content of the register after each shift?

7.37 Consider the serial adder shown in Figure 7.41. The carry flip-flop is initially cleared. Register A holds the binary number 0101, and register B holds 0111.
(a) List the binary value in A and Q after each shift.
(b) What are the contents of registers A and B at the termination of the addition process? What is Q?

7.38 Modify the circuit in Figure 7.41 so that the content of B can be subtracted from the content of A.

7.39 Derive the count sequence table and timing diagram for the ripple down counter of Figure 7.43.

7.40 Construct a 3-bit ripple up counter using positive edge-triggered JK flip-flops.
(a) Show a block diagram indicating which of the flip-flops' outputs (Q or Q') indicates the count.
(b) Show the count sequence table and a timing diagram.

7.41 Construct a 3-bit ripple up counter and a 3-bit ripple down counter using positive edge-triggered D flip-flops.

7.42 Construct a mod-12 counter using the MSI circuit of Figure 7.49. Show two alternatives.

7.43   Using two MSI circuits as specified in Figure 7.49, construct a binary counter that counts from 0 to 64.

7.44   Design a digital clock in block-diagram form. The clock uses the 60-Hz power line frequency to generate the count pulses and displays the time of the day in hours, minutes, and seconds.

7.45   Show that an n-bit binary counter connected to an $n \times 2^n$ decoder is equivalent to a standard-ring counter with 2^n flip-flops. Draw the block diagrams of both circuits for $n = 3$.

7.46   Decode the outputs of the mod-7 twisted-ring counter of Figure 7.55.

7.47   Construct a decade twisted-ring counter with positive edge-triggered D flip-flops. How many invalid states does it have?

7.48   Assume that the counter in Figure 7.57 is made up of three cascaded BCD counters. The unknown frequency of the input signal is in the range from 1 to 10kHz. The sampling interval can be selected as 1 second, 0.1 second, 10 millisesconds, or 1 millisecond. What would be the best sampling interval?

7.49   How would you generate the sampling interval required in Problem 7.48 from a 100-kHz pulse source? Show a block diagram.

Appendix: Logic Graphic Symbols

A.1 INTRODUCTION

The purpose of the appendix is to give a brief overview of the new IEEE standard of graphic symbols for logic functions.[†] The standard is a symbolic language used to describe the relationship between each input and each output of a digital circuit, without showing explicitly how the circuit is constructed internally.

The new standard promotes the use of *rectangular-outline* symbols as the preferred symbols, but also permits, to some extent, the use of the traditional, *distinctive-shape* symbols.

A.2 SYMBOL COMPOSITION

The general composition of a logic symbol is shown in Figure A.1. The dimensions of the rectangular outline are arbitrary. The normal direction of signal flow is from left to right, unless indicated otherwise by an arrowhead attached to the corresponding signal line. The graphic symbol is associated with one or more **qualifying symbols** classified as (1) general qualifying symbols and (2) qualifying symbols relating to inputs and outputs.

The *general qualifier* is used to define the logic function implemented by the device and is placed near the top center of the graphic symbol, as shown in Figure A.1. Examples of general qualifying symbols defined in the new standard are listed in Table A.1. The possible positions for

---

[†] *IEEE Standard Graphic Symbols for Logic Functions* (ANSI/IEEE Std 91-1984). The Institute of Electrical and Electronics Engineers, New York, 1984.

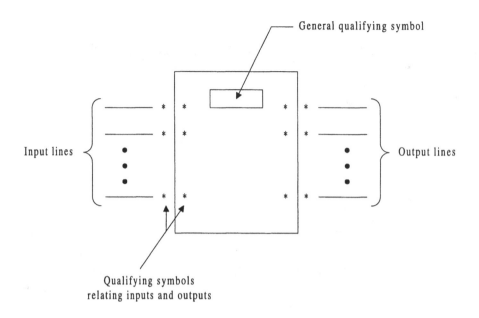

■ **Figure A.1** General logic symbol

qualifiers relating to inputs and outputs are designated by * in Figure A.1. As seen, these qualifiers can be placed both *outside* and *inside* the outline of the graphic symbol. Examples of such qualifiers are listed in Tables A.2 and A.3, respectively.

A circuit logic diagram can be drawn more compactly by having symbol outlines abutted or embedded (i.e., shared between elements), as shown in Figure A.2. In general, a common line that is in the direction of the signal flow lines, as in Figure A.2(a), indicates that there is no functional relation between the devices sharing that line. By contrast, a common line perpendicular to the direction of the signal flow lines, as in Figure A.2(b), indicates that there is at least one logical connection between the devices sharing that line. In this case, qualifying symbols can be placed on one or both sides of the common line, as shown in Figure A.2(c). [Note how the internal connection symbol (see Table A.2) is used in Figure A.2(c) to indicate the logical connections.]

If a graphic symbol depicts a number of the same devices, a single (common) general qualifier suffices to characterize the logic function implemented by all of the devices. For example, the symbol shown in

■ **Table A.1** General qualifying symbols

Symbol	Description	Comments
&	AND	
≥ 1	OR	Symbol indicates that, to perform an OR function, at least one *active* input is needed to produce an active output.
$= 1$	XOR	
$=$	Equivalence	
1	Single input	The one input must be active.
X / Y	Code converter	For example, decimal/BCD, BCD/excess 3, and binary/seven-segment code converters.
$2k$	Even element	An even number of inputs must be active to produce an active output—for example, an even-parity generator.
$2k + 1$	Odd element	An odd number of inputs must be active to activate the output—for example, an odd-parity generator.
MUX	Multiplexer	
COMP	Comparator	
Σ	Adder	
Π	Multiplier	
SRGm	Shift register	m designates the *length* of the register.
CTRm	Counter	m designates the number of bits. Hence, the count sequence consists of 2^m states.
RCTRm	Ripple counter	Count sequence length $= 2^m$.
CTR DIVm	Synchronous counter	Count sequence length $= m$.
ROM	Read-only memory	
RAM	Random-access read/write memory	

■ **Table A.2** Qualifying symbols for inputs and outputs

Symbol	Description	Comments
	Logic negation at input	An external 0 produces an internal 1.
	Logic negation at output	An internal 1 produces an external 0.
	Active-LOW input	Equivalent to ─o⌐ in positive logic.
	Active-LOW output	Equivalent to ⌐o─ in positive logic.
	Dynamic input	Input is active on the transition from (external) 0 to (external) 1.
	Dynamic input with negation	Input is active on the transition from (external) 1 to (external) 0.
	Active-LOW dynamic input	Input is active on the transition from (external) HIGH to (external) LOW.
	Internal connection	A 1 on the left side produces a 1 on the right side.
	Negated internal connection	A 1 on the left side produces a 0 on the right side.

Figure A.3(a) defines a hex inverter, a device (integrated-circuit) consisting of six inverters. Similarly, the element shown in Figure A.3(b) depicts a device consisting of four 2-input NOR gates. In either example, the single qualifier implies that each device in the package performs the same function.

Common inputs to more than one device in a circuit can be grouped in a *common control block* like the notched block shown in Figure A.4. As seen, the input *a* affects each of the elements below the common control block. On the other hand, when a common input affects only some of the

■ **Table A.3** Qualifying symbols for inside the outline

Symbol	Description	Comments
⌐—	Postponed output	The output changes when the input initiating (causing) the change returns to its initial state - for example, a pulse-triggered (master-slave) flip-flop.
—[EN	Enable input	
—[→ m	Shift-right input	$m = 1,2,3$, etc., designates the number of bit positions that will be right-shifted when the input is activated.
—[← m	Shift-left input	
—[+ m	Count-up input	$m = 1,2,3$, etc., designates the amount by which the content of the device is increased when the input is activated.
—[− m	Count-down input	
—[>	Greater-than input of a comparator	
—[<	Less-than input of a comparator	
—[=	Equal input of a comparator	
—[CT=m	Content-setting input	When the input is activated, the content of the element (for example, a counter or a register) will take on the value m.
$J, K, T,$ S, R, D	Data inputs	Inputs typically associated with bistable devices or storage elements.

elements below the common control block, this input must be qualified by *dependency notation* in a manner to be explained later.

We can also group common outputs into a *common output element*, as shown in Figure A.5. The common output element is designated by a double line at its top. A common output depends on all the elements above the common output block, but may also depend on inputs into the common output block. Consequently, just like any other element, the logic function of the common output element must be specified by a general qualifier. In Figure A.5, the common output element implements the AND function.

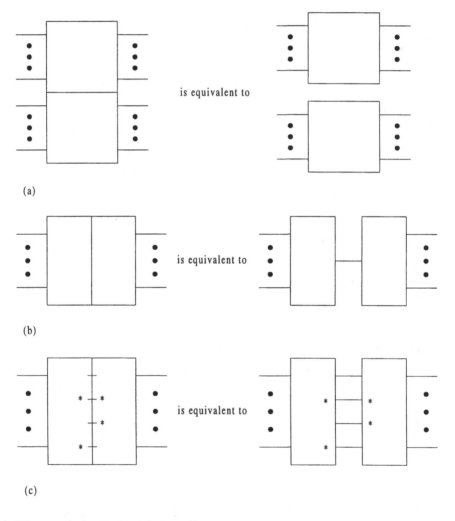

(a)

(b)

(c)

■ **Figure A.2** Embedded outlines

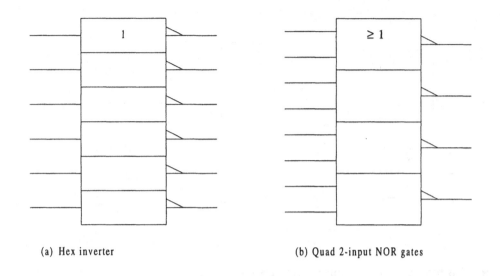

(a) Hex inverter (b) Quad 2-input NOR gates

■ **Figure A.3** Gate configurations

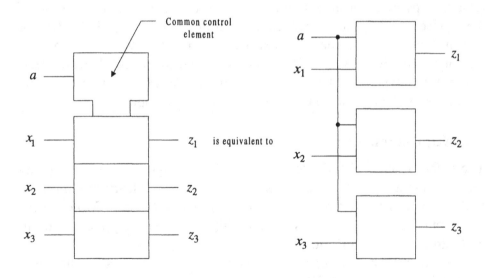

■ **Figure A.4** Common control block

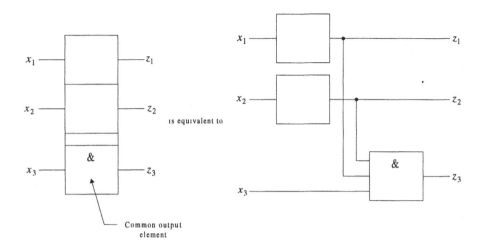

■   **Figure A.5**   Common outline element

A.3.   DEPENDENCY NOTATION

Dependency notation is fundamental to the new standard. It supplements the information provided by the qualifying symbols by specifying the relationships between inputs and outputs so that the logical operation of the device can be determined entirely from the logic symbol.   Of the eleven types of dependencies defined in the standard, we chose to introduce only three: G (AND), C (control), and M (mode) dependencies. These three types of dependencies will enable us to summarize the discussion of the new standard using a nontrivial example.

G (AND) Dependency

The letter G denotes AND dependency. In this type of dependency, any input or output designated with Gm (where m is an identifying number) is ANDed with any other input or output labeled with m.   Moreover, if two inputs or outputs are designated with Gm, then they are related to each other by an OR function.

An example of G dependency is shown in Figure A.6.   Since the input x_2 is labeled G1, it is ANDed with x_1 because the x_1 input is identified with the same label, $m = 1$. On the other hand, because the x_3 input is

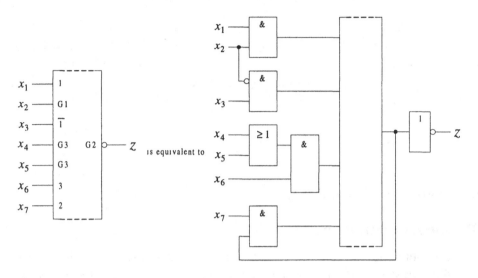

■ **Figure A.6**   G dependency

labeled with $\bar{1}$ [the bar indicates complement (negation)], x_3 is ANDed with the complement of x_2. Similarly, because the *internal* state of the output line is designated with G2, it affects x_7 since x_7 is also labeled with 2. Finally, because inputs x_4 and x_5 are designated with G3, they are first ORed and the result then ANDed with x_6, whose label is $m = 3$.

C (Control) Dependency

Control dependency is denoted by the letter C and is typically associated with a clock input. In this type of dependency, if the *internal* state of any input or output designated with Cm is 1, then any other input or output labeled with m is enabled. On the other hand, if the internal state of a Cm input or output is 0, then inputs or outputs affected by Cm are disabled and have no effect on the function of the device.

Figure A.7 shows an example of control dependency (combined with G dependency). We see that input x_3 selects which of inputs x_1 and x_2 affects the input D when x_4 becomes 1.

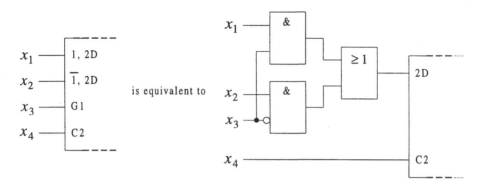

■ **Figure A.7** C dependency

M (Mode) Dependency

Mode dependency is denoted by the letter M and is used to indicate how the functions of affected inputs or outputs depend on the mode of operation of the device. M dependency affects inputs similar to C dependency: When the internal state of an Mm input or output is 1, any input or output labeled with m is enabled. On the other hand, if the internal state of the Mm input or output is 0, the affected inputs have no effect on the function of the element and the affected outputs can be ignored.

Figure A.8 shows an example of mode dependency. When $a = 1,$

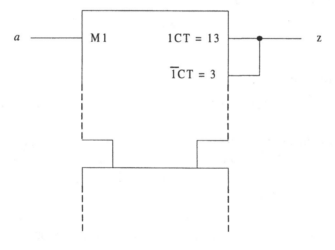

■ **Figure A.8** M dependency

mode 1 is established so that the output z will be 1 only if the content of the device (a register or a counter; see Table A.3) equals 13. On the other hand, if $a = 0$, the output z will be 1 only when the content of the element equals 3. (Note the $\overline{1}$ at the CT=3 label.)

A.4 CONCLUDING EXAMPLE

Consider the device shown in Figure A.9. The general qualifier, CTR DIV 16, indicates that the device is a divide-by-16 (4-bit) synchronous counter (see Table A.1). The four elements below the common control block are

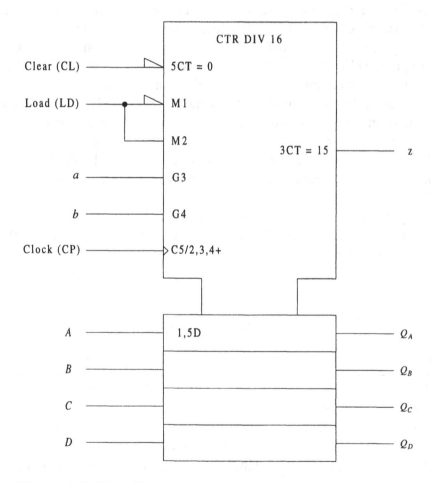

■ **Figure A.9** Four-bit synchronous counter

identical (the label 1,5D appears only once) and represent the four flip-flops, which are of the D type. The four inputs to the D flip-flops and the load (LD) line into the common control block indicate that the counter has a parallel load capability.

The dynamic indicator at the clock (CP) input implies that the C5 input is activated by a low-to-high transition of CP. In other words, the flip-flops are positive edge-triggered. The label /2,3,4+ following C5 means that the clock increments the counter (+ stands for count-up; see Table A.3) only when M2, G3, and G4 are HIGH.

Since the letter C denotes control dependency, any input labeled with a 5 prefix is dependent on (synchronized with) the clock. For example, the label 5CT=0 on the clear (CL) input means that the clear function is dependent on the clock; that is, it is a synchronous clear. Thus, when CL is LOW, the counter is reset to 0 (CT=0) on the positive-going edge of CP. Similarly, the 5D labels at the inputs of the flip-flops indicate that the flip-flops operate in synchronization with the triggering edge of the clock.

The counter has two modes of operation. When LD is LOW, the counter is in mode M1, in which the input data are loaded into the flip-flops on the positive-going edge of CP. (The number 1 appears in the 1,5D label.) When LD is HIGH, the counter is in mode M2 and will be able to advance through its count sequence provided that G3 and G4 are also HIGH.

Finally, the qualifier 3CT=15 means that the output z will be HIGH only when G3 is HIGH *and* the count is 15. Table A.4 summaries the operation of the counter.

■   **Table A.4**   Function table of the counter in Figure A.9

CL	LD	a	b	Function
LOW	×	×	×	Clear
HIGH	LOW	×	×	Load
HIGH	HIGH	Either a or b LOW		No change
HIGH	HIGH	HIGH	HIGH	Increment count

Bibliography

Almaini, A.E.A. *Electronic Logic Systems*. London: Prentice-Hall International, 1986.

Boole, G. *An Investigation of the Laws of Thought*. New York: Dover, 1958.

Brayton, R.K., G. Hachtel, C. McMullen, and A. Sangiovanni-Vincentelli. *Logic Minimization Algorithm for VLSI Synthesis*. Boston: Kluwer, 1984.

Comer, D.J. *Digital Logic and State Machine Design*. New York: Holt, Rinehart and Winston, 1984.

Dewey, A.M. *Analysis and Design of Digital Systems with VHDL*. Boston, M.A.: PWS Publishing Company, 1997.

Fletcher, W.I. *An Engineering Approach to Digital Design*. Englewood Cliffs, N.J.: Prentice-Hall, 1980.

Floyd, T.L. *Digital Fundamentals* (3rd ed.). Columbus, Ohio: Merrill, 1986.

Friedman, A.D. *Fundamentals of Logic Design and Switching Theory*. Rockville, MD.: Computer Science Press, 1986.

Gajski, D.D. *Principles of Digital Design*. Prentice-Hall, 1997.

Garrod, S.A.R., and R.J. Borns. *Digital Logic: Analysis, Application and Design*. Saunders, 1991.

Huffman, D.A. "The Synthesis of Sequential Switching Circuits." *Journal of the Franklin Institute* 257 (1954): 161-190 and 257-303.

Karnaugh, M. "The Map Method for Synthesis of Combinational Logic Circuits." *Transactions of AIEE* I, 72 (1953): 593-598.

Katz, R.H. *Contemporary Logic Design*. Benjamin/Cummings, 1995.

Kohavi, Z. *Switching and Finite Automata Theory* (2nd ed.). New York: McGraw-Hill, 1978.

Lala, P.K. *Practical Digital Logic Design and Testing*. Englewood Cliffs, N.J.: Prentice-Hall, 1996.

Mano, M.M. *Digital Design*. Englewood Cliffs, N.J.: Prentice-Hall, 1984.

Mano, M.M., and C.R. Kime. *Logic and Computer Design Fundamentals.* Upper Saddle River, N.J.: Prentice-Hall, 1997.

McCluskey, E.J. "Minimization of Boolean Functions." *Bell System Technical Journal* 35 (1956): 1417-1444.

_____. *Logic Design Principles.* Englewood Cliffs, N.J.: Prentice-Hall, 1986.

Mealy, G.H. "A Method for Synthesizing Sequential Circuits." *Bell System Technical Journal* 34 (1955): 1045-1079.

Moore, E.F. "Gedanken-experiments on Sequential Machines." *Automata Studies, Annals of Mathematical Studies*, no. 34, pp. 129-153. Princeton, N.J.: Princeton University Press, 1956.

Nagel, H.T., B.D. Carroll, and J.D. Irwin. *An Introduction to Computer Logic.* Englewood Cliffs, N.J.: Prentice-Hall, 1975.

Nelson, V.P., H.T. Nagle, B.D. Carroll, and J.D. Irwin. *Digital Logic Circuit Analysis and Design.* Englewood Cliffs, N.J.: Prentice-Hall, 1995.

Prosser, F.P., and D.E. Winkel. *The Art of Digital Design* (2nd ed.). Englewood Cliffs, N.J.: Prentice-Hall, 1987.

Quine, W.V. "The Problem of Simplifying Truth Functions." *American Mathematical Monthly* 59 (1952): 521-531.

_____. "A Way to Simplify Truth Functions." *American Mathematical Monthly* 62 (1955): 627-631.

Roth, C.H. *Fundamentals of Logic Design* (4th ed.). PWS, 1995.

_____. *Digital Systems Design Using VHDL*, Boston, M.A.: PWS Publishing Company, 1997.

Shannon, C.E. "A Symbolic Analysis of Relay and Switching Circuits." *Transactions of AIEE* 57 (1938): 713-723.

_____. "The Synthesis of Two-Terminal Switching Circuits." *Bell System Technical Journal* 28 (1949): 54-98.

Shaw, A.W. *Logic Circuit Design.* Saunders, 1993.

Unger, S.H. *Asynchronous Sequential Switching Circuits.* New York: Wiley, 1969.

Wakerly, J.F. *Digital Design: Principles and Practices*, (2nd ed.), Englewood Cliffs, N.J.: Prentice-Hall, 1994.

White, R. How Computer Work, Emeryville, CA.: Ziff-Davis Press, 1993

Wilkinson, B. *Digital System Design.* London: Prentice-Hall International, 1987.

Williams, G.E. *Digital Technology* (3rd ed.). Chicago: Science Research, 1986.

Answers to Selected Problems

Chapter 1

1.1 (a)

A	B	C	L
0	0	0	0
0	0	1	0
0	1	0	0
0	1	1	1
1	0	0	0
1	0	1	1
1	1	0	0
1	1	1	1

(b) $L(A,B,C) = C \cdot (A + B)$

1.3 (a) $F(A,B) = A' \cdot B' + A' \cdot B + A \cdot B'$ **(b)** $F(A,B) = A' \cdot B'$

1.4 (a)

A	B	L
0	0	0
0	1	1
1	0	1
1	1	0

(b)

A	B	C	L
0	0	0	1
0	0	1	0
0	1	0	1
0	1	1	1
1	0	0	0
1	0	1	1
1	1	0	1
1	1	1	1

1.9   **(a)** 500,000 pulses per second   **(b)** 250 pulses per second
         (c) 5 pulses per second

1.15 **(a)** $t_r \simeq 1.1$ microseconds   **(b)** $t_f \simeq 1.1$ microseconds
         (c) $t_p \simeq 6.1$ microseconds

Chapter 2

2.1 (a) $(101011)_2 = (43)_{10}$ **(b)** $(476)_8 = (318)_{10}$
         $(11011.1101)_2 = (27.8125)_{10}$ $(365.27)_8 = (245.359375)_{10}$
         $(.01101)_2 = (.40625)_{10}$ $(.7105)_8 = (.8918457)_{10}$
      (c) $(F23A)_{16} = (62010)_{10}$ **(d)** $(121.201)_3 = (16.7037037)_{10}$
         $(7A41.C8)_{16} = (31297.78125)_{10}$ $(1302.12)_4 = (114.375)_{10}$
         $(.FD21)_{16} = (.9887848)_{10}$ $(.354)_6 = (.6574074)_{10}$
                                                  $(298)_{12} = (404)_{10}$

2.2 $(120211.1111)_3$, $(1551.3)_6$, $(653.4)_8$, $(1AB.8)_{16}$

2.4 (a) $(01000111.00101000)_{BCD}$ **(b)** $(001001000010)_{BCD}$
      (c) $(0001100101100111.000101110001100001110101)_{BCD}$

2.5 $(11101011001.10011)_{BCD} = (10111101111.11111)_2$ (truncated after 5 bits)
                             $= (2F7.F8)_{16}$ (converted from binary)

2.8 15's complement: **FD85, 0749, F850.B3**
         16's complement: **FD86, 074A, F850.B4**

2.11 1's complement: **10010010, 0101000.0011, 01010010100**
         2's complement: **10010011, 0101000.0100, 01010010101**

2.12 $(0741)_{10}$, $(952315)_{10}$

2.14 (a)

	11010
2's complement of 01101:	+10011
	1 01101

Ignore end carry to yield 1101.

	11010
1's complement of 01101:	+10010
End-around carry:	1 01100
	+ ↘1
	1101

(b)

	00101
2's complement of 11001:	+00111
	01100

No end carry; hence, -(2's complement of 01100)=-10100.

	00101
1's complement of 11001:	+00110
	01011

No end carry; hence, -(1's complement of 01011)=-10100.

(c)

	11010
2's complement of 10000:	+10000
	1 01010

Ignore end carry to yield 1010.

	11010
1's complement of 10000:	+01111
	1 01001
End-around carry:	+ ↘1
	1010

2.16

	+547	−547
Sign-magnitude	0000001000100011	1000001000100011
1's complement	0000001000100011	1111110111011100
2's complement	0000001000100011	1111110111011101

2.18 (a) 1001000111   **(b)** 10110001001

2.19 **(a)** 111 (last remainder 011000)   **(b)** 10101 (last remainder 0000)

2.21

Decimal	Self-complementing 3231	Non-self-complementing 3231
0	0000	0000
1	0001	0001
2	0100	0100
3	0010	0010
4	0011	0011
*5	1100	0110
*6	1101	0111
*7	1011	1010
8	1110	1110
9	·1111	1111

2.23 **(a)** 642(−3): 01000000   **(b)** 4221: 01100000   **(c)** 5211: 01110000

2.25

Decimal	First Alternative	Second Alternative
0	0000	0000
1	0001	0010
2	0011	0011
3	0010	0001
4	0110	0101
5	0100	0111
6	0101	0110
7	0111	0100
8	1111	1100
9	1110	1101
10	1100	1111
11	1101	1110
12	1001	1010
13	1011	1011
14	1010	1001
15	1000	1000

2.26 **(a)** $(55)_{16}$, $(01010101)_2$ **(b)** $(5B)_{16}$, $(01011011)_2$

2.29

Decimal	Parity	642(−3)
0	1	000 0
1	1	010 1
2	0	001 0
3	1	100 1
4	0	010 0
5	0	101 1
6	1	011 0
7	0	110 1
8	1	101 0
9	1	111 1

Chapter 3

3.6 $x\Delta(x+y) = x(x+y)' + x'(x+y) = x(x'y') + x'x + x'y = x'y$

3.7 (a) $(x'y')' = x+y$
(b) $x+y' = 1$ if and only if $x+y = x$
(c) $x = 1$ if and only if $y = (x+y')(x'+y)$ for all y

3.10 (a) $x \oplus y$ (b) $x'yw'$ (c) 0 (d) $y(w+x)$ (e) $x'y' + y(x+z)$

3.14 $2^{2^{n-1}}$

3.15 (a) $f(x, f(y,0)) = (xy)'$ (NAND) (b) $f(1,y,z) = (y+z)'$ (NOR)

3.18 $x'y' + x'y + xy$

3.20 $(x+y'+z)(x+y'+z')(x'+y+z')$

3.25 $f(x,y,z) = \{[(xy)'(xy)']'z\}'$

3.27 (a) $F = \sum(1,3,5,7,9,11,13,15) = \prod(0,2,4,6,8,10,12,14)$
(b) $F = \sum(1,3,5,9,12,13,14) = \prod(0,2,4,6,7,8,10,11,15)$
(c) $F = \sum(3,5,6,7) = \prod(0,1,2,4)$

3.30 $B = A \oplus \text{FLAGIN}; \text{FLAGOUT} = A + \text{FLAGIN}$

3.31 $y_1 = x_1 \oplus 0$
$y_2 = x_2 \oplus x_1$
$y_3 = x_3 \oplus (x_1 + x_2)$

3.33 $z_0 = x_0 y_0$
$z_1 = x_1 y_0 (x_0 y_1)' + x_0 y_1 (x_1 y_0)'$
$z_2 = x_1 y_1 (x_0 y_0)'$
$z_3 = x_0 x_1 y_0 y_1$

3.35 $x = \sum (1,2,3,5)$
$y = \sum (3,4,5,6,7)$

3.38 $\text{N/S} = (A + B)C'D' + AB(C' + D'); \text{E/W} = (\text{N/S})'$

3.41 2^{n+1}

3.42 (a) $S_{1,2,5}(A', B, C, D, E)$
(b) $S_{0,1,4,5}(A, B, C, D, E)$
(c) $S_{0,1,4}(B, C, D, E)$

3.43 $H(w,z) = w'z + wz'; \ f(w,x,y,z) = Hy + H'x' = G[H(w,z),x,y]$

3.44 $H(x_1,x_4) = x_1 \odot x_4; \ f(x_1,x_2,x_3,x_4) = Hx_2'x_3' + H'x_2 x_3$

3.45 (a) $\{-2,2,2;1\}$ (b) $\{1,-1,-2;-3/2\}$ (c) Not a threshold function

3.46 $\sum (2,4,6,14)$

3.47 $\sum (1,4,5,7,9,11,12,13,15)$

Chapter 4

4.1   (b) $(xy)'(y+z')+(xz+y)w$

4.5   (b) $x_2'(x_3'+x_1)$

4.7   (a) $x_2'+x_1x_3$ (c) $\sum(2,3,5,7)$

4.9 $KB_1(P'+B_2)TD$

4.11 (a) $L=x_1x_2+x_1x_3+x_2x_3$
  (b) $f = x_1x_2x_3 + x_1x_2x_4 + x_1x_2x_5 + x_1x_3x_4 + x_1x_3x_5 + x_1x_4x_5 + x_2x_3x_4$
     $+ x_2x_3x_5 + x_2x_4x_5 + x_3x_4x_5$
  (c) $F = (x_1'+x_2')(x_1'+x_3')(x_2'+x_3')$

4.13 Let $f(x_1,x_2,x_3)=1$ when oil flows. Let $x_i=1$ if the switch S_i is open.
Then, $f = x_1'x_2'x_3' + x_1x_2'x_3' + x_1x_2x_3$.

4.23 (a) $x_1x_2x_4 + x_1x_2x_3 + x_2x_3x_4$ (b) $x_1'x_2 + x_2'x_3$
  (c) $x_2x_3' + x_1'x_2'x_4 + x_1x_2'x_3x_4'$ (e) $x_2x_4 + x_2'x_4' + x_1'x_2$
  (g) x_2' (h) $x_2'+x_1'x_4'$ (j) $x_1x_4' + x_2'x_4' + x_3x_4'$
  (l) $x_1'x_2'x_3' + x_1'x_2x_4 + x_1x_2x_3 + x_1x_2'x_4'$
     or $x_2'x_3'x_4' + x_1'x_3'x_4 + x_2x_3x_4 + x_1x_3x_4'$

4.25 (a) $x_2x_5 + x_1x_4'x_5 + x_1'x_2'x_5'$ (i) $x_4x_5'x_6 + x_1x_2'x_5'x_6 + x_1x_2x_4x_6$

4.26 (a) $F = x_1'x_4 + x_3x_4 = x_4(x_1'+x_3)$
  (b) $F = x_1x_2x_3' + x_1'x_3x_4 = (x_3'+x_4)(x_1'+x_3')(x_1'+x_2)(x_1+x_3)(x_1+x_4)$
  (c) $F = x_4'$
  (d) $F = x_1'x_3 + x_2'x_4' = x_1'(x_3+x_4')(x_2'+x_3)$
  (e) $F = x_2'x_4' + x_1'x_4 = (x_1'+x_4')(x_2'+x_4)$
  (f) $F = x_1x_2 + x_2x_3' + \begin{cases} x_1'x_2'x_3 \\ \text{or} \\ x_1'x_2'x_4 \end{cases}$

(g) $F = x_3'x_4 + x_2'x_3' + x_1'x_3x_4'$

(h) $F = x_1'x_2'x_4' + x_2x_3'x_4' + x_2'x_3'x_4 + x_2x_3x_4$

(i) $F = x_3'x_4' + x_3x_4$

(j) $F = x_1 + x_2x_3 + x_2x_4$

(l) $F = x_1'x_2 + x_5'$

(n) $F = x_2'x_4' + x_2x_4 = (x_2 + x_4')(x_2' + x_4)$

4.28 To eliminate the 1-hazards, implement $x_2'x_4' + x_1x_3 + x_1'x_3'$

4.32 $F = x_1x_2'x_4' + x_1'x_2x_3 + \begin{cases} x_1'x_3x_4' \\ \text{or} \\ x_2'x_3x_4' \end{cases}$

(b) $F = x_1x_2' + x_2'x_3'x_4 + x_1'x_2x_4' + x_2x_3x_4$

(c) $F = x_2'x_3x_4 + x_1'x_2' + x_1'x_3'x_4 + x_2'x_3'x_4'$

(d) $F = x_1x_2x_3' + x_1x_3x_4 + x_1'x_2x_3 + x_1'x_2'x_4$

(e) $F = x_1x_2' + x_1'x_2 + \begin{cases} x_1x_3+x_2'x_3'x_4' \\ \text{or} \\ x_2x_3+x_2'x_3'x_4' \\ \text{or} \\ x_1x_3+x_1'x_3'x_4' \\ \text{or} \\ x_2x_3+x_1'x_3'x_4' \end{cases}$

(f) $F = x_2'x_3' + x_3x_4' + x_1'x_2x_4$

(h) $F = x_2x_3x_4x_5' + x_2'x_3'x_5 + x_1x_2x_5' + \begin{cases} x_1'x_3'x_4x_5+x_1'x_2'x_3x_5 + x_1'x_2x_4'x_5 \\ \text{or} \\ x_1'x_2'x_4x_5+x_1'x_3x_4'x_5 + x_1'x_2x_3'x_5 \end{cases}$

(i) $F = x_1x_2x_3x_4x_5x_6x_7 + x_1'x_3x_5x_6'x_7' + x_2x_3'x_4'x_5x_6$

(j) $F = x_1'x_2'x_3'x_4x_5x_6' + x_1'x_2x_3'x_4'x_5 + x_3x_5'x_6 + \begin{cases} x_1'x_2x_4'x_5x_6 \\ \text{or} \\ x_1'x_2x_3x_4'x_6 \end{cases}$

(k) $F = x_2'x_3 + x_3x_4 + x_1x_4$

(l) $F = x_1'x_2 + x_1x_2' + \begin{cases} x_1'x_4 \\ \text{or} \\ x_2'x_4 \end{cases}$

(m) $F = x_1x_4 + x_2'x_3x_4 + x_2x_3x_4' + x_1'x_2'x_3'x_4'$

4.41 $F = L(x_1', x_2', x_3', x_4) + L(x_1, x_2, x_3, x_4') + L(x_1, x_2', x_3, x_4)$

4.45 $F_2 = \sum(0,1,2,3,8,9,10,11) + \sum_d(4,7,12,14,15)$

4.47 $F_1 = x_1'x_3x_4' + x_2x_3'x_4 + x_1'x_2'x_3'x_4'$
$F_2 = x_1x_3x_4 + x_1x_3'x_4 + x_1'x_2'x_3'x_4'$
$F_3 = x_1'x_3'x_4'$

Chapter 5

5.1 $F = [(x_2'x_4')'(x_1x_2')'(x_1x_3)']'$

5.10

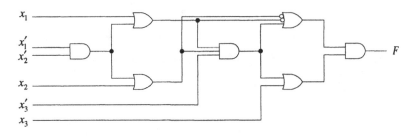

5.11 $F = x_3'x_4(x_1' + x_2) + x_3x_4'(x_2 + x_1)$

5.15 $F = x_1'x_2 + x_1x_2'$ AND-OR
$= [(x_1'x_2)'(x_1x_2')']'$ NAND-NAND
$= [(x_1 + x_2')(x_1' + x_2)]'$ OR-NAND
$= (x_1 + x_2')' + (x_1' + x_2)'$ NOR-OR

$$= (x_1'x_2' + x_1x_2)' \qquad \text{AND-NOR}$$
$$= (x_1'x_2')'(x_1x_2)' \qquad \text{NAND-AND}$$
$$= (x_1 + x_2)(x_1' + x_2') \qquad \text{OR-AND}$$
$$= [(x_1 + x_2)' + (x_1' + x_2')']' \qquad \text{NOR-NOR}$$

5.18 Addition table with the sum written in base 3 and in binary code:

| | Binary Sum | | | Corrected Sum in Base 3 | | |
Base 3	K	T_2	T_1	C	S_2	S_1
0	0	0	0	0	0	0
1	0	0	1	0	0	1
2	0	1	0	0	1	0
10	0	1	1	1	0	0
11	1	0	0	1	0	1
12	1	0	1	1	1	0

To correct, add 1 to the binary sum when $C = 1$; otherwise, add 0. Thus, $C = K + T_2T_1$, and we obtain the circuit shown in Figure A5.18.

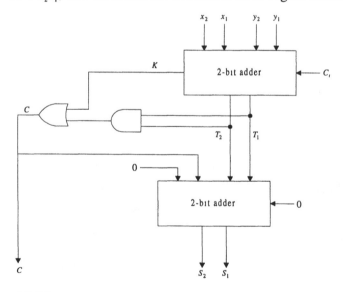

■ **Figure A5.18**

5.20 $P_1 = x_1 y_1$
$P_2 = x_2 y_1 y_2' + x_1 y_1' y_2 + x_1' x_2 y_1 + x_1 x_2' y_2$
$P_3 = x_1' x_2 y_2 + x_1 y_1' y_2$
$P_4 = x_1 x_2 y_1 y_2$

5.21 BCD value $=$ excess 3 value $- (3)_{10}$
$\qquad\qquad\qquad = $ excess 3 value $+ $ 2's complement of $(3)_{10}$
$\qquad\qquad\qquad = $ excess 3 value $+ (1101)_2$

Answer is the same as Figure 5.17 except that $B = 1101$.

5.26 Add one 4-bit comparator to the circuit in Figure 5.26.

5.27 The equality comparator can be designed the same as E was designed in Figure 5.27 with four equivalence gates and one 4-input AND gate.

5.29 **(b)** I_0 through I_7 are: 1, 1, 0, x_1', x_1', 0, 0, and x_1.

5.36 I_0 through I_3 for F_1, F_2, and F_3 respectively are: x_1', x_1', x_1, 0; x_1', x_1', 1, 1; and x_1', x_1', x_1, x_1.

5.37 I_0 through I_7 for F_1, F_2, and F_3 respectively are: 0, x_4', x_4', 0, 0, 1, 1, 0; 0, 0, 1, 0, 0, 1, x_4, 0; and x_4, 1, 0, 0, 0, 1, x_4', 0.

5.40

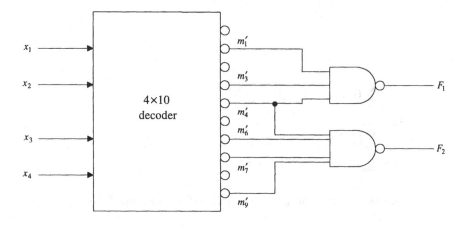

5.41 Designate the input BCD code by x_1, x_2, x_3, and x_4 (x_1 msb) and the output excess 3 code by F_1, F_2, F_3, and F_4 (F_1 msb). Then,

$$F_1 = \sum(5,6,7,8,9)$$
$$F_2 = \sum(1,2,3,4,9)$$
$$F_3 = \sum(0,3,4,7,8)$$
$$F_4 = \sum(0,2,4,6,8) = x_4'$$

The decoder implementation is similar to that of Problem 5.40.

5.46 **(a)** 5 address lines, 8 output lines
(b)

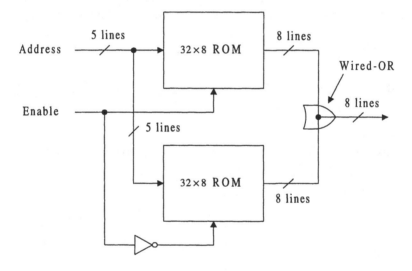

5.47 From Table 2.8, we get

$$g_3 = \sum(8,9,10,11,12,13,14,15)$$
$$g_2 = \sum(4,5,6,7,8,9,10,11)$$
$$g_1 = \sum(2,3,4,5,10,11,12,13)$$
$$g_0 = \sum(1,2,5,6,9,10,13,14)$$

To implement the circuit, use a 4×4 ROM as in Figure 5.46(b).

5.51 (a) $F_1 = x_1 + x_2 x_3 + x_2 x_4$

$F_2 = x_1' x_3 + x_2' x_4 + x_2 x_3' x_4'$

$F_3 = x_3 x_4 + x_3' x_4'$

$F_4 = x_4'$

(b) PLA table:

Product Terms	Inputs				Outputs			
	x_1	x_2	x_3	x_4	F_1	F_2	F_3	F_4
x_1	1	-	-	-	1	-	-	-
$x_2 x_3$	-	1	1	-	1	-	-	-
$x_2 x_4$	-	1	-	1	1	-	-	-
$x_2' x_3$	-	0	1	-	-	1	-	-
$x_2' x_4$	-	0	-	1	-	1	-	-
$x_2 x_3' x_4'$	-	1	0	0	-	1	-	-
$x_3 x_4$	-	-	1	1	-	-	1	-
$x_3' x_4$	-	-	0	0	-	-	1	-
x_4'	-	-	-	0	-	-	-	1

However, if we were to implement F_2' instead of F_2, then F_1 and F_2 could share two product terms and the number of product terms would be reduced by 2.

5.54 Assigning 0 to all the don't-care conditions, we obtain the following expressions:

$$a = A'C + A'BD + B'C'D' + AB'C'$$
$$b = A'B' + A'C'D' + A'CD + AB'C'$$
$$c = A'B + A'D + B'C'D' + AB'C'$$
$$d = A'CD' + A'B'C + B'C'D' + AB'C' + A'BC'D$$
$$e = A'CD' + B'C'D'$$
$$f = A'BC' + A'C'D' + A'BD' + AB'C'$$
$$g = A'CD' + A'B'C + A'BC' + AB'C'$$

All functions except d can be implemented by the PAL since none have more than four product terms. The expression for d has five product terms, but can be rewritten as:

$$d = A'CD' + A'B'C + B'C'D' + E$$

where

$$E = AB'C' + A'BC'D$$

Hence, we can use the extra PAL output to generate E and connect E to one of the PAL inputs as shown in the following figure.

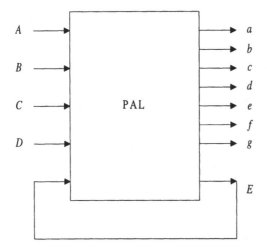

Chapter 6

6.1

	NS		Output $(z_1 z_2)$	
PS	$x = 0$	$x = 1$	$x = 0$	$x = 1$
A	A	B	00	01
B	B	C	01	10
C	C	D	10	11
D	D	A	11	00

6.6   **(a)**

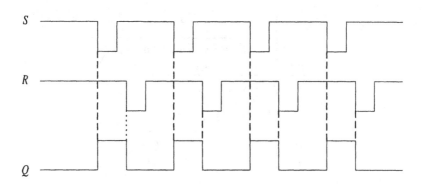

(b)

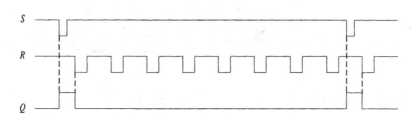

(c)

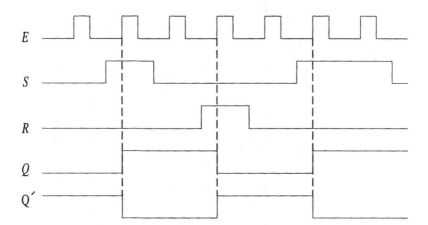

6.7 Primitive flow table:

	NS (Y)			
	Input Combination $(x_1 x_2)$			
PS	00	01	11	10
a	ⓐ	b	-	e
b	a	ⓑ	d	-
c	a	ⓒ	d	-
d	-	c	ⓓ	e
e	a	-	d	ⓔ

6.10 (c)

	NS/Output			
	Input Combination $(x_1 x_2)$			
PS	00	01	11	10
a	ⓐ/⓪	ⓐ/⓪	ⓐ/⓪	$b/-$
b	ⓑ/①	$a/-$	ⓑ/①	ⓑ/①

6.12 Equivalence classes: $\{2,8\}$, $\{5,12\}$, $\{3,10\}$, $\{4,11\}$

	NS				Outputs
	Input Combination $(x_1 x_2)$				
PS	00	01	11	10	$z_1 z_2$
1	①	7	-	4	11
2	②	5	-	4	01
3	-	7	③	4	10
4	2	-	3	④	00
5	6	⑤	9	-	11
6	⑥	7	-	4	01
7	1	⑦	14	-	10
9	-	7	⑨	13	01
13	2	-	14	⑬	11
14	-	5	⑭	4	00

6.15 Equivalence classes: $\{1, 9, 12\} = a$, $\{2, 11\} = b$, $\{3\} = c$, $\{4, 13\} = d$, $\{5, 14\} = e$, $\{6\} = f$, $\{7\} = g$, $\{8\} = h$, $\{10, 15\} = i$

Maximal merger groups: $(a, d) = A$, $(b, c) = B$, $(e, f) = C$, $(g, h) = D$, $(i) = E$

Reduced table:

	NS/output			
	Input Combination $(x_1 x_2)$			
PS	00	01	11	10
A	Ⓐ/01	B/–	Ⓐ/00	E/–
B	C/–	Ⓑ/11	A/–	Ⓑ/11
C	Ⓒ/10	Ⓒ/11	D/–	E/–
D	Ⓓ/01	C/–	Ⓓ/00	B/–
E	A/–	C/–	A/–	Ⓔ/11

6.17 Assign $a = 00$, $b = 01$, $c = 11$, and $d = 10$. Note that the transition from row c to row a constitutes a noncritical race.

6.19

	NS				Output
	Input Combination (TC)				
PS	00	01	11	10	Q
a	ⓐ	b	-	c	0
b	a	ⓑ	d	-	0
c	a	-	d	ⓒ	0
d	-	b	ⓓ	e	0
e	f	-	g	ⓔ	1
f	ⓕ	h	-	e	1
g	-	h	ⓖ	c	1
h	f	ⓗ	g	-	1

6.21

	NS				Output
	Input Combination $(x_1 x_2)$				
PS	00	01	11	10	z
a	ⓐ	b	-	d	0
b	a	ⓑ	c	-	0
c	-	f	ⓒ	d	1
d	e	-	c	ⓓ	0
e	ⓔ	f	-	d	1
f	e	ⓕ	c	-	1

6.24

	NS				Output
	Input Combination $(x_1 x_2)$				
PS	00	01	11	10	z
a	ⓐ	ⓐ	b	c	0
b	a	ⓑ	ⓑ	d	1
c	a	ⓒ	ⓒ	ⓒ	0
d	ⓓ	ⓓ	b	ⓓ	1

6.27 $S_1 = x_1(y_2' + x_2), R_1 = x_1'x_2$
$S_2 = x_1'y_1 + x_2, R_2 = x_2'(x_1 \odot y_1)$

6.28 (a) $S = AB, R = A'$
$z = Ay$
(b) $D = Ay + B$
$z = Ay$

6.31 $Y = x_1x_2 + x_2'y$
$z = y$

6.33 To prevent the occurrence of the essential hazard, y_2 must be delayed as it enters the AND gate whose second input is x'.

Chapter 7

7.1

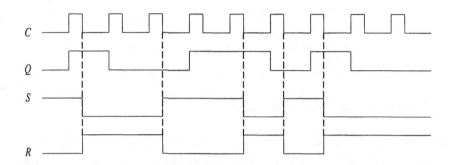

7.3

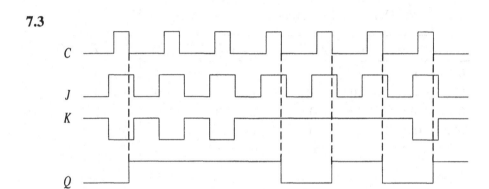

7.5 **(a)** SR flip-flop implementation: $T = Q'S + QR$
   (b) D flip-flop implementation: $T = Q'D + QD' = Q \oplus D$
   (c) JK flip-flop implementation: $T = Q'J + QK$

7.6 **(a)** SR flip-flop implementation: $J = S, K = R$
   (b) D flip-flop implementation: $J = D, K = D'$
   (c) T flip-flop implementation: $T = K = T$

7.7 **(a)** SR flip-flop implementation: $D = R'Q + S$
   (b) JK flip-flop implementation: $D = JQ' + K'Q$

(c) T flip-flop implementation: $D = T \oplus Q$

7.10

PS	NS/Output (z_1) Input Combination $(x_1 x_2)$			
z_2	00	01	11	10
$a = 0$	0/0	0/1	1/0	0/1
$b = 1$	0/1	1/0	1/1	1/0

The circuit functions as a full-adder.

7.14 Using an arbitrary state assignment, $a = 000$, $b = 100$, $c = 011$, $d = 001$, and $e = 010$, we obtain the following circuit excitation table.

PS	NS $(Y_1 Y_2 Y_3)$		Output (z)		$x = 0$			$x = 1$		
$y_1 y_2 y_3$	$x = 0$	$x = 1$	$x = 0$	$x = 1$	T_1	T_2	T_3	T_1	T_2	T_3
000	011	100	0	1	0	1	1	1	0	0
001	001	100	0	1	0	0	0	1	0	1
010	010	000	0	1	0	0	0	0	1	0
011	001	010	0	1	0	1	0	0	0	1
100	010	011	0	0	1	1	0	1	1	1

Utilizing the unused states as don't-cares, we obtain:

$$T_1 = y_1 + y_2'x$$
$$T_2 = y_1 + y_2'y_3'x' + y_2(y_3 \oplus x)$$
$$T_3 = x(y_1 + y_3) + y_1'y_2'y_3'x'$$
$$z = y_1'x$$

The circuit is self-starting.

7.18 $J = x_3$ and $K = x_3'$

7.21

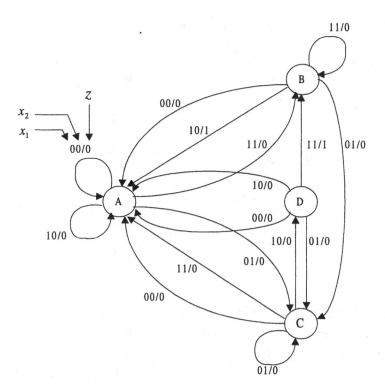

State assignment: $A = 00,\ B = 01,\ C = 11,\ D = 10$

Excitation equations: $D_1 = x_1'x_2 + y_1y_2x_1x_2'$

$$D_2 = (x_1y_1y_2)'x_2$$

Output equation: $z = y_1'y_2x_1x_2' + y_1y_2'x_1x_2$

7.24 Excitation equations: $A_1 = xy_2 + x'y_1y_2',\ B_1 = 1$

$$A_2 = x',\ B_2 = 1$$

The output equation remains the same as in Example 7.1.

7.29 Example of an input/output sequence:

$$x:\ 000\ 101\ 111\ 010$$
$$z:\ 0000101011110101$$

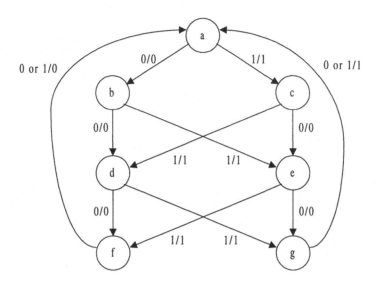

State assignment: $a = 000$, $b = 001$, $c = 010$, $d = 011$, $e = 100$, $f = 101$, and $g = 110$. The unused state is taken as don't-care.

Excitation equations: $T_2 = y_0 x + y_1 x' + y_2 y_1 + y_2 y_0$

$$T_1 = y_2 y_1 + y_1 x' + y_2' y_0 x' + y_2 y_0' x' + y_2' y_1' y_0' x$$
$$T_0 = y_0 x + y_2 y_0 + y_2' y_1 x + y_2' y_1 y_0' + y_2 y_1' x + y_2' y_0' x'$$

(The subscript 2 designates the most significant bit.)

Output equation: $z = y_0' x + y_2' x + y_2 y_1$

7.31 **(a)** The MAR requires 13 flip-flops; the MBR requires 32 flip-flops.
(b) $2^{15} = 32,768$ words

7.36

	SI	1101
t_1	1	1110
t_2	0	0111
t_3	1	1011
t_4	1	1101
t_5	0	0110
t_6	1	1011

7.39

Count	A_3	A_2	A_1	A_0	
0	1	1	1	1	(Counter is initially set)
1	1	1	1	0	
2	1	1	0	1	
3	1	1	0	0	
4	1	0	1	1	
5	1	0	1	0	
6	1	0	0	1	
7	1	0	0	0	
8	0	1	1	1	
9	0	1	1	0	
10	0	1	0	1	
11	0	1	0	0	
12	0	0	1	1	
13	0	0	1	0	
14	0	0	0	1	
15	0	0	0	0	

7.43

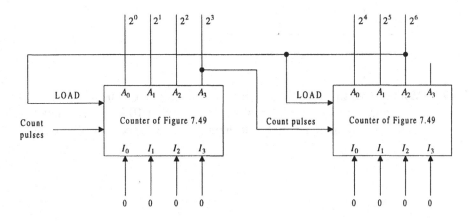

When count $= 2^6 = (1000000)_2$, LOAD is enabled and the counter is reset.

7.46

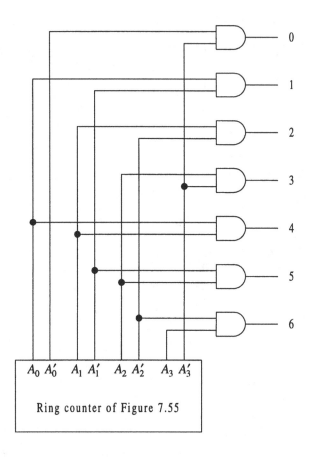

Ring counter of Figure 7.55

7.48   With three BCD counters, the total capacity of the counter is 999. Selecting the sampling interval as 0.1 second, a 10 kHz frequency would produce a count of 1000, thus utilizing the full capacity of the counter.

Index